A PRACTICAL GUIDE TO UNDERSTANDING THE NMR OF POLYMERS

A PRACTICAL GUIDE TO UNDERSTANDING THE NMR OF POLYMERS

Peter A. Mirau

A JOHN WILEY & SONS, INC., PUBLICATION

Copyright © 2005 by John Wiley & Sons, Inc. All rights reserved.

Published by John Wiley & Sons, Inc., Hoboken, New Jersey.
Published simultaneously in Canada.

No part of this publication may be reproduced, stored in a retrieval system, or transmitted in any form or by any means, electronic, mechanical, photocopying, recording, scanning, or otherwise, except as permitted under Section 107 or 108 of the 1976 United States Copyright Act, without either the prior written permission of the Publisher, or authorization through payment of the appropriate per-copy fee to the Copyright Clearance Center, Inc., 222 Rosewood Drive, Danvers, MA 01923, 978-750-8400, fax 978-646-8600, or on the web at www.copyright.com. Requests to the Publisher for permission should be addressed to the Permissions Department, John Wiley & Sons, Inc., 111 River Street, Hoboken, NJ 07030, (201) 748-6011, fax (201) 748-6008.

Limit of Liability/Disclaimer of Warranty: While the publisher and author have used their best efforts in preparing this book, they make no representations or warranties with respect to the accuracy or completeness of the contents of this book and specifically disclaim any implied warranties of merchantability or fitness for a particular purpose. No warranty may be created or extended by sales representatives or written sales materials. The advice and strategies contained herein may not be suitable for your situation. You should consult with a professional where appropriate. Neither the publisher nor author shall be liable for any loss of profit or any other commercial damages, including but not limited to special, incidental, consequential, or other damages.

For general information on our other products and services please contact our Customer Care Department within the U.S. at 877-762-2974, outside the U.S. at 317-572-3993 or fax 317-572-4002.

Wiley also publishes its books in a variety of electronic formats. Some content that appears in print, however, may not be available in electronic format.

Library of Congress Cataloging-in-Publication Data:

Mirau, Peter A.
 A practical guide to understanding the NMR of polymers / Peter A. Mirau.
   p. cm.
 Includes bibliographical references and index.
 ISBN 0-471-37123-8 (Cloth)
 1. Polymers—Spectra. 2. Nuclear magnetic resonance spectroscopy. I. Title.
 QC463.P5M57   2004
 547′.7046—dc22
2004009888

Printed in the United States of America.

10 9 8 7 6 5 4 3 2 1

CONTENTS

Preface xi

1 Introduction to NMR 1

    1.1 Introduction, 1
    1.2 Basic Principles of NMR, 2
        1.2.1 Introduction, 2
        1.2.2 Magnetic Resonance, 3
        1.2.3 The Rotating Reference Frame, 7
        1.2.4 The Bloch Equations, 9
        1.2.5 Pulsed NMR, 11
        1.2.6 The Fourier Transform, 15
        1.2.7 The Product Operator Formalism, 16
    1.3 Chemical Shifts and Polymer Structure, 20
        1.3.1 Molecular Structure and Chemical Shifts, 21
            1.3.1.1 Chemical Structure Effects, 21
            1.3.1.2 Inductive Effects, 23
            1.3.1.3 Anisotropic Shielding, 24
            1.3.1.4 Chemical Exchange, 26
        1.3.2 Proton Chemical Shifts, 29
        1.3.3 Carbon Chemical Shifts, 30
        1.3.4 Other Nuclei, 34
            1.3.4.1 Fluorine, 34
            1.3.4.2 Silicon, 34
            1.3.4.3 Phosphorus, 35
            1.3.4.4 Nitrogen, 36

1.4 Spin-Spin Coupling, 36
    1.4.1 Introduction, 36
    1.4.2 Nomenclature for Spin-Spin Coupling, 38
    1.4.3 Spin-Spin Coupling Patterns, 39
        1.4.3.1 Strong Coupling, 40
        1.4.3.2 Scalar Coupling and nD NMR, 41
    1.4.4 Proton–Proton Coupling, 41
    1.4.5 Proton–Carbon Coupling, 44
        1.4.5.1 Other Nuclei, 46
        1.4.5.2 Fluorine Couplings, 46
        1.4.5.3 Phosphorus Couplings, 48
        1.4.5.4 Silicon Couplings, 49
        1.4.5.5 Nitrogen Couplings, 49
    1.4.6 Homonuclear Couplings in Insensitive Nuclei, 50
1.5 NMR Relaxation, 51
    1.5.1 Introduction, 51
    1.5.2 Relaxation Mechanisms, 54
        1.5.2.1 Dipole–Dipole Interactions, 54
        1.5.2.2 Quadrupolar Interactions, 54
        1.5.2.3 Chemical Shift Anisotropy, 55
        1.5.2.4 Paramagnetic Relaxation, 55
        1.5.2.5 Other Relaxation Mechanisms, 56
    1.5.3 Spin-Lattice Relaxation, 56
        1.5.3.1 Heteronuclear Spin-Lattice Relaxation, 59
        1.5.3.2 Homonuclear Spin-Lattice Relaxation, 61
    1.5.4 Spin-Spin Relaxation, 61
    1.5.5 The Nuclear Overhauser Effect, 63
        1.5.5.1 Heteronuclear Nuclear Overhauser Effects, 64
        1.5.5.2 Homonuclear Nuclear Overhauser Effects, 65
1.6 Solid-State NMR, 66
    1.6.1 Chemical Shift Anisotropy, 66
    1.6.2 Magic-Angle Sample Spinning, 70
    1.6.3 Dipolar Broadening and Decoupling, 73
    1.6.4 Cross Polarization, 75
    1.6.5 Quadrupolar NMR, 77
1.7 Multidimensional NMR, 79
    1.7.1 Magnetization Transfer in nD NMR, 85
        1.7.1.1 Through-Bond Magnetization Transfer, 85
        1.7.1.2 Through-Space Magnetization Transfer, 91
    1.7.2 Solution 2D NMR Experiments, 92
        1.7.2.1 COSY, 92
        1.7.2.2 TOCSY, 93
        1.7.2.3 Heteronuclear Multiple Quantum Coherences, 94
        1.7.2.4 2D Exchange NMR, 95
        1.7.2.5 J-Resolved NMR, 98

  1.7.3 Solid-State 2D NMR Experiments, 98
    1.7.3.1 2D Exchange NMR, 99
    1.7.3.2 Wideline Separation Spectrsocopy, 100
    1.7.3.3 Heteronuclear Correlation, 101

2 Experimental Methods   104

2.1 Introduction, 104
2.2 The NMR Spectrometer, 105
  2.2.1 The Magnet, 105
  2.2.2 Shim Coils, 106
  2.2.3 RF Console, 107
  2.2.4 NMR Probes, 107
  2.2.5 Computer, 108
2.3 Tuning the NMR Spectrometer, 108
  2.3.1 Adjusting the Homogeneity, 108
  2.3.2 Adjusting the Gain, 108
  2.3.3 Tuning the Probe, 109
  2.3.4 Adjusting the Pulse Widths, 110
2.4 Solution NMR Methods, 112
  2.4.1 Sample Preparation, 112
  2.4.2 Data Acquisition, 114
  2.4.3 Decoupling, 115
  2.4.4 Data Processing, 117
    2.4.4.1 Baseline Corrections, 117
    2.4.4.2 Digital Resolution and Zero-Filling, 117
    2.4.4.3 Window Functions, 118
    2.4.4.4 Phasing, 119
    2.4.4.5 Quadrature Detection, 119
    2.4.4.6 Referencing, 119
  2.4.5 Quantitative NMR, 120
  2.4.6 Sensitivity Enhancement, 122
  2.4.7 Spectral Editing, 123
2.5 Solid-State NMR Methods, 126
  2.5.1 Magic-Angle Spinning, 126
  2.5.2 Cross Polarization, 128
  2.5.3 Decoupling, 131
  2.5.4 Wideline NMR, 131
  2.5.5 Solid-State Proton NMR, 133
2.6 NMR Relaxation, 137
  2.6.1 NMR Relaxation in Solution, 138
    2.6.1.1 Spin-Lattice Relaxation, 138
    2.6.1.2 Spin-Spin Relaxation, 140
    2.6.1.3 Nuclear Overhauser Enhancement, 141
  2.6.2 NMR Relaxation in Solids, 142

            2.6.2.1 Spin-Lattice Relaxation, 142
            2.6.2.2 Rotating-Frame Spin-Lattice Relaxation, 143
    2.7 Multidimensional NMR, 144
        2.7.1 Data Acquisition, 145
            2.7.1.1 Digital Resolution and Acquisition Times in nD NMR, 145
            2.7.1.2 Inverse Detection, 146
            2.7.1.3 Phase Cycling, 147
            2.7.1.4 Quadrature Detection, 147
            2.7.1.5 Pulse-Field Gradients, 148
            2.7.1.6 Decoupling, 152
        2.7.2 Data Processing, 152
            2.7.2.1 Apodization, 152
            2.7.2.2 Phasing, 154
            2.7.2.3 Baselines and t_1 Noise, 156
            2.7.2.4 Linear Prediction and Zero-Filling, 157

3 The Solution Characterization of Polymers   **160**

    3.1 Introduction, 160
        3.1.1 Polymer Microstructure, 161
            3.1.1.1 Regioisomerism, 161
            3.1.1.2 Stereochemical Isomerism, 162
            3.1.1.3 Geometric Isomerism, 163
            3.1.1.4 Branching and Endgroups, 164
            3.1.1.5 Chain Architecture, 165
            3.1.1.6 Copolymers, 166
        3.1.2 Spectral Assignments in Polymers, 167
            3.1.2.1 Model Compounds and Polymers, 168
            3.1.2.2 Polymer Chain Statistics, 169
            3.1.2.3 Chemical Shift Calculations, 173
            3.1.2.4 The γ-Gauche Effect, 175
            3.1.2.5 Spectral Editing, 178
            3.1.2.6 Multidimensional NMR, 179
    3.2 Stereochemical Characterization of Polymers, 182
        3.2.1 The Observation of Stereochemical Isomerism, 183
        3.2.2 Resonance Assignments for Stereosequences, 189
            3.2.2.1 Assignments of Stereosequences Using Model Compounds, 189
            3.2.2.2 Assignments of Stereosequences Using Polymerization Statistics, 192
            3.2.2.3 Assignments of Stereosequences Using Chemical Shift and Conformational Calculations, 193
            3.2.2.4 Assignments of Stereosequences Using nD NMR, 196
    3.3 Regioisomerism in Polymers, 211

3.4 Defects in Polymers, 214
    3.4.1 Branching, 214
    3.4.2 Endgroups, 218
3.5 Polymer Chain Architecture, 223
3.6 Copolymer Characterization, 225
    3.6.1 Random Copolymers, 227
    3.6.2 Alternating Copolymers, 232
    3.6.3 Block Copolymers, 234
3.7 The Solution Structure of Polymers, 236
    3.7.1 Polymer Chain Conformation, 237
    3.7.2 Intermolecular Interactions in Polymers, 243

4 The Solid-State NMR of Polymers   248

4.1 Introduction, 248
4.2 Chain Conformation in Polymers, 250
    4.2.1 Semicrystalline Polymers, 251
        4.2.1.1 Solid-Solid Phase Transitions, 261
    4.2.2 Amorphous Polymers, 263
    4.2.3 Elastomers, 266
    4.2.4 Reactivity and Curing in Polymers, 273
4.3 The Structure and Morphology of Polymers, 278
    4.3.1 Introduction, 278
    4.3.2 Spin Diffusion and Polymer Morphology, 280
        4.3.2.1 Spin Diffusion and Interfaces, 285
        4.3.2.2 Spin-Diffusion Coefficients, 286
        4.3.2.3 Polarization Gradients for Measuring Spin Diffusion, 287
        4.3.2.4 Proton Relaxation and Morphology, 294
    4.3.3 Semicrystalline Polymers, 295
    4.3.4 Block Copolymers, 302
    4.3.5 Multiphase Polymers, 310
    4.3.6 Polymer Blends, 317

5 The Dynamics of Polymers   336

5.1 Introduction, 336
5.2 Chain Motion of Polymers in Solution, 337
    5.2.1 Modeling the Molecular Dynamics of Polymers in Solution, 338
    5.2.2 Relaxation Mechanisms for Polymers in Solution, 345
    5.2.3 NMR Relaxation in Solution, 348
    5.2.4 The Relaxation of Polymers in Solution, 349
5.3 NMR Relaxation in the Solid State, 359
    5.3.1 Introduction, 359
    5.3.2 NMR Relaxation in Solid Polymers, 360
    5.3.3 Spin Exchange in Solid Polymers, 367

5.3.4 Polymer Dynamics and Lineshapes, 370
    5.3.4.1 Wideline Deuterium NMR, 371
    5.3.4.2 Chemical Shift Anisotropy and Polymer Dynamics, 385
    5.3.4.3 Dipolar Lineshapes and Polymer Dynamics, 390

Index   **397**

PREFACE

NMR spectroscopy has emerged as an important analytical method for the characterization of polymers and other materials, both in solutions and in the solid state. NMR has the ability to provide information about the structure and dynamics of polymers over a wide range of length scales and time scales. NMR spectroscopy is a discipline in its own right, and one of the more elegant laboratory demonstrations that quantum mechanics really works. The intense interest in NMR has led to the development of both practical methods as well as complex and sometimes esoteric applications. This makes understanding and appreciating the power of NMR somewhat daunting to students and researchers from other fields. The overarching goal of this book is to provide a framework for understanding the power and practical applications of the NMR in polymer science at an introductory level.

An appreciation of NMR in polymer science requires an understanding of several factors, including the NMR phenomenon, the NMR spectrometer, experimental methods and polymer structure. This book is organized to provide such a basic framework. Chapter 1 introduces the basic NMR phenomena, including chemical shifts, spin-spin coupling, NMR relaxation, multidimensional (nD) NMR and solid-state NMR. While the focus of Chapter 1 is an introduction to NMR methods, special attention is given to NMR parameters (such as the chemical shift) that are particularly important for the NMR of polymers.

The focus of Chapter 2 is experimental methods for polymer characterization. In addition to an introduction to the spectrometer, Chapter 2 contains many practical tips for the preparation of NMR samples, as well as for data collection and analysis. Many of the methods used for polymer analysis, including quantitative analysis, spectral editing and nD NMR require careful setup of the spectrometer and specialized data processing. Chapter 2 contains the most commonly used pulse sequences that typically come standard on a modern NMR spectrometer.

Chapter 3 is concerned with the solution characterization of polymers, and begins with a review of polymer microstructure with a particular emphasis on how changes in structure lead to identifiable features in the NMR spectra. One of the largest applications of NMR in polymer science is materials characterization. The first step in this analysis is assigning the peaks, and the many methods used to establish the resonance assignments are reviewed. I have tried to choose a variety of examples (using different nuclei and methods) to illustrate that there are often different ways to establish the assignments. These examples illustrate the level to which polymers can be analyzed by solution NMR.

Solid-state NMR has emerged as a routine and powerful method for the analysis of polymers, and this topic is considered in Chapter 4. One of the unique features of solid-state NMR is that the chemical shifts are often not averaged by molecular motion, so the NMR spectrum can provide information about polymer chain conformation. Solid-state NMR can also be used to characterize insoluble meterials and to monitor the reactivity and curing of polymers. Proton spin diffusion provides information about the mixing of polymers in blends and on the length scale of phase separation on a length scale (.5–20 nm) that is difficult to measure by other means.

The dynamics of polymers are covered in Chapter 5 for polymers both in solution and the solid state. The dynamics have a fundamental affect on the appearance of the NMR spectra, and it is important to understand the relationships between polymer dynamics and the relaxation times which influence the peak intensities in quantitative NMR experiments. Chapter 5 also considers some of the models used to interpret relaxation data in an attempt to understand the relationship between polymer structure and dynamics. The lines are often very broad in solids and the means by which the lineshapes are averaged provides detailed information about the amplitude and geometry of molecular motions in solid polymers.

I attempted to write this book on a number of levels in an attempt to satisfy the needs of students and researchers from other fields that do not have a deep understanding of polymer NMR. On the lowest level this book provides an introduction to the NMR phenomenon and the power of NMR in polymer science. Since the goal in many cases is to illustrate the general principles, rather than a complete review of the applications of NMR, many of the examples were chosen from our experiments to show the fundamental principles. Many sections contain additional details (such as the product operator formalism) that more advanced students can use to pursue a deeper understanding of the phenomena. Many of the examples were chosen to show the level to which information about the structure and dynamics can be obtained from the NMR spectra. In addition the book contains many references to other applications for those interested in a particular topic. This book was completed during my move from Bell Laboratories to the Air Force Research Laboratories while waiting for the NMR laboratory to be renovated.

1

INTRODUCTION TO NMR

1.1 INTRODUCTION

Nuclear magnetic resonance (NMR) spectroscopy has emerged as one of the premier methods for the characterization of polymers, both in solutions and in the solid state. The popularity of NMR as a method is due to the fact that many molecular-level features can be measured from the NMR spectra, including polymer microstructure, chain conformation, and dynamics. The NMR spectra of solid polymers are sensitive to these same features, but also to the length scale of mixing in blends and phase-separated materials, and the domain sizes in semicrystalline polymers. Since the spectral features and relaxation times are affected by local interactions, they provide information about the structure of polymers on a length scale (0.2–20 nm) that is difficult to measure by other methods.

The first NMR studies of polymers were reported (1) only about a year after nuclear resonance was discovered in bulk matter (2,3). In these studies it was reported that the proton linewidth for natural rubber at room temperature is more like that of a mobile liquid than of a solid, but that the resonance broadens near the glass transition temperature (T_g). This was recognized as being related to a change in chain dynamics above and below T_g. NMR methods developed rapidly after these initial observations, first for polymers in solution and, more recently, for polymers in the solid state.

Solution NMR is frequently used for polymer characterization because of its high resolution and sensitivity. The chemical shifts are sensitive to polymer microstructure, including polymer stereochemistry, regioisomerism, and the presence of branches and defects. These observations led to an improved understanding of both polymer microstructure and polymerization mechanisms (4). With the advent of higher magnetic fields and improved NMR methods and spectrometers, it has become possible to

A Practical Guide to Understanding the NMR of Polymers, by Peter A. Mirau
ISBN 0-471-37123-8 Copyright © 2005 John Wiley & Sons, Inc.

characterize even very low levels of defects in high molecular-weight polymers. The chemical shift assignments in the early studies were established by comparing the spectra with those from model compounds. The developments of spectral editing and multidimensional NMR methods have made it possible to assign the spectra without recourse to model compounds. The assignments can be established not only for carbons and protons, but also for any silicon, nitrogen, phosphorus, or fluorine atoms that may be present. The detailed microstructural characterization for polymers in solution has led to a deeper understanding of polymer structure–property relationships.

The solid-state NMR analysis of polymers emerged after the solution studies because the specialized equipment required for high-resolution studies was not routinely available. Solid-state NMR has become such an important analytical method that most modern spectrometers can be adapted easily for the solid-state studies. The interest in the NMR of solid polymers is due in part to the fact that most polymers are used in the solid state, and in some cases the NMR properties can be directly related to the macroscopic properties. Solid-state NMR can also provide information about the structure and dynamics of polymers over a range of length scales and time scales. Polymers have a restricted mobility in solids, and the chemical shifts can in favorable cases be directly related to the chain conformation (5).

Solid-state NMR is also an efficient way to monitor the reactivity of polymers, since the chemical changes that occur during curing often give rise to large spectral changes. The relaxation times in solids depend not only on the chain dynamics, but also on the morphology over a length scale of 0.2–20 nm. NMR has been extensively used to measure the length scale of mixing in blends and multiphase polymers, and the domain sizes in semicrystalline polymers. Solid-state NMR methods have been greatly expanded with the introduction of multidimensional NMR (6). These studies have led to a molecular-level understanding of the dynamics traditionally observed by dielectric and dynamic–mechanical spectroscopy, and a better understanding of the relationship between polymer morphology and macroscopic properties.

1.2 BASIC PRINCIPLES OF NMR

1.2.1 Introduction

Magnetic resonance is a complex phenomenon with many chemical and physical implications. One of the fundamental properties of magnetic resonance is that the separation between energy levels are quantized, as expected from quantum mechanics. The implication of this is that the resonance frequency contains important information about the chemical structure and the local magnetic environment. While the rigorous quantum mechanical treatment is required for a complete understanding of the NMR, there are many situations where magnetic resonance phenomena can be understood with classic vector diagrams. In this text we use the vector diagrams whenever they provide a physical insight into the magnetic resonance phenomena.

There are examples from classic physics that provide an introduction into the behavior of magnetic moments, the magnetic resonance phenomenon, and the vector diagrams that provide some physical insight into the experimental observations. One

1.2 BASIC PRINCIPLES OF NMR

of the most fundamental properties is that magnetic moments tend to align in a magnetic field. The classic analogy of this phenomenon is a compass needle that aligns in the Earth's magnetic field. In magnetic resonance experiments we can gain information about a spin system by perturbing it from equilibrium. In the compass example this would involve applying a force to move the needle from its equilibrium position. The torque, T, that acts on the needle is given by

$$T = \frac{\partial J}{\partial t} = r \times F \tag{1.1}$$

where J is the angular momentum, which is the equivalent for rotational motions to linear momentum, where r is the radius, and F is the force. In NMR the torque that we apply to the magnetization is in the form of radio frequency (rf) pulses. Much of what we can learn comes from monitoring a spin system as it returns to equilibrium following some perturbation. In most cases the spins interact with their neighbors, and we can often perturb one spin and measure the effect on its neighbors to gain information about the chemical structure and molecular dynamics.

A second classic example that allows some insights into the magnetic resonance phenomenon is that of a gyroscope. A simple gyroscope can be constructed as a wheel connected to a shaft. If we stand such a gyroscope up, it will simply fall down. However, if we spin it, it will remain standing as long as it is spinning. If we closely examine the spinning gyroscope we would see that it is not exactly vertical, but rather makes some angle ϕ with respect to vertical. Furthermore, the gyroscope slightly wobbles, or *precesses*, about the vertical axis. The rate at which the shaft precesses (ω) depends on a number of factors, including the spinning rate, the weight of the wheel, and the length of the shaft.

NMR-active nuclei are similar to gyroscopes in that they possess spin. When placed in a magnetic field they precess about the field direction at a rate that depends on the strength of the magnetic field and the nuclear properties. The rate of precession also depends on the local electronic structure, so the precession frequency also contains important chemical information.

1.2.2 Magnetic Resonance

The NMR phenomenon is possible because NMR-active isotopes possess spin, or angular momentum. Since a spinning charge generates a magnetic field, there is a magnetic moment associated with this angular momentum. According to a basic principle of quantum mechanics, the maximum experimentally observable component of the angular momentum of a nucleus possessing a spin (or of any particle or system having angular momentum) is a half-integral or integral multiple of $h/2\pi$, where h is Planck's constant. This maximum component is I, which is called the spin quantum number or simply "the spin." The magnitudes of the nuclear magnetic moments (μ) are often specified in terms of the ratio of the magnetic moment to the angular momentum. This parameter γ is known as the magnetogyric ratio and is given by

$$\gamma = \frac{2\pi \mu}{Ih} \tag{1.2}$$

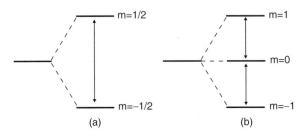

FIGURE 1.1 The energy-level diagram for (a) $I = 1/2$ and (b) $I = 1$ spin systems. Transitions are allowed only for $\Delta m = \pm 1$.

In general, there are $2I + 1$ possible orientations or states of the nucleus. For spin $1/2$ nuclei the possible magnetic quantum numbers are $+1/2$ and $-1/2$, corresponding to spins aligned with or against the magnetic field. Figure 1.1 shows the energy-level diagram for spin systems with $I = 1/2$ and $I = 1$.

When placed in a magnetic field, nuclei having spin undergo precession about the field direction as in the preceding gyroscope example. This is illustrated in Figure 1.2. The frequency of this so-called Larmor precession is designated as ω_0 in radians per second or υ_0 in hertz (Hz), cycles per second ($\omega_0 = 2\pi \upsilon_0$). The Larmor precession frequency or resonance frequency is given by

$$\omega_0 = \gamma B_0 \tag{1.3}$$

where B_0 is the magnetic field strength. The two quantities that determine the observation frequency for NMR signals are the magnetogyric ratio, γ, and the magnetic field strength, B_0. Table 1.1 lists some of the nuclear properties of spins that are of interest to polymer chemists. Our ability to detect the NMR signals depends both on the magnetogyric ratio and the natural abundance of the NMR-active nuclei. Protons have the highest sensitivity because they have a large magnetogyric ratio and natural abundance near 100%. At a magnetic field strength of 11.7 T (1 tesla = 10^5 gauss) the proton NMR signals are observed at 500 MHz. Fluorine is the second most sensitive nuclei, but it is not a common element in polymers. Most polymers of interest contain carbon, and Table 1.1 shows that the sensitivity is very low compared to that of

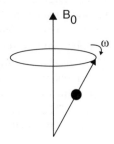

FIGURE 1.2 The precession of a magnetic moment in a magnetic field.

1.2 BASIC PRINCIPLES OF NMR

TABLE 1.1 The Properties of NMR-Active Nuclei of Interest in Polymer Science.

Isotope	Abundance (%)	Spin	$(\gamma \times 10^{-8})^a$	Sensitivity[b]	Frequency (MHz)[c]
^{1}H	99.98	1/2	2.6752	1.0	500.0
^{19}F	100.0	1/2	2.5167	0.83	470.2
^{29}Si	4.7	1/2	−0.5316	0.078	99.3
^{31}P	100.0	1/2	1.0829	0.066	202.3
^{13}C	1.1	1/2	0.6726	0.0159	125.6
^{2}H	0.015	1	0.4107	0.00964	76.7
^{15}N	0.365	1/2	−0.2711	0.001	50.6

[a] The magnetogyric ratio in SI units.
[b] The sensitivity relative to protons.
[c] The resonant frequency in a 11.7 T magnetic field.

protons. However, the sensitivity of modern NMR spectrometers is such that carbon spectra can be routinely observed. Nitrogen is also a common element in polymers, but it is difficult to study because of its low magnetogyric ratio and natural abundance. ^{15}N NMR studies are possible, but usually only after isotopic labeling. The sensitivities of silicon and phosphorus are intermediate between those of protons and carbons.

The splitting of the energy levels in the presence of the magnetic field leads to a population difference between the upper and lower spin states that is determined by the Boltzmann distribution. Since the energy difference between the upper and lower states is very small, the population difference between the upper and lower states is also very small ($\sim 1/10^5$). This makes NMR a relatively insensitive technique compared to optical spectroscopy, and concentrations on the order of 10^{-3} M are usually required.

When placed in a magnetic field the nuclear spins tend to align with (the α spin state) or against (the β spin state) the magnetic field and the populations in the upper (N^+) and lower (N^-) spin state is given by the Boltzmann distribution as

$$\frac{N^+}{N^-} = 1 + \frac{2\pi\mu}{kT} \qquad (1.4)$$

In an NMR experiment, a large number of spins are placed in the magnetic field, and we observe the net magnetization by perturbing the spin system. The equilibrium state of a collections of spins in a magnetic field is illustrated in Figure 1.3a, which shows the precession of a collection of spins aligned with or against the magnetic field. From the Boltzmann considerations there are slightly more spins in one of the states, and the sum of the individual vectors is represented by the simple vector diagram in Figure 1.3b.

The equilibrium state of the NMR-active nuclei in the magnetic field cannot be directly detected, but we can perturb the system and monitor its return to equilibrium. This can be accomplished by the application of a second magnetic field, typically at a right angle to the main magnetic field, as illustrated in Figure 1.4. If the frequency

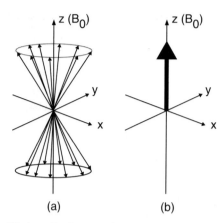

FIGURE 1.3 The equilibrium distribution of spins. The drawing shows (a) the random distribution for the individual spins and (b) the net magnetization.

of the applied field B_1 is close to ω_0, then the magnetization will begin to be affected by the B_1 field. The result is that the magnetization now precesses about the effective field B_{eff}. The application of a second magnetic field perturbs the spin system from equilibrium and allows us to measure the resonance frequency.

Following the application of a second magnetic field, the spin system relaxes toward equilibrium. Two kinds of NMR relaxation are important, *longitudinal* (along the z axis in Figure 1.4), and *transverse* (in the xy plane) relaxation. Longitudinal relaxation is called spin-lattice relaxation and is designated by the symbol T_1. The relaxation times provide information about the molecular dynamics of polymers since they depend on the rates and amplitudes of atomic fluctuations. The time scale of motion is related to the *rotational correlation time* τ_c that will be discussed in some detail in Chapter 5.

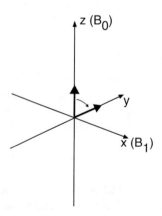

FIGURE 1.4 The application of an rf pulse along the x direction to the equilibrium magnetization.

1.2 BASIC PRINCIPLES OF NMR

The transverse relaxation is also called spin-spin relaxation and is given the symbol T_2. The spin-spin relaxation times determine the resolution in an NMR experiment because the relaxation time is inversely related to the linewidth $\Delta v_{1/2}$ by

$$\Delta v_{1/2} = \frac{1}{\pi T_2} \tag{1.5}$$

As the chain motion is restricted, the linewidths become broader, limiting the resolution that can be obtained. As with the spin-lattice relaxation, the spin-spin relaxation depends on the local atomic motions, and faster molecular motions lead to sharper lines and a higher resolution.

1.2.3 The Rotating Reference Frame

It is easier to understand many basic NMR concepts and to visualize the effect of rf pulses on the magnetization using the *rotating reference frame*. The actual NMR frequencies are determined by the magnetogyric ratios and the magnetic field strength as give by Equation (1.3). For protons in an 11.7-T magnet, the natural frequency is 500 MHz, and to understand the effects of pulses and chemical shift evolution it is necessary to understand how the high-frequency radiation interacts with the magnetic moments. The idea behind the rotating reference frame is to consider the effect of rf pulses and the chemical shift evolution from the perspective of a spin in the magnetic field.

The so-called laboratory frame of reference defined by the magnet and the rf coils inside the NMR probe. In the laboratory frame, the z axis is along the magnetic field direction and the x and y axes are perpendicular. At equilibrium the spin precesses about the magnetic field direction at a rate given by Equation (1.3) (500 MHz for protons), and the effect of rf pulses on the magnetization can be understood by considering how an rf pulse oscillating at 500 MHz interacts with a vector that is precessing at 500 MHz.

The response of the magnetization to rf pulses is more easily visualized in a reference frame that rotates at the resonance frequency of the nuclei of interest. In such a moving reference frame the spin appears stationary and the effects of pulses and chemical shift evolution can be more easily visualized. The moving reference frame is illustrated in Figure 1.5 for the example of a person observing buses from two perspectives. To a person standing by the road, the busses appear to be moving very fast (40 and 50 mph in the laboratory frame). However, when we place the observer on one bus (the moving reference frame), the other bus appears to be moving slowly.

This same idea can be applied to the precessing moments in a magnetic field. In the laboratory reference frame (Figure 1.6a) the precession is observed at the resonance frequency. However, if the observer is moving with a rotational frequency of 500 MHz, the magnetization now appears stationary. The rotating reference frame is a much better way to visualize the effect of rf pulse, scalar coupling, and chemical shift evolution and will be used in the remainder of this book.

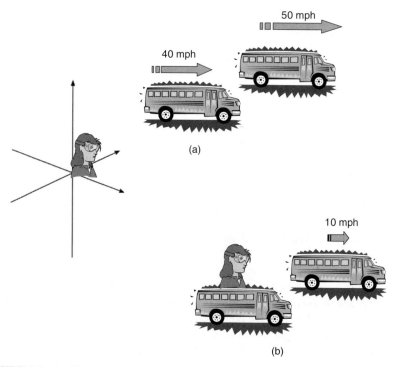

FIGURE 1.5 An illustration of the rotating reference frame showing the difference in perspective of a person standing on the road as a bus passes vs. an observer on the bus.

The rotating reference frame is particularly useful for visualizing the effect of rf pulses on the magnetization. In the rotating frame we can visualize rf pulses as a torque applied to the magnetization. This makes it easy to visualize rf pulses, such as the so-called 90° pulse that rotates the magnetization from along the z axis into the xy plane. It is also easier to visualize the effects of small differences in chemical

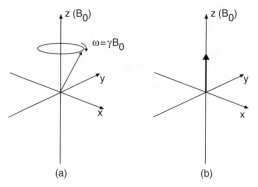

FIGURE 1.6 The rotating reference frame showing the precessing magnetic moment in the (a) laboratory frame and (b) the rotating reference frame.

1.2 BASIC PRINCIPLES OF NMR

shifts, such as those that arise from differences in chemical structure for the same nuclei, or from the effects of through-bond scalar couplings (Section 1.4).

1.2.4 The Bloch Equations

To get a better understanding of the behavior of nuclear spins and the effects of external fields, it is instructional to use the approach of Bloch (7) and consider the assembly of nuclei in macroscopic terms. In this treatment we define M as the total of the magnetic moments for all of the nuclei. We also assume that $I = 1/2$ and that the direction of the static field is along the z axis. Under these conditions, the magnetization precesses around the z axis as shown in Figure 1.2. At equilibrium the projection along the z axis M_z is constant and is given by

$$\frac{dM_z}{dt} = 0 \tag{1.6}$$

The magnitudes of the x and y components will vary with time and are given by

$$\frac{dM_x}{dt} = \gamma M_y \cdot B_0 = \omega_0 M_y \tag{1.7}$$

and

$$\frac{dM_y}{dt} = -\gamma M_x \cdot B_0 = -\omega_0 M_x \tag{1.8}$$

In addition to the static field, B_0, we also consider a much smaller field, B_1 (the rf pulse), that is applied along the x or y axis. The field B_1 rotates the magnetization away from the z axis. This field is not constant, but is usually applied for a brief period of time (a few μs). The effect of this B_1 field can be described as the tipping of magnetic vectors by a magnetic field and gives

$$\frac{dM_x}{dt} = \gamma \left[M_y B_0 - M_z (B_1)_y \right] \tag{1.9}$$

$$\frac{dM_y}{dt} = -\gamma \left[M_x B_0 + M_z (B_1)_x \right] \tag{1.10}$$

$$\frac{dM_z}{dt} = \gamma \left[M_x (B_1)_y - M_y (B_1)_x \right] \tag{1.11}$$

where $(B_1)_x$ and $(B_1)_y$ are the components of B_1 along the x and y axes, and are given by

$$(B_1)_x = B_1 \cos(\omega t) \tag{1.12}$$

and

$$(B_1)_y = -B_1 \sin(\omega t) \tag{1.13}$$

The spin system is perturbed from equilibrium by the application of the B_1 field, and relaxes along the x, y, and z axes. The relaxation of the z component is given by

$$\frac{dM_z}{dt} = \frac{M_z - M_0}{T_1} \tag{1.14}$$

where M_0 is the equilibrium magnetization and T_1 is the spin-lattice relaxation time. The transverse relaxation is given by

$$\frac{dM_x}{dt} = \frac{M_x}{T_2} \tag{1.15}$$

and

$$\frac{dM_y}{dt} = \frac{M_y}{T_2} \tag{1.16}$$

where T_2 is the spin-spin relaxation time. It is important to note that the T_1 relaxation restores the magnetization along the z axis, while the transverse magnetization decays to zero. Combining Equations (1.6)–(1.8) and Equations (1.9)–(1.16), we obtain the complete Bloch equations

$$\frac{dM_x}{dt} = \gamma[M_y B_0 - M_z B_1 \sin(\omega t)] - \frac{M_x}{T_2} \tag{1.17}$$

$$\frac{dM_y}{dt} = -\gamma[M_x B_0 - M_z B_1 \cos(\omega t)] - \frac{M_y}{T_2} \tag{1.18}$$

and

$$\frac{dM_z}{dt} = \gamma[M_x B_1 \sin(\omega t) + M_y B_1 \cos(\omega t)] - \frac{M_z - M_0}{T_1} \tag{1.19}$$

that describe the chemical shift evolution and relaxation that follows the application of an rf pulse. We can gain a clearer insight into the behavior of the spin system by changing to the rotating reference frame where we resolve the projections along the x and y axes as u and v. This gives

$$M_x = u\cos(\omega t) - v\sin(\omega t) \tag{1.20}$$

and

$$M_y = u\sin(\omega t) - v\cos(\omega t) \tag{1.21}$$

1.2 BASIC PRINCIPLES OF NMR

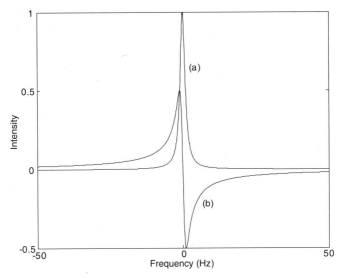

FIGURE 1.7 The (a) absorption and (b) dispersion lineshapes calculated from the Bloch equations.

Under conditions where $\gamma^2 B_1^2 T_1 T_2$ is very small, we can use Equations (1.17)–(1.19), and (1.20) and (1.21) to solve for u and v and obtain

$$u = M_0 \frac{\gamma B_1 T_2^2 (\omega_0 - \omega)}{1 + T_2^2 (\omega_0 - \omega)^2} \tag{1.22}$$

and

$$v = -M_0 \frac{\gamma B_1 T_2^2}{1 + T_2^2 (\omega_0 - \omega)^2} \tag{1.23}$$

where ω is the frequency and ω_0 is the exact resonance. These equations are of particular importance because they show the lineshapes for the signals detected along the x and y axes in the rotating frame, which closely parallels how the actual experiments are performed. The u and v lineshapes correspond to the dispersion and absorption lineshapes shown in Figure 1.7. For the highest resolution we measure the absorption-phase lineshapes. The linewidth $\Delta v_{1/2}$ in the absorption-phase signal is inversely related to the T_2 relaxation time as given by Equation (1.5). Therefore, the longer the T_2, the narrower the linewidth and the better the resolution.

1.2.5 Pulsed NMR

The NMR signals in the earliest studies were detected by sweeping the magnetic field and monitoring a change in the absorption of rf energy. It is much more efficient to keep the magnetic field constant and to sample the NMR signals with rf pulses. These pulses are rf waves oscillating at the resonance frequency of the nuclei and tip the magnetic moments from along the z axis into the xy plane for the observation of NMR signals. We can do experiments at different frequencies to observe the proton, carbon, silicon, or other nuclei in polymers. More complex information can be extracted from a series of pulses and delays. Simple pulse sequences can, for example, be used to measure the spin-lattice and spin-spin relaxation times. More complex pulse sequences are used in multidimensional NMR to provide information about the spin system, including information about the structure and dynamics.

As noted in Section 1.2.4, pulses can be applied to perturb the spin system from its equilibrium state. The effect that the pulses have depends on the length of the pulse, the strength of the B_1 field, the frequency, and phase of the pulse. In most instances, the lengths of the pulses are on the order of μs.

The amount that the magnetization is rotated by the B_1 field is called the tip angle θ and is given by

$$\theta = \gamma B_1 t_w \qquad (1.24)$$

where t_w is the pulse length.

One very useful pulse is the so-called 90° (or $\pi/2$) pulse that rotates the magnetization by 90° in the rotating-frame coordinate system. The exact effect of this pulse depends on the state of the spin system before the pulse is applied and the phase of the pulse. If the B_1 field is applied along the x axis to the magnetization at equilibrium, the magnetization is rotated from the z axis to the y axis, as shown in Figure 1.8a. The signal then can be observed by measuring the signal along the y axis. The signal is observed in the absence of applied rf fields and is called the *free induction decay* (FID).

We could also perturb the equilibrium magnetization by applying a pulse along any of the other axes, and Figure 1.8b shows the application of a 90° pulse along the

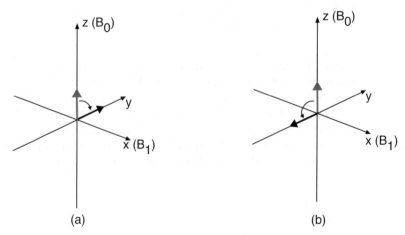

FIGURE 1.8 The effect of 90° rf pulses on the equilibrium magnetization for (a) a $90°_x$ pulse and a $90°_{-x}$ pulse.

1.2 BASIC PRINCIPLES OF NMR

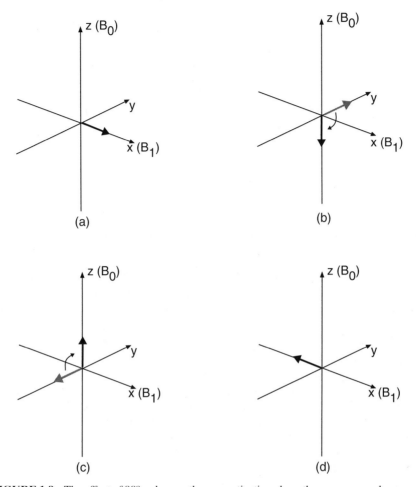

FIGURE 1.9 The effect of 90°_x pulses on the magnetization along the x, y, $-y$, and $-x$ axes.

$-x$ axis. The signal after the 90°_{-x} pulse is opposite in phase of that following the 90°_x pulse, and if the signals from the two experiments were added they would cancel. In a similar way, we can apply pulses along the y and $-y$ axes.

As noted in Section 1.2.4, the effect of rf pulses on the magnetization follows the mathematics of vectors. This means we can use simple mathematics to understand the effects of pulses from different phases for any given starting state. This is illustrated in Figure 1.9, which shows the effect of 90°_x pulses on magnetization starting along the x, y, $-y$ and $-x$ axes. For the y and $-y$ cases, the magnetization is rotated by $90°$ along the desired axis. Note that if we apply a 90°_x pulse to magnetization along the $-y$ direction, it is returned to its equilibrium position along the z axis. Also note that the application of a 90°_x pulse to magnetization along the x or $-x$ axis has no effect, since the rf field must be orthogonal to the magnetization to have any effect, as expected from the mathematics of vectors.

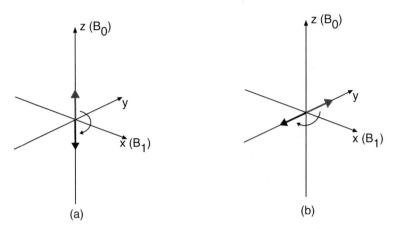

FIGURE 1.10 The effect of a 180_x° pulse on (a) equilibrium z-magnetization and (b) magnetization along the y axis.

One of the other more useful pulses is the 180° (or π) pulse. Figure 1.10 shows the application of a 180_x° pulse to magnetization along the z and y axes. The effect of the 180_x° to the z magnetization is to rotate it 180° to the $-z$ axis. The magnetization has been inverted and the 180° pulse is very useful for measuring the spin-lattice relaxation times. Application of the 180_x° pulse to magnetization along the y axis rotates it to along the $-y$ axis. This pulse is very useful for measuring the spin-spin relaxation time.

In the discussion thus far we have assumed that the rf pulses were applied exactly on resonance. One of the reasons that NMR is such a useful method is that the resonant frequencies (or chemical shifts) are very sensitive to the chemical structure and conformation. Thus, it may not be possible to apply pulses that are on resonance to all of the nuclei, and we must consider the *excitation profile* of the pulses.

The excitation profile, or tip angle as a function of frequency, excited by a pulse depends on the rf power, offset and the pulse length. The excitation bandwidth is inversely proportional to the pulse length, so short pulses, high-powered pulses give the broadest excitation. The relationship between the pulse length and the excitation profile is illustrated in Figure 1.11. The shape of the time domain rf signal and the

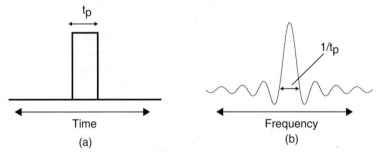

FIGURE 1.11 The relationship between the (a) pulse shape and (b) the frequency excitation profile.

1.2 BASIC PRINCIPLES OF NMR

frequency spectrum are approximately related by the Fourier transform (FT). The FT of a square wave is the sinc $(\sin(x)/x)$ function shown in Figure 1.11b. The bandwidth is the frequency difference between lobes of the sinc function, which is approximately $1/2t_p$. To get uniform excitation, this value should be much larger than the range of chemical shifts. In proton NMR, for example, the 90° pulse widths are often on the order of 10 μs, giving a bandwidth of 50 kHz. This is much larger than the chemical shift dispersion of 5 kHz observed on an 11.7-T magnet.

For nuclei with a large range of chemical shifts, such as carbon, it is often not possible to use such high power that the bandwidth for the 90° pulse is much larger than the chemical shift dispersion. This is important for quantitative applications, because the tip angle (and signal intensity) can depend on the frequency offset from the center of the spectrum. In such cases a shorter pulse (with a tip angle of 30° or 45°) can be used. The signal intensities following the shorter pulse would be smaller, but the signals would be uniformly excited.

For some applications, such as relaxation and multidimensional (nD) NMR experiments, it is important to give uniform 90° and 180° pulses to all of the spins. In those cases we often use *compensated* pulses that give a more uniform excitation profile (8). In other cases we would like to use *selective excitation* to only excite some of the spins in the system. In those cases we can use long pulses (ms) to excite only those spins near the center of the spectrum. This also can be accomplished with *shaped pulses* rather than square-wave pulses (8).

There are many advantages to pulsed NMR, including the ability to easily prepare the spin system in a nonequilibrium state by applying a series of pulses, and the ability to signal average to increase the signal-to-noise ratio (SNR). Many NMR signals are too weak to be detected in a single acquisition, and the SNR can be increased by adding the results from many acquisitions. The SNR from acquiring n scans is related to the SNR after one scan by

$$\left(\frac{S}{N}\right)_n = \sqrt{n} \left(\frac{S}{N}\right)_1 \tag{1.25}$$

It can be seen from Equation (1.25) that the SNR increases as the $\sqrt{n}$.

1.2.6 The Fourier Transform

The preceding section shows that there are many advantages to using pulsed NMR. The signals from pulse NMR experiments are recorded as free induction decays, which is a plot of voltage as a function of time. These data are difficult to interpret, and it is much more informative to view the frequency spectrum. The widespread use of NMR is due in part to the FT, a mathematical procedure that converts the time domain signals (FIDs) into the frequency domain spectrum.

The FT was developed by Joseph Fourier (1768–1830). His insight was that any function $y = f(x)$ could be represented as a series of the form

$$y = \frac{a_0}{2} + \sum_{n=1}^{\infty} a_n \cos(nx) + \sum_{n=1}^{\infty} b_n \sin(nx) \tag{1.26}$$

which is now known as the Fourier series. This is very useful because the NMR signal is of the form $v = f(t)$, where v is the voltage. The equation that we use to transform the NMR data is

$$F(\omega) = \int_{-\infty}^{+\infty} f(t) e^{-i\omega t} dt \qquad (1.27)$$

where $f(t)$ is the time domain data (the FID), $F(\omega)$ is the frequency spectrum, and

$$e^{-i\omega t} = \cos(\omega t) - i \sin(\omega t) \qquad (1.28)$$

The magic of the FT is that it transforms a very difficult to interpret time domain signal into an easier to interpret plot of peaks as a function of frequency. One of the major advantages of the FT is that it is a linear transform, and the amplitude of the sine and cosine waves contained in the FID are directly related to the peak intensities in the frequency spectrum. It is this property that allows us to use NMR as a quantitative method. The transform is reversible, and the frequency spectrum can be converted to the FID using the inverse FT.

The FT is implemented on computers using the so-called *fast fourier transform* (FFT) algorithm introduced by Cooley and Tukey in 1965 (9). This algorithm is most efficient if performed on 2^n data points, and data sets of this size (1024, 2048, 4096, etc.) are commonly recorded in an NMR experiment.

Figure 1.12 shows the FID and the transformed spectrum. For a multispin system the FID contains a number of frequency components that decay from spin-spin relaxation during the acquisition time. Each of these components has its own amplitude, frequency, and spin-spin relaxation time in the FID. Upon FT, each frequency component corresponds to a peak at a specific frequency and amplitude and its own linewidth.

1.2.7 The Product Operator Formalism

The time-dependent behavior of spins during the pulses and delays of a pulse sequence follows the dictates of quantum mechanics and, at least for relatively small collections of spins, can be calculated explicitly. Although NMR is one of the most concrete examples of the applications of quantum mechanics, the results are sometimes difficult to appreciate using the density matrix approach. In simple cases the vector picture often leads to a better understanding of the spin physics, but the vectors are often difficult to visualize for collections of coupled spins. A greater insight into the effects of the pulses and delays in complex, multispin systems can be gained by using the *product operator formalism* (POF) (10). The simple rules governing the POF allow us to understand exactly how the pulses, their phases, and the delays between pulses affects the spins. In some nD NMR experiments we generate multiple-quantum coherences that cannot be directly observed, but they can be recognized in the POF formalism, and it becomes obvious how pulses and delays affect them. In two-dimensional (2D)

1.2 BASIC PRINCIPLES OF NMR

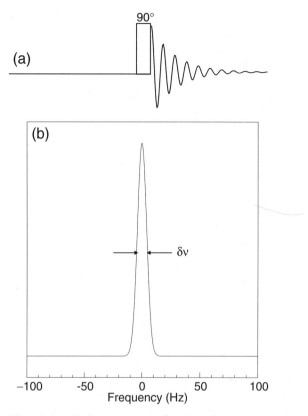

FIGURE 1.12 The relationship between (a) the free induction decay and (b) the signal.

correlated spectroscopy (COSY) we use the POF to understand how magnetization can be transferred between coupled spins, something that is rather difficult to understand using vector diagrams. In addition, we will be able to recognize the product operators that are generated by certain pulse sequences and recognize the operators that are associated with certain types of peaks in the nD spectra.

In the following sections we give a brief overview of the POF. For a more complete discussion, the reader should see the original papers (10). The goal is not a complete treatment, but rather an introduction to the fundamentals, so that we can understand how pulse sequences work, and to recognize that the product operators are important for nD NMR.

In its simplest form, the POF uses two spins, I and S, to illustrate the effects of pulses and delays on the evolution of the product operators. The spins may be from different nuclei, such as carbons and protons, or the same nuclei (protons) with well-resolved chemical shift differences. The POF accounts for the forces (Hamiltonians) acting on the spins, including pulses, chemical shifts, couplings, and relaxation. The terms in POF can be either single operators, such as I_z (the magnetization of spin I along the z axis), or products of operators, such as $I_x S_z$.

There are six single-spin operators in the two-spin treatment of the POF, I_x, I_y, and I_z, and S_x, S_y, and S_z. The terms I_z and S_z represent magnetization for the two spins aligned along the $+z$ axis in the rotating frame. Experiments typically begin with these spins in their equilibrium state. The remaining terms correspond to magnetization for the two spins along the x and y axes. Since magnetization is typically detected along the y direction, I_y and S_y correspond to absorption phase magnetization, while I_x and S_x correspond to dispersive magnetization.

The utility of the POF can be illustrated with a simple one-pulse experiment for a single spin, where the signal is observed after a $90°_x$ pulse. Pulses act on the spin operators, as expected from the mathematics of vectors. As we saw earlier, a $90°_x$ pulse rotates magnetization from along the z axis to the y axis. This can be written as

$$M_z \xrightarrow{90°_x} M_y \tag{1.29}$$

and the signal is detected along the y axis. Under the POF we show explicitly the effect of the pulse and the evolution of the chemical shift during the detection period. Starting again from equilibrium z magnetization, the pulse of tip angle θ is given by

$$I_z \xrightarrow{\theta I_x} I_z \cos\theta - I_y \sin\theta \tag{1.30}$$

which for $\theta = 90°$ gives only $-I_y$. Note that the sign convention in the POF is opposite that of the traditional vector picture, and a $90°_x$ pulse converts z magnetization into $-y$ magnetization. Equation (1.30) is valid for any pulse angle, and when $\theta = 180°$, z magnetization is converted into $-z$ magnetization, as expected for a $180°_x$ pulse.

The two-spin notation introduced here can be used to describe both homonuclear and heteronuclear NMR experiments. In homonuclear experiments the pulses are applied to both the I and S spins, while in heteronuclear experiments the pulses can be selectively applied to either the I or S spins, or to both.

As expected from the general principles of NMR, magnetization in the xy plane evolves under the influence of the chemical shifts. In the POF notation this is given by

$$I_y \xrightarrow{\omega t I_z} I_y \cos(\omega t) - I_x \sin(\omega t) \tag{1.31}$$

where ω is the resonance frequency. Both the I_y and I_x components are recorded and subjected to a complex FT to give a spectrum with a peak that is offset from the center of the spectrum by $\omega/2\pi$. It is important to remember that only the single-spin terms give rise to observable magnetization and that many complex terms could be generated by a pulse sequence that are simply not observable.

The POF is perhaps most useful for understanding the effects of spin-spin coupling in nD experiments. As discussed in more detain in Section 1.4, scalar or through-bond couplings split the signals into two or more peaks that are separated from each other in frequency by the spin coupling constant J. In the vector model the scalar coupling

1.2 BASIC PRINCIPLES OF NMR

causes a divergence of the signals at a rate of $\frac{1}{2} J_{IS}$ as the spin system evolves. In the POF this is represented by

$$I_y \xrightarrow{(\pi J_{IS}\tau)2I_z S_z} I_y \cos(\pi J_{IS}\tau) - 2I_x S_z \sin(\pi J_{IS}\tau) \quad (1.32)$$

where $(\pi J_{IS}\tau) 2I_z S_z$ is the spin-spin coupling operator. If the delay time is chosen such that $\tau = 1/(2J_{IS})$, then the first term is zero and the second term becomes $2I_x S_z$. These types of terms cannot be directly detected, but are important because they can evolve into observable magnetization. During a delay time the pair of coupled spins would evolve under the influence of both the chemical shift and the spin-spin coupling. This is accomplished by performing one operation followed by the other. Thus the 90° pulse and observation of a two-spin system with scalar coupling would be given by

$$I_z \xrightarrow{90° I_x} \xrightarrow{(\pi J_{IS}t)2I_z S_z} I_y \cos(\pi J_{IS}t) \cos(\omega t) - 2I_x S_x \sin(\pi J_{IS}t) \cos(\omega t)$$
$$- I_x \cos(\pi J_{IS}t) \sin(\omega t) + 2I_y S_z \sin(\pi J_{IS}t) \sin(\omega t) \quad (1.33)$$

While the algebra may at first sight appear complex, it contains a complete description of the spin system. For the purposes of understanding where the magnetization comes from in complex experiments, it is often sufficient to simplify Equation (1.33) to

$$I_z \xrightarrow{90° I_x} \xrightarrow{(\pi J_{IS}t)2I_z S_z} I_y - 2I_x S_x - I_x + 2I_y S_z \quad (1.34)$$

where it can now be easily observed that there are only four product operator terms. Two of these terms are observable magnetization and two have the potential to be converted into observable magnetization with further pulses and delays.

In the majority of solution NMR experiments, we are concerned with only three things: the effects of pulses, the evolution under the chemical shifts, and the effects of spin-spin coupling. We evaluate pulse sequences by sequentially calculating the effects of pulses and delays. The effects of the different operations are summarized in Table 1.2 for a two-spin system. The left-hand column contains the spin operators and the top row shows the operations that are performed on these operators. This table easily shows that a $90°_x$ pulse (I_x) operates on z magnetization (I_z) to generate magnetization along the $-y$ axis ($-I_y$), and that evolution under the coupling operator ($2I_z S_z$) of $2I_x S_z$ magnetization generates an observable signal (I_y). For a full calculation, the sine and cosine terms must be added as previously described. During delay periods, the spins evolve under the influence of both the chemical shift (I_z) and coupling ($2I_z S_z$) operators and the product operators can be applied in either order. The effect of pulses on the two-spin operators follows the rules of single-spin operators, and the application of a pulse to the I spins only affects the I spins, such as

$$2I_z S_x \xrightarrow{I_x} -2I_y S_x \quad (1.35)$$

TABLE 1.2 The Effect of Various Spin Operators on the Spin System.

	I_x	I_y	I_z	S_x	S_y	S_z	$2I_zS_z$
I_x	$E/2$	$-I_z$	I_y	$E/2$	$E/2$	$E/2$	$2I_yS_z$
I_y	I_z	$E/2$	$-I_x$	$E/2$	$E/2$	$E/2$	$-2I_xS_z$
I_z	$-I_y$	I_x	$E/2$	$E/2$	$E/2$	$E/2$	$E/2$
S_x	$E/2$	$E/2$	$E/2$	$E/2$	$-S_z$	S_y	$2I_zS_y$
S_y	$E/2$	$E/2$	$E/2$	S_z	$E/2$	$-S_x$	$-2I_zS_x$
S_z	$E/2$	$E/2$	$E/2$	$-S_y$	S_x	$E/2$	$E/2$
$2I_zS_z$	$-2I_yS_z$	$2I_xS_z$	$E/2$	$-2I_zS_y$	$2I_zS_x$	$E/2$	$E/2$
$2I_xS_z$	$E/2$	$-2I_zS_z$	$2I_yS_z$	$-2I_xS_y$	$2I_xS_x$	$E/2$	I_y
$2I_yS_z$	$2I_zS_z$	$E/2$	$-2I_xS_z$	$-2I_yS_y$	$2I_yS_x$	$E/2$	$-I_x$
$2I_zS_x$	$-2I_yS_x$	$2I_xS_x$	$E/2$	$E/2$	$-2I_zS_z$	$-2I_zS_y$	S_y
$2I_zS_y$	$-2I_yS_y$	$2I_xS_y$	$E/2$	$2I_zS_z$	$E/2$	$-2I_zS_x$	$-S_x$
$2I_xS_x$	$E/2$	$-2I_zS_x$	$2I_xS_y$	$E/2$	$-2I_xS_z$	$2I_xS_y$	$E/2$
$2I_xS_y$	$E/2$	$-2I_zS_y$	$2I_yS_y$	$2I_xS_z$	$E/2$	$-2I_xS_x$	$E/2$
$2I_yS_x$	$2I_zS_x$	$E/2$	$-2I_xS_x$	$E/2$	$-2I_yS_z$	$2I_yS_y$	$E/2$
$2I_yS_y$	$2I_zS_y$	$E/2$	$-2I_xS_y$	$2I_yS_z$	$E/2$	$-2I_yS_x$	$E/2$

The unity matrix is E.
Note: The spin operators (top) operate on the spin state (left column) to produce new spin states.

1.3 CHEMICAL SHIFTS AND POLYMER STRUCTURE

In the discussions thus far, we have treated the chemical shift as a single value, when in reality the chemical shifts depend on the local magnetic environment. From Equation (1.3) we saw that the resonance frequency depends on the magnetogyric ratio and the static magnetic field. Thus, for a given magnet we would expect to observe different frequencies for different atoms with different magnetogyric ratios. In an 11.7-T magnet, for example, the carbons and protons would be observed at 125 and 500 MHz, a very large and easily observable difference. However, it is also true that the field around a nucleus is affected by the local magnetic environment in a way that depends on the electronic structure and conformation. The local magnetic field around a nuclei will be given by

$$B_{\text{loc}} = B_0 (1 - \sigma) \tag{1.36}$$

where σ is the chemical shift tensor that is related to the electron density ρ at distance r from the nucleus, as given by Lamb's equation

$$\sigma = \frac{4\pi e^2}{3mc^2} \int_{-\infty}^{\infty} r\rho(r)\,dr \tag{1.37}$$

There are three principal components to the chemical shift tensor, σ_{11}, σ_{22}, and σ_{33}, that are important for solid-state NMR studies. For solution studies chain motion

1.3 CHEMICAL SHIFTS AND POLYMER STRUCTURE

averages the chemical shift anisotropy and the isotropic value σ_{iso} is observed.

$$\sigma_{iso} = \frac{1}{3}(\sigma_{11} + \sigma_{22} + \sigma_{33}) \tag{1.38}$$

The NMR frequencies are reported in parts per million (ppm) relative to some reference compound, as

$$\delta = \frac{(\sigma_{ref} - \sigma_{sample})}{10^6} \tag{1.39}$$

To the extent that the differences in the local field are large enough that a peak will be shifted from its neighbors, we will observe different peaks for atoms with the same magnetogyric ratios, but with different magnetic environments. Much of the power of NMR derives from the fact that very subtle differences in structure, such as tacticity, can give rise to observable differences in the chemical shifts.

1.3.1 Molecular Structure and Chemical Shifts

The screening constant σ depends most strongly on the hybridization, and the peaks from aromatic atoms are often well resolved from aliphatic ones. Furthermore, the methine, methylene, and methyl signals are often well resolved from each other. The chemical shifts depend on the local electron density, so any substituents that donate or withdraw electrons can also influence the chemical shifts, as can hydrogen bonding and other types of intra- or intermolecular interactions. Electrons in π orbitals can also generate local magnetic fields that can affect the chemical shifts, depending on the intermolecular distance and orientation of the anisotropic group relative to the nuclei of interest. In addition, nuclei that undergo chemical exchange can experience large changes in their local magnetic environment that cause noticeable effects in the spectra. These chemical exchange effects depend both on the difference in chemical shift between the two environments and the rate of chemical exchange.

1.3.1.1 Chemical Structure Effects The factor leading to the largest change in chemical shifts is the chemical structure, where the difference in resonance frequencies are often much larger than the linewidths. The shift due to chemical structure is much larger for nuclei with a large range in chemical shifts (carbon, fluorine, silicon, or nitrogen) compared to those with a smaller shift range (protons). Figure 1.13 shows the effect of chemical structure on the carbon and proton chemical shifts for aromatic, methine, methylene, and methyl groups. In both cases methyl groups are the most shielded and appear at the lowest chemical shift (highest field). The methylene, methine, olifinic, and aromatic groups are more deshielded and appear at progressively larger chemical shifts. Note that each of the peaks can appear over a range of chemical shifts. This is because the chemical shifts are determined by other factors in addition to the chemical structure, and structural change as many as five atoms away can affect the local electron density and chemical shift. Other factors, such as branching, are also known to affect the chemical shifts, and the chemical shifts will be very sensitive to the degree of branching several atoms away. This makes the NMR spectrum very

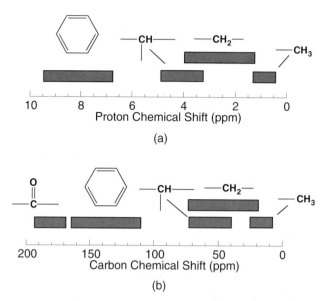

FIGURE 1.13 The effect of chemical structure on the chemical shifts for aromatic, olefinic, and aliphatic (a) protons and (b) carbons.

sensitive to gross chemical changes that accompany regioisomerism, where the methine groups in a vinyl polymer in a head-to-tail arrangement with two methylene neighbors can be easily resolved from a methine in a head-to-head configuration that has a methine and methylene neighbor.

The large range in chemical shift for the olefinic and aromatic groups is due to the fact that the electron density in the π electrons systems is very sensitive to substitution. This means that the introduction of strong electron donating or withdrawing groups leads to large (several ppm) chemical shift changes. These effects in aromatic groups can be very strong even when the substitution is made several atoms away. This leads to well-resolved spectra for compounds like chlorobenzene compared to benzene.

The chemical structure effects on the chemical shift lead to well-resolved spectra because the induced shifts from a change in structure are much larger than the polymer linewidths. The proton chemical shifts span a range of ca. 10 ppm and the difference in chemical shift for the aromatic and methyl protons is on the order of 6 ppm. In an 11.7 T field where the proton resonance frequency is 500 MHz, this corresponds to a frequency difference of 3000 Hz. Since the intrinsic linewidths are on the order of 1 Hz for many polymers, these two lines are very well resolved. The carbon chemical shifts span a much broader range (200 ppm) and the methyl and aromatic carbons appear at ca. 10 and 120 ppm. In this same magnetic field the carbon resonance frequency is 125 MHz and these signals are separated by 13,000 Hz. The effect of

1.3 CHEMICAL SHIFTS AND POLYMER STRUCTURE

chemical structure on the proton and carbon chemical shifts will be considered in detail in sections 1.3.1 and 1.3.2.

1.3.1.2 Inductive Effects The chemical shifts are influenced by changes in electron density that arise from the presence on nearby electron-donating and withdrawing groups. As may be expected, groups with the greatest electronegitivity have the largest effect on the chemical shift. Inductive effects are attenuated by the intervening bonds, but they can still be observable for groups with as many as five bonds removed. Other interactions that lead to a change in electron density, such as hydrogen bonding or ionic interactions, can also lead to large changes in chemical shift. For water-soluble polymers, the chemical shifts are very sensitive to the degree of ionization.

Electron-withdrawing groups draw electron density from nearby carbon and result in downfield shifts. This is illustrated in Figure 1.14, which shows a plot of the carbon chemical shift for the methylene and methyl carbons in a series of substituted pentanes. Fluorine is the most electronegative substituent and has the largest effect on the chemical shift. This plot shows that there is a clear correlation between the electronegativity and the induced shift. The effect is attenuated with an increasing number of bonds to the substituted carbon, and although the induced shifts for the methyl carbon are smaller than for the substituted methylene, they are still measurable.

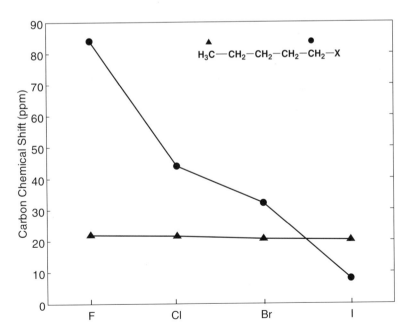

FIGURE 1.14 The effect of electronegativity on the chemical shifts for the methylene and methyl carbons in substituted pentanes. The chemical shifts are plotted for the substituted methylene (•) and methyl carbons (▲).

These effects are additive, so a carbon with two electron-withdrawing groups would be shifted further downfield than a carbon with a single substituent.

Hydrogen bonding can also affect the electron density of both the proton directly in the center of the hydrogen bond, and nearby carbon and nitrogen atoms. This was observed early on when Arnold and Packard (11) reported that the proton resonance for the hydroxyl protons in ethanol shifted upfield by 1.5 ppm between −117 and 78°C. There also are many reports of the effect of dilution on the chemical shifts of phenols and alcohols (12). The change in chemical shift is observed because the lower temperature or higher concentration promotes dimer formation, and a larger fraction of the molecules are hydrogen bonded. Hydrogen bonding withdraws electrons and shifts the proton signals downfield. In the absence of hydrogen bonding (at high temperature or low concentration), the hydroxyl protons appear around 1 ppm.

Although the protons in hydrogen bonds are central to the formation of hydrogen bonds, they are often difficult to study because they can exchange with the solvent, leading to shifting resonance positions and line broadening (Section 1.3.1.4). It is more common to study hydrogen bonding in polymers by monitoring the chemical shifts of the nearby carbon and nitrogen atoms. Carbonyl carbons in hydrogen bonding polymers can often be used to monitor hydrogen-bond formation, since hydrogen bonding can lead to chemical shift changes as large as 5 ppm. This change is large enough that the carbon chemical shifts can be used to monitor hydrogen-bond formation in solid polymers.

The ionization state also affects the electron density and gives rise to large changes in the NMR spectrum. The protons in carboxylic acids are extremely deshielded and often appear at chemical shifts below 15 ppm. These acidic protons can rapidly exchange with water and are often difficult to observe by proton NMR. The protonation can be indirectly observed via the carbon chemical shifts of nearby atoms. The carbon peak can change by 5–10 ppm with protonation, and the pK_a can be determined from a plot of the chemical shift vs. pH.

1.3.1.3 Anisotropic Shielding
The chemical shifts are sensitive not only to the chemical structure, but also to local fields from π electron clouds in aromatic, acetylinic, olefinic, and carbonyl groups. This effect is illustrated in Figure 1.15,

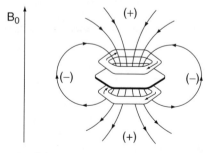

FIGURE 1.15 A schematic diagram of an aromatic ring with its π electron clouds in a magnetic field.

1.3 CHEMICAL SHIFTS AND POLYMER STRUCTURE

which shows a schematic drawing of a benzene ring with two doughnut-like π-electron clouds on each face of the ring. These circulating electron clouds produce short-range magnetic fields that can shift a resonance upfield or downfield, depending on the orientation of the ring relative to the nuclei of interest. An atom that is above the electron cloud is shifted upfield, while an atom in the plane of the ring experiences a downfield shift.

The magnitude of the induced shift depends both on the orientation relative to the group and the distance. This can be defined more quantitatively by considering the diamagnetic susceptibility for a group aligned parallel and perpendicular to the magnetic field direction. If a group is aligned with its principal axis along the field, then the observed magnetic susceptibility would be $\chi_\parallel$, and the magnetic susceptibility for a group orientated perpendicular to the field would be $\chi_\perp$. The induced shift is given by

$$\sigma_g = \frac{(\chi_\parallel - \chi_\perp)(1 - 3\cos^2\theta)}{3r^2} \tag{1.40}$$

This model assumes that the anisotropic group can be approximated as a point dipole and r is the distance between the point dipole and the atom of interest. The angle θ is defined by the line connecting the atom to the point dipole and the principal axis system. The induced shifts, sometimes called "ring current" shifts, are largest for protons. Figure 1.16 shows a plot of the ring current shifts for a proton in the

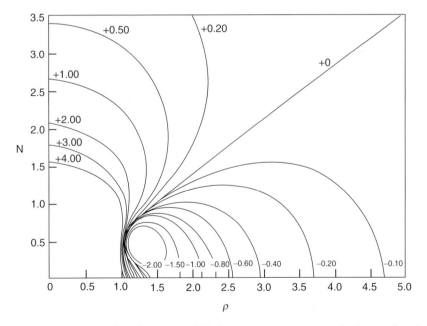

FIGURE 1.16 A plot of the ring current shifts for a proton near an aromatic ring as a function of distance and orientation.

vicinity of an aromatic ring. Note that those atoms in the plane of the benzene ring will be shifted downfield, while those above the ring will be shifted upfield. The principle of ring current shifts can be directly demonstrated with molecules such as 1,4-polymethylenebenzenes (13). In such a molecule the central methylenes are constrained to be directly above the benzene ring, and they are observed at 0.3 ppm rather than at the 1.3 ppm expected for methylenes in linear alkanes.

1,4 Decamethylene Benzene

Any π electron system, including carbonyls, acetylinic and olefinic groups, can generate induced shifts that depend on the distance and orientation of a nuclei relative to the π electron system. The induced shifts for carbonyl and acetylinc groups are usually smaller than for aromatic rings.

1.3.1.4 Chemical Exchange

The chemical exchange between sites with different magnetic environments can have a large effect on the appearance of the NMR spectrum. The spectral changes depend both on the magnitude of the chemical shift and the exchange rate. If the exchange is slow, then two separate peaks would be observed, while if the exchange is very fast, then a single peak would be observed midway between the chemical shifts for the two sites.

The effect of the exchange can be incorporated into the Bloch equations of Section 1.2.4, so that the peak positions and lineshapes can be calculated as a function of the chemical shift difference and the exchange rate. For this discussion, we will consider a two-spin system (A and B) with unequal chemical shifts ($\delta_A \neq \delta_B$), equal spin-spin relaxation rates ($T_{2A} = T_{2B} = T_2$), and exchange site lifetimes of τ_A and τ_B. Modification of the Bloch equations for two-site chemical exchange gives (14)

$$\frac{dM_{Ax}}{dt} = \gamma \left(M_{Ax} B_0 + M_{Az} B_1 \sin(\omega t)\right) - \frac{M_{Ax}}{T_2} - \frac{M_{Ax}}{\tau_A} + \frac{M_{Bx}}{\tau_B} \quad (1.41)$$

$$\frac{dM_{Ay}}{dt} = \gamma \left(M_{Az} B_1 \cos(\omega t) - M_{Ax} B_0\right) - \frac{M_{Ay}}{T_2} - \frac{M_{Ay}}{\tau_A} + \frac{M_{By}}{\tau_B} \quad (1.42)$$

$$\frac{dM_{Bx}}{dt} = \gamma (M_{By} B_0 + M_{Bz} B_1 \sin(\omega t)) - \frac{M_{Bx}}{T_2} - \frac{M_{Bx}}{\tau_B} + \frac{M_{Ax}}{\tau_A} \quad (1.43)$$

and

$$\frac{dM_{By}}{dt} = \gamma \left(M_{Bz} B_1 \cos(\omega t) - M_{Bx} B_1\right) - \frac{M_{By}}{T_2} - \frac{M_{By}}{\tau_B} + \frac{M_{Ay}}{\tau_A} \quad (1.44)$$

1.3 CHEMICAL SHIFTS AND POLYMER STRUCTURE

The general solution to the modified Bloch (or McConnell) equations is

$$f(\omega) = \frac{1}{2}\gamma M_0 \frac{\left(1 + \tau/T_2\right) + QR}{P^2 + R^2} \tag{1.45}$$

where

$$\tau = \frac{\tau_A \tau_B}{(\tau_A + \tau_B)} \tag{1.46}$$

$$P = \frac{\tau}{T_2^2} - \left[\frac{1}{2}(\omega_A + \omega_B) - \omega\right]^2 + \frac{1}{4}(\omega_A - \omega_B)^2 - \frac{1}{T_2} \tag{1.47}$$

$$Q = \tau\left[\frac{1}{2}(\omega_A + \omega_B) - \omega - \frac{1}{2}(p_A - p_B)(\omega_A - \omega_B)\right] \tag{1.48}$$

$$R = \left[\frac{1}{2}(\omega_A + \omega_B) - \omega\right]\left(1 + \frac{2\tau}{T_2}\right) + \frac{1}{2}(p_A - p_B)(\omega_A - \omega_B) \tag{1.49}$$

where p_A and p_B are the fractional populations of sites A and B.

Figure 1.17 shows the effect of exchange rate on a two-spin system where the difference in resonance frequency for the two peaks is 10 Hz. Exchange has a large effect on the spectral appearance and depends on the exchange rate relative to the inverse of the peak separation. The three chemical exchange regimes that are commonly discussed are the fast ($1/\tau \gg \Delta\upsilon$), intermediate ($1/\tau \approx \Delta\upsilon$), and slow ($1/\tau \ll \Delta\upsilon$) exchange regimes.

In the slow exchange regime, the exchange is slow compared to the peak separation and two separate peaks are observed, and as the exchange rate increases, broadening is observed. For the special case of an equally populated site ($p_A = p_B$), the linewidth in the slow exchange regime is given by

$$\frac{1}{T_{2\text{obs}}} = \frac{1}{T_2} + k \tag{1.50}$$

Thus the first effects of exchange are line broadening when the exchange rate becomes comparable to $1/T_2$.

Broad spectral features are observed in the intermediate exchange regime where the line broadening is almost as large as the separation between the peaks. If the peak separation is large, then very broad peaks will be observed. In some cases these peaks become so broad that they are not easily observed in the high-resolution spectrum.

The peaks become sharper in the fast exchange limit, and the peak frequency is an average of the peak frequencies for A and B. For two-site exchange with equal populations the spin-spin relaxation rate is given by

$$\frac{1}{T_{2\text{obs}}} = \frac{1}{T_2} + \frac{(\omega_A - \omega_B)}{8k} \tag{1.51}$$

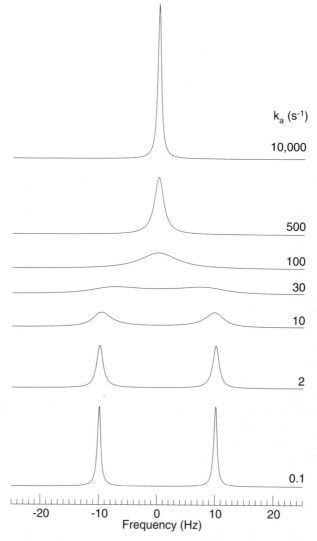

FIGURE 1.17 The effect of chemical exchange on the appearance of the spectra as a function of the exchange rate.

Chemical exchange is most often studied by recording the spectrum as a function of temperature. Exchange processes usually have high activation energies, so large changes are observed as a function of temperature. The frequencies and lineshapes can be measured as a function of temperature to extract the exchange rates, activation energies, and site populations. The line broadening observed in the intermediate regime is sometimes a nuisance, and the temperature can be changed to obtain a spectrum in the fast or slow exchange limit. Exchange rates that are too slow to cause noticeable changes in the linewidths and can be studied by 2D exchange NMR (15).

1.3 CHEMICAL SHIFTS AND POLYMER STRUCTURE

The concepts behind chemical exchange are important for understanding other NMR phenomena, such as decoupling and magic-angle sample spinning. In both cases the idea is to cause the spin system to be at the fast exchange limit, either by applying pulses to rapidly rotate the magnetization or by spinning the sample, so that sharp lines are observed.

1.3.2 Proton Chemical Shifts

Protons are the most widely studied nuclei because they are ubiquitous and they have a high sensitivity. The range of chemical shifts for most organic compounds is around 10 ppm, and it is only protons attached to strongly electron-withdrawing groups that are observed at a lower field. In an 11.7-T magnet the proton frequency is 500 MHz, so the proton signals are spread over 5 kHz. Mobile polymers may have homogeneous proton linewidths on the order of 1 Hz, so very high-resolution spectra can be observed.

As with all chemical shifts, the proton chemical shifts depend on the hybridization and the presence of nearby electron-donating and -withdrawing groups. Figure 1.18 shows a plot of the proton chemical shifts for a variety of chemical structures. The highest field peaks (0 ppm) are from methyl groups attached to silicon atoms. Methyl groups bonded to carbons typically resonante between 0.5 and 1.5 ppm, while methyl groups attached to nitrogens appear at a lower field (2.1–2.9 ppm). Methylene protons are observed between 1.5 and 2.5 ppm, while the methine protons appear between 4 and 5 ppm. Olefinic protons appear in the 5–6-ppm range, while aromatic protons are observed in the 6.5–8-ppm range. Aldehydes appear near 10 ppm, and very acidic protons can appear at a lower field (12–15 ppm). The frequencies and linewidths

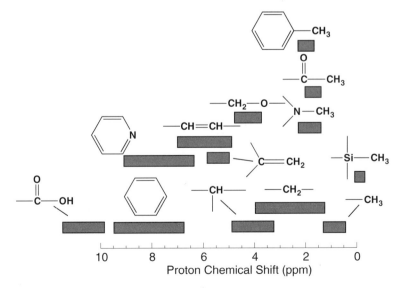

FIGURE 1.18 A plot of proton chemical shifts for various structures.

for the exchangeable protons depends on the rate of chemical exchange between sites.

1.3.3 Carbon Chemical Shifts

Carbon NMR is often used to study polymers in part because of the large chemical shift range and resolution. Although ^{13}C is not very sensitive compared to protons, the carbon spectrum is spread over a much larger range, so there is a greater chance that the carbon spectrum will be resolved. In an 11.7-T magnet the carbon frequency is 125 MHz, and the chemical shift range of 200 ppm corresponds to 25 kHz. Since the linewidths for mobile carbons can be as low as 1 Hz, there is a good chance that the peaks from rather subtle structural features can be resolved. Figure 1.19 shows a plot of the chemical shifts expected for different chemical structures. The relatively broad range of chemical shifts for each structure is an indication that the shifts are extremely sensitive to the local structure.

Carbon-13 chemical shifts are also of interest in polymer studies because they are very sensitive to molecular structure and conformation, quite apart from the influence of chemical structure and substituent groups. This is illustrated by comparing the ^{13}C and ^{1}H spectra of paraffinic hydrocarbons, a class to which many important polymers belong. The paraffinic signals are dispersed over a range of only about 2 ppm in the proton spectrum, while in the carbon chemical spectrum they are spread over is more than 40 ppm. Since the linewidths for the carbons and protons are roughly comparable, there is a much better chance that structural features will be better resolved in the carbon spectrum.

This sensitivity of the carbon chemical shift to the structure and conformation makes ^{13}C spectroscopy one of the most powerful tools for studying polymers. The

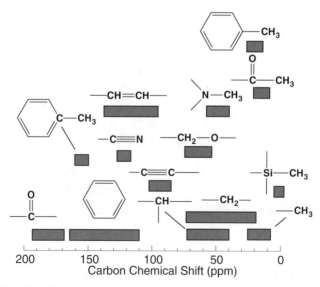

FIGURE 1.19 A plot of the carbon chemical shifts for various structures.

1.3 CHEMICAL SHIFTS AND POLYMER STRUCTURE

TABLE 1.3 The α-Effect on the ^{13}C Chemical Shifts.

Structure	δ_C (ppm)	α Effect (ppm)
CH$_3$—H	−2.1	—
CH$_3$—CH$_3$	5.9	8.0
CH$_2$—(CH$_3$)$_2$	16.1	10.2
CH—(CH$_3$)$_3$	25.2	9.1
C—(CH$_3$)$_3$	27.9	2.7

correlation between carbon chemical shift and molecular structure has been extensively investigated and empirical correlations between the structure and chemical shift have been reported (16). One important class of correlations is for hydrocarbons, where a correlation has been observed between the chemical shift and the hydrocarbon structure. The so-called α, β, and γ effects lead to large changes in chemical shift for structures separated by one, two, and three bonds from the carbon of interest. Correlations have also been reported for structures removed by four or five bonds (the δ- and ε-effects), but these are smaller effects. Quantitative correlations also exist for substituted hydrocarbons as well as those with olefinic or aromatic groups (16).

The effect of the next-nearest-neighbor structure on the chemical shift in hydrocarbons (the α-effect) is illustrated in Table 1.3. In this table we compare the carbon chemical shift for methane through neopentane, where the chemical shift changes from −1.2 ppm to 27.9 ppm. For each added methyl carbon we see a progressively downfield shift, giving a deshielding of about 9 ± 1 ppm for each added carbon, except in neopentane, where crowding apparently reduces the effect.

Table 1.4 shows the so-called β-effect for hydrocarbons. As with the α-effect, the shifts from structures two bonds removed from the carbon of interest can be very large, sometimes as large as 9.7 ppm. Comparison of the substituted ethanes, propanes, and butanes shows that the magnitude of the β-effect depends strongly on the chemical structures. Again, steric crowding in highly substituted structures reduces the magnitude of the induced shift.

Finally, we consider the so-called γ effect (Table 1.5), which, although smaller than α- and β-effects, is of particular interest for polymers, because it depends both

TABLE 1.4 The β-Effect on ^{13}C Chemical Shifts.

Structure	δ_C (ppm)	β Effect (ppm)
CH$_3$—CH$_3$	5.9	—
CH$_3$—CH$_2$—CH$_3$	15.6	9.7
CH$_3$—CH—(CH$_3$)$_2$	24.3	8.7
CH$_3$—C—(CH$_3$)$_3$	31.5	7.2
CH$_3$—**CH$_2$**—CH$_3$	16.1	—
CH$_3$—**CH$_2$**—CH$_2$—CH$_3$	25.0	8.9
CH$_3$—**CH$_2$**—CH—(CH$_3$)$_2$	31.8	6.8
CH$_3$—**CH$_2$**—C—(CH$_3$)$_3$	36.7	4.9

TABLE 1.5 The γ-Effect on the ^{13}C Chemical Shifts.

Structure	δ_C (ppm)	β Effect (ppm)
CH$_3$—CH$_2$—CH$_3$	15.6	—
CH$_3$—CH$_2$—CH$_2$—CH$_3$	13.2	−2.4
CH$_3$—CH$_2$—CH—(CH$_3$)$_2$	11.3	−1.9
CH$_3$—CH$_2$—C—(CH3)$_3$	8.8	−2.5
CH$_3$—**CH$_2$**—CH$_2$—CH$_3$	25.0	—
CH$_3$—**CH$_2$**—CH$_2$-CH$_2$—CH$_3$	22.6	−2.4
CH$_3$—**CH$_2$**—CH$_2$-CH—(CH$_3$)$_2$	20.7	−1.9
CH$_3$—**CH$_2$**—CH$_2$—C—(CH$_3$)$_3$	18.8	−1.9

on the structure and conformation (5). The data in Table 1.5 show that the shifts from the γ-effect are smaller in magnitude and in the opposite direction of the α- and β-effects. The γ-effect are also found for elements other than carbon, which are three bonds removed from the observed carbon, and to show a correlation with the electronegativity of this element (*vide infra*).

For a better understanding of the γ-effect (or the *γ-gauche* effect), consider the staggered conformers of propane, butane, and isopentane shown in Figure 1.20. Propane has no carbons in the γ position and does not experience a *γ-gauche* effect. It has been experimentally observed that the carbon of interest in butane appears

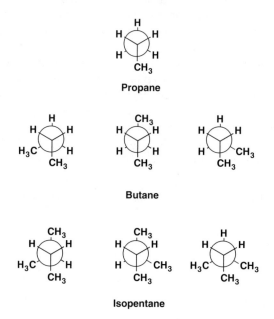

FIGURE 1.20 A schematic drawing of the conformers of propane, butane, and isopentane, showing the number of *gauche* interactions.

1.3 CHEMICAL SHIFTS AND POLYMER STRUCTURE 33

at 2.3 ppm to higher field than expected. The explanation for the induced shift is that it arises from through-space interaction with the γ neighboring carbon in the *gauche* conformation. Two of the three butane conformers has a γ-*gauche* interaction, so if we divide the induced shift by the *gauche* conformer content (ca. 0.45, from energy calculations), we calculate a γ-*gauche* effect of -5.3 ppm per *gauche* interaction. In isopentane, two of the conformers have a γ-*gauche* interaction, while the third conformer has two. The induced chemical shift is obtained by multiplying the fraction of each conformer times the number of interactions and the value of -5.3 ppm. This treatment assumes that this shielding occurs only when the three-bond four-carbon system is in or near a *gauche* state. The magnitude of the shift depends on the dihedral angle and increases to about 8 ppm for the eclipsed conformation, and decreases to zero for dihedral angles in the range of $90°-100°$. Shielding differences of this magnitude can be readily observed and measured by solid-state NMR and directly correlated with the crystal structures derived from X-ray diffraction.

Empirical studies have shown that the induced shift of -5.3 ppm for each γ-*gauche* interaction quite accurately explains the dependence of carbon chemical shifts on stereochemical configuration in many chiral compounds and macromolecules (5). For the more sterically crowded compounds (i.e., 2-methylbutane or 2,2-dimethylbutane) the effect appears somewhat smaller.

The rules and regularities embodied in Tables 1.3, 1.4, and 1.5 and in the γ-*gauche* effect enable one to *predict* as well as interpret carbon chemical shifts, and thereby to test proposed structures and conformations. One of the fundamental problems in polymer NMR is assigning all the peaks in the spectrum. It is often observed that the carbon spectrum is very well resolved and that many peaks from stereochemical isomerism can be measured. In many cases, these peaks can be assigned from the combination of conformational calculations and the γ-*gauche* effect. This topic is considered in more detail in Section 3.1.2.3.

Inductive effects are also of great importance in ^{13}C chemical shifts of nonparaffinic compounds and macromolecules. This was illustrated in Figure 1.14, which showed the effect of substituents that differ in electronegititivy on the ^{13}C chemical shifts in pentanes. These effect can be very large, and the difference between the ^{13}C chemical shifts for the methylenes bonded to fluorines and iodines is 76 ppm. The inductive effects are attenuated by the intervening bonds, and theory suggests that the deshielding influence of electronegative elements or groups should be propagated down the chain with alternating effect, decreasing as the third power of the distance (17,18). This is qualitatively, but not quantitatively correct. It is true that the shielding of C_1 shows the expected dependence on the electronegativity of the attached atom. Good empirical correlations exist between the electronegativity and the induced chemical shift as a function of distance from the substitution (16).

We can observed from Figure 1.19 that the olefinic and aromatic carbons are strongly deshielded compared to those of saturated carbon chains—an effect parallel to that for protons (Figure 1.18), but much larger. Theoretical explanations involve the paramagnetic screening term rather than diamagnetic shielding or ring-current considerations (16,19,20). Again, the wide range in chemical shifts for the olefinic and aromatic carbons is an indication that the chemical shifts are very sensitive to the local structures. The introduction of electron-donating or -withdrawing groups into an aromatic ring also leads to a large dispersion in chemical shifts. Carbonyl carbons are among the most deshielded and are often easily resolved from any other resonances.

1.3.4 Other Nuclei

A variety of nuclei other than protons and carbons have been used for the solution and solid-state NMR analysis of polymers. The main limitation in these studies is the sensitivity and natural abundance of the nuclei under study. As shown in Table 1.1, the sensitivity of fluorine is near that of protons, and the spectra of fluorinated polymers can be recorded with a high sensitivity. The only difficulty is that fluorines and protons are close in frequency and a special probe must be used to observe the fluorines while decoupling the protons (and vice versa).

The sensitivity of silicon and phosphorus is also very good. Both of these nuclei have a wide range in chemical shifts, and the spectra are quite sensitive to polymer microstructure. Nitrogen also has a wide range in chemical shifts, but the sensitivity is very low. The nitrogen spectra are most often recorded on isotopically enriched samples.

1.3.4.1 Fluorine Fluorine is an excellent nucleus for the analysis of polymers because it has good sensitivity and a wide range of chemical shifts. In an 11.7-T magnet (500 MHz for protons), the fluorines are observed at 473 MHz. The high sensitivity for fluorine NMR is a consequence of both the natural abundance of the NMR-active nuclei (100%) and the large magnetogyric ratio.

The effect of chemical structure on the fluorine chemical shifts is shown in Figure 1.21. The CHF and CF fluorines are observed at the highest field, in the -175 to -200-ppm range relative to CCl_3F. CF_2 groups can be observed in the -160 to -80-ppm range, and aromatic fluorines appear in the -115 to -170-ppm range. CF_3 groups can be observed in the -40 to -90-ppm range. The wide range in chemical shift for the various fluorinated groups is again an indication that the chemical shifts are extremely sensitive to the local chemical structure.

1.3.4.2 Silicon Silicon NMR is important for the characterization of silane and siloxane polymers, as well as other main-chain polymers that are derivatized with silicon-containing groups. Silicon-29 has a natural abundance of 4.7% and a magnetogyric ratio similar to that of carbon. In an 11.7-T magnet the silicons are observed at 99 MHz. The silicon spectrum has a higher sensitivity than carbons because of the higher natural abundance. Some silicon spectra are difficult to analyze quantitatively because the spin-lattice relaxation times can be extremely long. Special methods are sometimes required because background signals from the NMR tubes and parts of the probe can interfere with the quantitative measurement of peaks near -100 ppm.

Figure 1.22 shows a plot of silicon chemical shifts for some chemical structures found in polymers. The silicon chemical shift is very sensitive to the number of

1.3 CHEMICAL SHIFTS AND POLYMER STRUCTURE

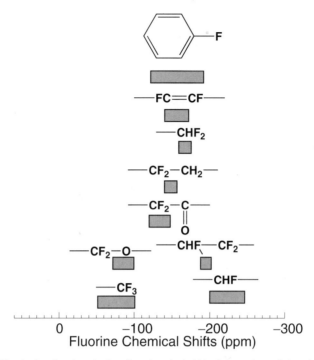

FIGURE 1.21 A plot showing the fluorine chemical shifts for a variety of chemical structures.

attached organic groups. Silicons bonded to four oxygens appear in the -80 to -110-ppm range, and the peaks move progressively upfield as each of the oxygens is replaced with an organic moiety. The polysilanes are observed in the -15 to -35-ppm range, and polysiloxanes are observed between 10 and 40 ppm. The silicon chemical shifts are reported relative to tetramethylsilane at 0 ppm.

1.3.4.3 Phosphorus Phosphorus NMR is used for the characterization of polyphosphazines and other polymers that are derivatized with phosphate groups. The

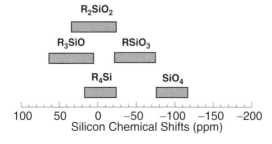

FIGURE 1.22 A plot showing the silicon chemical shifts for a variety of chemical structures.

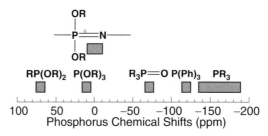

FIGURE 1.23 A plot showing the phosphorus chemical shifts for a variety of chemical structures found in polymers.

sensitivity of phorphorus is very good, since it has a large magnetogyric ratio and 100% natural abundance. In an 11.7-T magnet the phosphorus signals are observed at 203 MHz.

Figure 1.23 shows a plot of the phosphorus chemical shifts for several of the structures found in polymers. The chemical shift range for phosphorus is on the order of 300 ppm, and the chemical shifts are usually referenced to 85% phosphoric acid at 0 ppm.

1.3.4.4 Nitrogen The low natural abundance (0.365%) and magnetogyric ratio of nitrogen make the ^{15}N characterization of polymers very challenging. However, since many important polymers contain nitrogen, several studies that use isotopic labeling have been reported. In an 11.7-T magnet the nitrogen peaks are observed at 50 MHz.

Figure 1.24 shows a plot of nitrogen chemical shifts for a variety of functional groups found in polymers. The nitrogen chemical shifts have a very broad range and are sensitive to the local structure.

1.4 SPIN-SPIN COUPLING

1.4.1 Introduction

One of the more useful properties of spin systems for resonance assignments and structure determination is the scalar, or spin-spin, of coupling between NMR-active nuclei. By coupling we mean that these spins are interacting, and this interaction

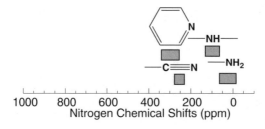

FIGURE 1.24 A plot showing the nitrogen chemical shifts for a variety of chemical structures.

1.4 SPIN-SPIN COUPLING

provides a mechanism for spin exchange that we can use to map out the local environment. There are two types of couplings that are important in high-resolution NMR studies: dipolar couplings and scalar couplings. The dipolar couplings are through-space interactions that can affect the lineshapes and relaxation times. For polymers in solutions, the chain motion averages out the dipolar interactions, and high-resolution spectra can be observed. The dipolar couplings are of interest in solid-state NMR since they may be much larger than the chemical shift range and they contain information about the molecular dynamics. The dipolar couplings in solids are considered in much more detail in Section 1.6.3.

The spin-spin or scalar couplings are mediated through bonds and cause the resonances to be split into multiple peaks. This coupling is due to the Fermi-contact interaction, which is related to the coupling between the nucleus and the electrons. Electrons that occupy an orbital that has some density at the nucleus (such as an s-orbital) have an electron spin that is antiparallel to the nuclear spin. If this atom is bonded to another, then the orbitals for the two atoms form an s-orbital that is connected to two electrons with antiparallel spins. The second nucleus can now have its nuclear spin parallel or antiparallel to the second electron spin, depending on whether it is in the α or β state. The antiparallel arrangement has a lower energy and gives a peak at a lower frequency. The magnitude of this coupling is called the scalar coupling and is denoted as $^n J_{XY}$, where n is the number of intervening bonds, and X and Y are the two nuclei that are coupled. Thus $^1 J_{CH}$ is the one-bond carbon-proton coupling and $^3 J_{HH}$ is the three-bond proton-proton coupling. Scalar (or J) couplings do not depend on the magnetic field strength, but they depend very strongly on the nuclei involved, the chemical structure, and the bond lengths. In favorable cases, scalar couplings can be observed through as many as five bonds, although the magnitude of the couplings decrease with the number of intervening bonds (21).

The peak multiplicities from scalar couplings depend on the number of coupled spins and their nuclear properties, and can provide important structural information about the number of NMR-active nuclei in close proximity. The multiplicity of the peak is given by $2nI + 1$, where n is the number or coupled spins and I is the spin quantum number. Thus, the carbon spectrum of chloroform ($CHCl_3$) has two lines, since $I = 1/2$ for protons. The carbon spectrum of deuterated chloroform gives rise to triplet, since deuterium is a spin-1 nucleus.

The number of peaks in a multiplet and their relative intensities provide information about the number of coupled nuclei. The relative intensities of the peaks in the multiplet can be calculated from simple arguments relating to the number of possible spin states. Two spins give rise to a 1:1 doublet because there is a nearly equal probability of the coupled nuclei being in the α or the β spin state (i.e., aligned with or against the field). The situation is slightly more complex in the carbon spectrum of a methylene group (CH_2), where the carbon peak is spit by two proton spins, and each of these protons could be in the α or β state. The four possible combinations are $\alpha\alpha$, $\alpha\beta$, $\beta\alpha$, and $\beta\beta$. Since each of these states has a nearly equal probability and the $\alpha\beta$ and $\beta\alpha$ states have the same energy, a 1:2:1 triplet is observed. In a similar way, the relative intensities for larger spin systems can be calculated. A simple mnemonic device to calculate the relative intensities is Pascal's triangle shown in Figure 1.25.

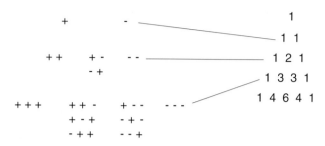

FIGURE 1.25 A diagram of Pascal's triangle used to calculate the relative intensities for multiples coupled to spin-$\frac{1}{2}$ nuclei.

The magnitudes of the J couplings depend on the structure and the nuclei involved, as well as the natural abundance of the coupled nuclei. In the chloroform example given earlier, we saw that the carbon signal would be split into a 1:1 doublet from the one-bond carbon–proton scalar coupling. When we measure the carbon spectrum, we are observing only the 1.1% of the carbon nuclei that have a ^{13}C nucleus, while in the proton spectrum, we are observing all of the protons, because the NMR-active ^{1}H is present at near 100% natural abundance. This leads to a singlet in the proton spectra from 98.9% of the protons that are bound to ^{12}C nuclei. In very high SNR spectra, we can observe small (1.1%) satellites due to the one-bond carbon–proton coupling from those protons in chloroform molecules bound to a ^{13}C nucleus.

The magnitudes of the J couplings depend both on the electronic structure, through the s-character of the bonds, and the magnetogyric ratios of the coupled nuclei. This is clearly demonstrated in one-bond carbon–proton couplings for sp^3, sp^2, and sp carbons where the magnitudes of the coupling constants are approximately 140 Hz, 160 Hz, and 250 Hz, respectively. Furthermore, the coupling constants are directly related to the magnetogyric ratios. If a deuterium is substituted for a proton in a molecule, the corresponding proton–deuterium coupling constant is reduced by a factor of 6.154 (γ_H/γ_D) relative to the proton–proton coupling.

1.4.2 Nomenclature for Spin-Spin Coupling

The peak multiplicities resulting from scalar coupling depend on the number of coupled spins and the magnetogyric ratios of the nuclei. However, the appearance of the spectra also depends on the chemical shift difference between the coupled spins. Given the complex chemical structures possible in polymers, many different types of coupling patterns could be imagined. These patterns are named based on the chemical shift difference between the coupled spins, and the number of spins involved. The simplest couplings are observed for pairs of spins in which the chemical shift difference is much greater than J. These coupling patterns are given the designation AX. The AX designation is equally well applied to heteronuclear or homonuclear couplings. The nomenclature for the A spin coupled to three X spins is AX_3, to four X spins, AX_4, and so forth. These systems are said to be weakly coupled.

1.4 SPIN-SPIN COUPLING

If the chemical shift separation is on the same order as the J coupling, then the spin system is said to be strongly coupled and the peak positions and intensities in the multiplets can be distorted from those predicted by Pascal's triangle. Both the peak positions and intensities will depend on the ratio of $J/\Delta v$, and these strongly coupled spin systems are given the AB designation.

Spin systems in which the chemical shift separation is intermediate between the AX and AB cases are given the M designation. Thus, for complex spin systems we could have, for example, coupling patterns corresponding to the AMX, $ABMX$, or AMX_3 designations.

1.4.3 Spin-Spin Coupling Patterns

Pascal's triangle can be used to calculate the relative peak intensities in multiplet only for the coupling between magnetically equivalent spins, such as the protons in a methyl group. In other cases the effect of the couplings must be considered individually. This is illustrated in Figure 1.26, which shows the spectrum expected for an AMX spin system under conditions where the coupling of spin A to M and X has the same and different coupling constants. If $J_{AX} \neq J_{MX}$, then a 1:1:1:1 quartet would be observed for the A signal, while equal coupling constants ($J_{AX} = J_{MX}$) give rise to a 1:2:1 triplet. Thus, the observation of a 1:2:1 triplet is not indicative of an AX_2 spin system.

Similar effects are expected for the other peaks in the AMX system. If $J_{AX} \neq J_{MX}$, then the AMX system would give rise to 12 peaks from the three 1:1:1:1 quartets from the A, M, and X spins, while if $J_{AX} = J_{MX}$, then 9 peaks would be observed from the three 1:2:1 triplets.

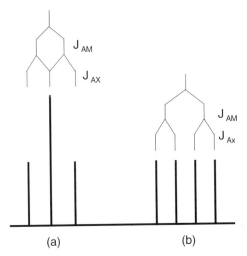

FIGURE 1.26 The coupling pattern for an AMX spin group in which (a) $J_{AM} = J_{AX}$, and (b) $J_{AM} \neq J_{AX}$.

The coupling patterns for a large number of spin systems have been observed and classified (12). The coupling patterns for polymers can be especially complex, since the equivalence between protons can be lost due to stereochemistry or chain defects. This is illustrated by considering the methine protons in a vinyl polymer, which are adjacent to two methylene groups. If the geminal protons are nonequivalent and each neighboring methylene has a different chemical shift, the coupling pattern for the methine proton could be very complex.

1.4.3.1 Strong Coupling The simple rules to calculate the multiplet intensities using Pascal's triangle are only applicable to couplings in the weak-coupling limit. In cases where $J/\Delta v > 0.1$, the spin states for the two nuclei become mixed and we can no longer discuss the lines in terms of arising from one particular nucleus. The effect of strong coupling can be demonstrated by comparing the peak positions and intensities expected for the AX spin system with the AB spin systems shown in Figure 1.27. For an AX spin group the peaks for A and X are 1:1 doublets. For the AB system, the peak position and line intensities depend on the ratio of $J/\Delta v$. For a ratio of 0.2 the spectrum appears as an AX pattern with slightly distorted intensities. The pattern becomes more distorted as the ratio increases and for a value of 1.0, the pattern appears as a double t with weak satellite peaks. Since the patterns can become very complex as more spins are involved (e.g., AB_3X), it is generally not possible to determine the spin topology from inspection of the coupling constant pattern. Such complex patterns are often present in polymers, since stereochemistry and some defects can make neighboring protons in polymer chains slightly nonequivalent.

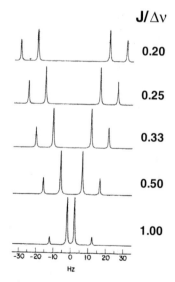

FIGURE 1.27 The effect of strong coupling on the appearance of the AB spin system.

1.4 SPIN-SPIN COUPLING

1.4.3.2 Scalar Coupling and nD NMR

The magnitudes and multiplicities of the coupling constants have important implications for the nD NMR of polymers. As we will see later (Section 1.7), nD NMR experiments using scalar couplings for magnetization transfer are important for establishing resonance assignments in polymers and for structure determination. A general feature of these nD NMR experiments is a delay period on the order of $1/2J$ for maximum magnetization transfer. One complication is that spin-spin relaxation also occurs during this period, so the best SNR is observed when $1/2J \ll T_2$. Many heteronuclear couplings are on the order of 100 Hz, so this condition is easily fulfilled. Homonuclear couplings are much smaller (2–15 Hz), and so losses from T_2 relaxation are much larger.

The scalar coupling multiplicity also affects the SNR in nD NMR experiments. In many correlation experiments antiphase cross peaks are generated in which parts of the multiplet cross peak are out of phase with each other. When the linewidths become significant compared to the coupling constant, the positive and negative lobes of the multiplet can cancel, decreasing the SNR. This effect can be partially mitigated using experiments that generate in-phase cross peaks (Section 1.7.1.1).

1.4.4 Proton–Proton Coupling

Proton–proton scalar couplings are important for resonance assignments and for the determination of polymer chain conformation. The couplings of most interest in polymers are the two-bond geminal couplings and the three-bond vicinal couplings $^2J_{HH}$ and $^3J_{HH}$. As is generally the case, the magnitude of the proton–proton scalar coupling decreases with the number of intervening bonds, so the two-bond couplings are always larger than the three-bond couplings. The magnitude of the three-bond coupling depends on the dihedral angle, and thus provides important conformational information.

The proton–proton coupling constants for a variety of aliphatic hydrocarbons are shown in Figure 1.28. The largest of these are the two-bond geminal couplings that

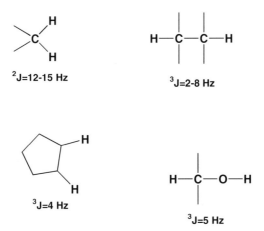

FIGURE 1.28 The magnitude of the proton–proton scalar couplings in a variety of structures.

are in the range 12–15 Hz. The two-bond coupling constant depends on the H—C—H angle, and in cyclopropane derivatives this coupling is on the order of 4–6 Hz. The vicinal couplings in aliphatic hydrocarbons vary between 2 and 8 Hz, depending on the chain conformation. Thus they are important not only for resonance assignments but also for the determination of chain conformation. For freely rotating bonds, as in butane, there is rapid interconverison of the *gauche* and *trans* conformers. The coupling constant is larger for the *trans* conformation than for the *gauche* conformation. For most polymers the energy barrier for interconversion of the *gauche* and *trans* conformers is low, so the exchange between conformers is fast on the NMR time scale. Under these conditions the coupling constant is a population-weighted average of the values for the *gauche* and *trans* states given by

$$J_{obs} = X_g J_g + X_t J_t \tag{1.52}$$

where X is the fraction of molecules in the *gauche* and *trans* states. The *trans* coupling constant $^3J_{trans}$ is typically on the order of 8 to 10 Hz, while the coupling constant for a *gauche* $^3J_{gauche}$ arrangement is on the order of 2 to 4 Hz. The conformational dependence of the magnitude of the coupling constant has been extensively studied, and the relationship between the conformation and the coupling constant is known as the Karplus relationship and is given by

$$^3J_{HH} = \begin{cases} 8.5\cos^2\varphi - 0.28 & 0° \leq \varphi \leq 90° \\ 9.5\cos^2\varphi - 0.28 & 90° \leq \varphi \leq 180° \end{cases} \tag{1.53}$$

where φ is the dihedral angle between the two protons. Figure 1.29 shows a plot of the Karplus relationship.

In aliphatic rings and other conformationally constrained structures it may be that not all conformations are allowed, and the full range of dihedral angles is not sampled. Rings, for example, can often sample only a few conformations. If the exchange between conformers is fast on the NMR time scale, the coupling constants will be a population-weighted average of the values in these conformations. Conformational averaging may be energetically unfavorable for very constrained structures, and the coupling constants will be directly related to the conformation. In such cases it is possible to have a coupling constant near zero for those cases where the dihedral angle is near 90°. Very small couplings ($J < 2$ Hz) are often difficult to observe in polymers.

In favorable cases the three-bond coupling to hydroxyl protons can be observed. The magnitude of this coupling is on the order of 5 Hz. The hydroxyl protons are very susceptible to exchange with other exchangeable protons (including residual water) and the couplings are averaged by chemical exchange and often difficult to observe.

The coupling constants for olefinic protons range between 1 and 19 Hz, as shown in Figure 1.30. The coupling constant for the two-bond geminal couplings are in the 1–4-Hz range, a large reduction relative to the aliphatic geminal couplings. Comparison of the aliphatic and olefinic geminal couplings shows that the values are very sensitive to the hybridization. As with the alphatic compounds, the three-bond couplings are

1.4 SPIN-SPIN COUPLING

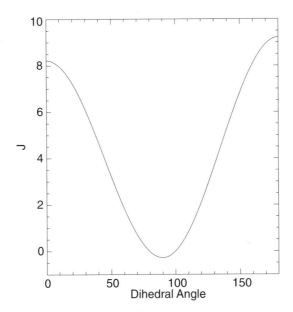

FIGURE 1.29 A plot of the dependence of the three-bond scalar coupling constant on the dihedral angle.

sensitive to the relative orientation of the olefinic protons. The *trans* orientation gives rise to coupling constants in the 12–19-Hz range, while the coupling constants for the *cis* orientation is on the order of 7–11 Hz.

The proton–proton scalar couplings have been measured for a large number of aromatic groups and some typical coupling constants are shown in Figure 1.31. The largest values (8 Hz) are observed for *ortho* coupling constants. The coupling constants decrease with an increasing number of intervening bonds to 2 Hz for *meta* couplings and 0.5 Hz for *para* couplings. The introduction of heteroatoms into aromatic groups

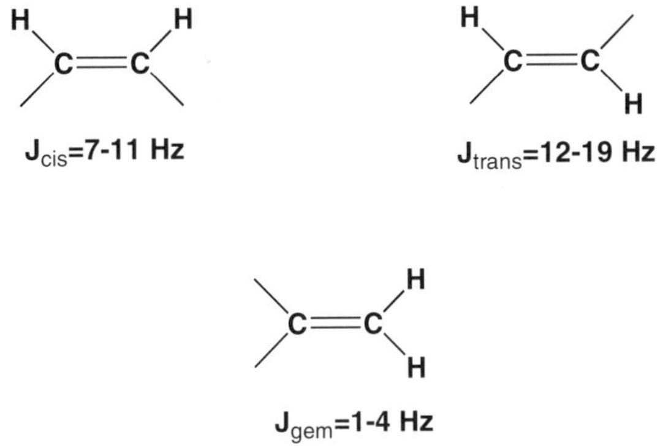

J_{cis}=7-11 Hz J_{trans}=12-19 Hz

J_{gem}=1-4 Hz

FIGURE 1.30 The proton–proton scalar couplings for olefinic structures.

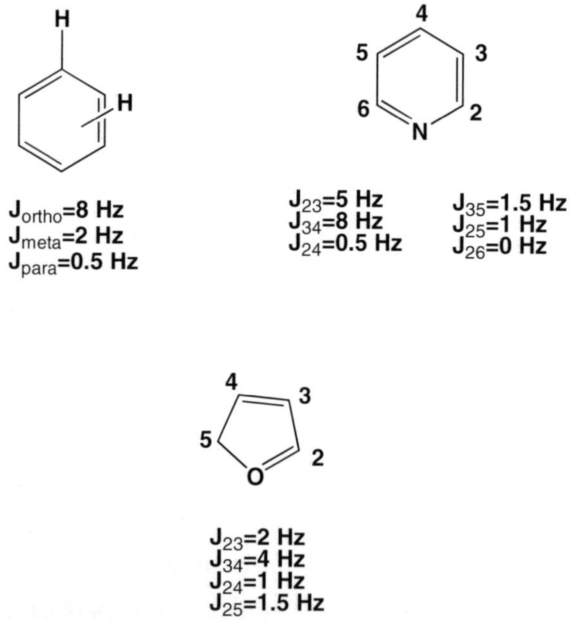

J_{ortho}=8 Hz
J_{meta}=2 Hz
J_{para}=0.5 Hz

J_{23}=5 Hz
J_{34}=8 Hz
J_{24}=0.5 Hz

J_{35}=1.5 Hz
J_{25}=1 Hz
J_{26}=0 Hz

J_{23}=2 Hz
J_{34}=4 Hz
J_{24}=1 Hz
J_{25}=1.5 Hz

FIGURE 1.31 The proton–proton scalar coupling constants for representative aromatic compounds.

only slightly perturbs the value of the couplings. The couplings are sensitive to the bond angles as the size of the ring is changed.

1.4.5 Proton–Carbon Coupling

The one-bond carbon–proton scalar couplings are much larger than the proton–proton couplings. Although the one-bond couplings do not depend on the conformation, they do provide important information about the number of directly bonded protons that can be used to assign the carbon type. The coupling patterns become increasingly complex as the number of directly attached protons increases, leading to extensive overlap for complex polymers. For the highest resolution spectra the couplings are removed by proton decoupling (Section 2.4.3)

The one-bond carbon–proton couplings $^1J_{CH}$ for a variety of structures are shown in Figure 1.32. The coupling constants are between 120 and 130 Hz for most aliphatic structures, and they are sensitive to both the hybridization and the bond angles in strained structures. The coupling constant depends on the hybridization, and is approximately given by

$$^1J_{CH} = 500\rho_{CH} \tag{1.54}$$

where ρ_{CH} is the fraction of s-character in the bonding orbitals. The couplings for the sp^3, sp^2, and sp carbons are on the order of 125, 165, and 250 Hz, respectively. Larger

1.4 SPIN-SPIN COUPLING

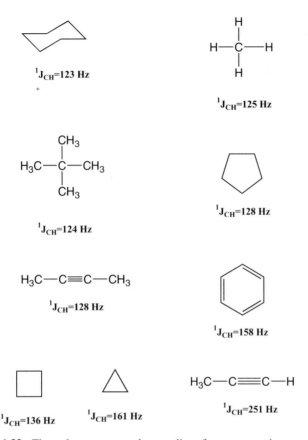

FIGURE 1.32 The carbon–proton scalar couplings for representative compounds.

coupling constants are observed for strained ring structures, and the values increase to 136 and 161 Hz for cyclobutane and cyclopropane.

The multiple-bond carbon–proton coupling constants are also large enough that they can sometimes be observed by high-resolution NMR (Figure 1.33). The magnitudes of the couplings are very dependent on the local chemical structures, but these couplings are often in the range of 3–6 Hz in aliphatic compounds. In aromatic compounds relatively long-distance couplings can be observed that do not monotonically decrease with the number of intervening bonds, as illustrated by the comparison of the $^2J_{CCH}$ and $^3J_{CCH}$ couplings in chlorobenzene.

The multiple-bond couplings are important for nD NMR studies using heteronuclear correlation of the carbon and proton spectra. If methods are used to select magnetization transfer via multiple-bond coupling constants, it is possible to correlate the proton spectra with carbons without directly bonded protons. In this way, for example, it is possible to assign carbonyl groups. The SNR is often low in these experiments because the coupling constants are smaller and more magnetization is lost via T_2 relaxation during the $1/2J$ delay for carbon–proton magnetization transfer.

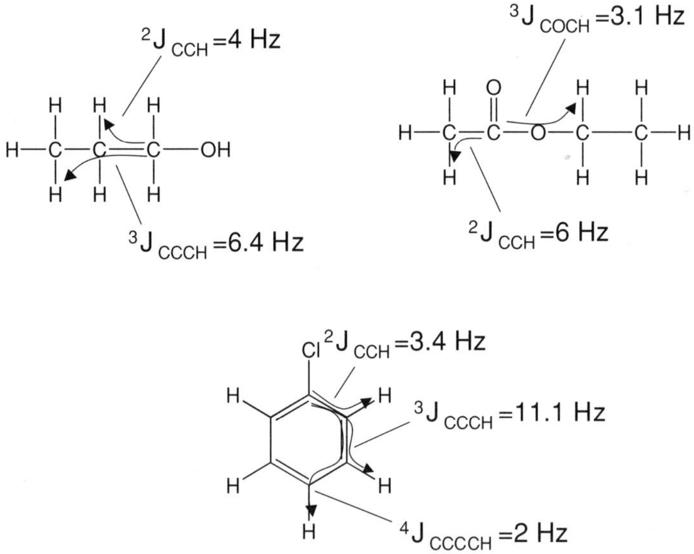

[Figures showing coupling constants in various organic compounds:

- Propene derivative (H-C-C=C-OH): $^2J_{CCH} = 4$ Hz, $^3J_{CCCH} = 6.4$ Hz
- Vinyl acetate-like structure: $^3J_{COCH} = 3.1$ Hz, $^2J_{CCH} = 6$ Hz
- Chlorobenzene: $^2J_{CCH} = 3.4$ Hz, $^3J_{CCCH} = 11.1$ Hz, $^4J_{CCCCH} = 2$ Hz]

FIGURE 1.33 The two-bond carbon–proton coupling constants in representative organic compounds.

1.4.6 Other Nuclei

The scalar couplings from a variety of other nuclei provide important information for resonance assignments in polymers. Among the NMR-active nuclei that have been studied are fluorine, phosphorus, silicon, and nitrogen. It is important to realize that there is nothing exotic about the study of these other nuclei, and with a suitable spectrometer many of the nD NMR methods used to study the carbon and proton spectra can be directly applied to these other nuclei as long as we modify the experiments to account for the magnitudes of the coupling constants.

1.4.6.1 Fluorine Couplings Fluorine is a high-sensitivity nuclei that is often coupled to other fluorines, protons or carbons. Figure 1.34 shows some fluorinated structures and the associated two, three, four, and five-bond fluorine–fluorine coupling constants. In contrast to the proton couplings constants, there is no a simple relationship between the magnitude of the coupling and the number of intervening bonds. The three-bond coupling constants in fluorinated aliphatic compounds vary between 4 and 15 Hz, while the four-bond coupling constants are often in the 7 Hz range. The coupling constants are larger in olefinic compounds. The geminal coupling is in the 70–80-Hz range, while the couplings for *cis* and *trans* olefinic are in the range of 50–60 and 110–120 Hz. The coupling constants for fluorinated aromatics are quite large for *ortho* and *para* couplings, but relatively small for *meta* couplings. The relatively large fluorine–fluorine coupling constants make fluorine relatively easy to study by nD NMR.

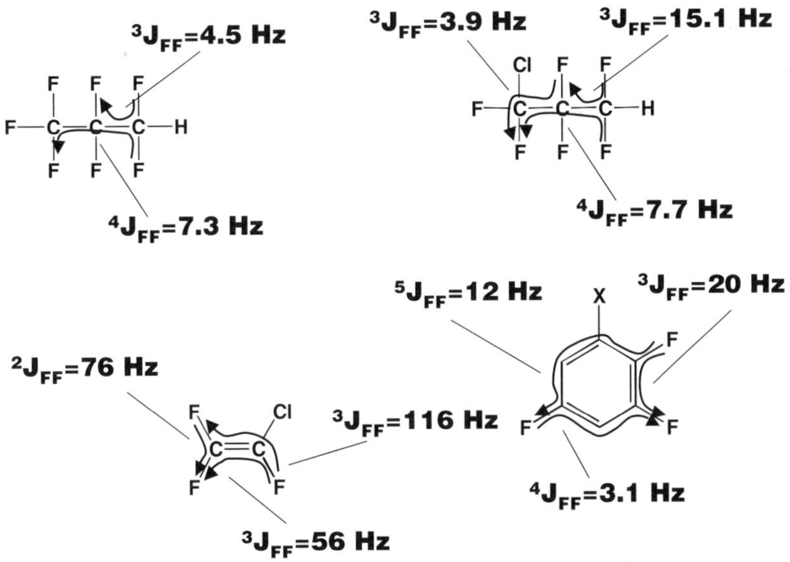

1.4 SPIN-SPIN COUPLING

FIGURE 1.34 The fluorine–fluorine scalar couplings in representative organic compounds.

The fluorines in polymers are often in close proximity to protons, and the proton–fluorine couplings can be used in heteronuclear correlation spectra for establishing the resonance assignments. The magnitudes of the proton–fluorine couplings for representative aliphatic, olefinic, and aromatic structures are shown in Figure 1.35. The two-bond couplings in aliphatic compounds are on the order of 40–50 Hz, while the three-bond couplings are in the 8–14-Hz range. In favorable cases, the four-bond fluorine–proton couplings can also be observed. The two-bond coupling is larger in olefinic compounds (70–80 Hz), while the three-bond couplings are smaller. The proton–fluorine couplings in aromatic compounds are somewhat larger than the comparable proton–proton coupling constants.

The carbon spectrum will also be split due to couplings with fluorine, and Figure 1.36 shows the coupling constants for representative aliphatic, olefinic, and aromatic structures. The one-bond coupling constants are approximately twice the value of those observed for carbon–proton couplings. The two-bond carbon–fluorine couplings are also large enough to be observed in the high-resolution spectrum.

The carbon spectrum of fluorinated polymers is especially complex because the lines will be split both by the one- and two-bond carbon–proton and carbon–fluorine couplings. We can improve the resolution with decoupling to collapse the carbon multiplets into singlets. However, in this case we must perform a triple resonance experiment with both proton and fluorine decoupling. It is also necessary to decouple the fluorines to observe the highest resolution proton spectrum. The resonance frequencies of protons and fluorines are relatively close and a special probe with rf filters is required.

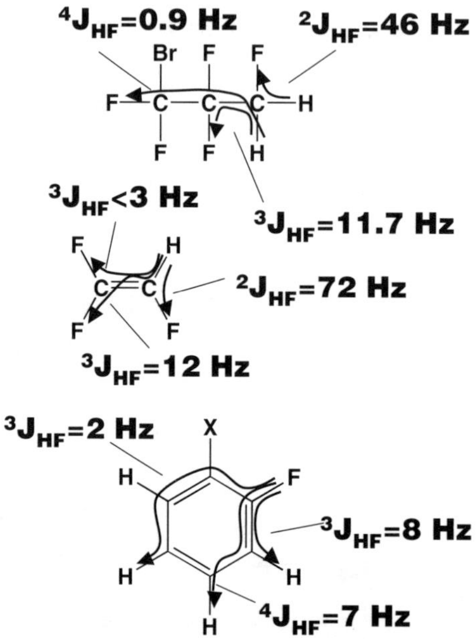

FIGURE 1.35 The fluorine–proton scalar couplings in representative organic compounds.

1.4.6.2 Phosphorus Couplings Phosphorus is a relatively high-sensitivity nucleus that shows scalar couplings to both carbons, protons, and other phosphorus atoms. This is illustrated in Figure 1.37 which shows the coupling constants observed for a variety of structures. Large (200-Hz) one-bond phosphorus couplings are observed, and the magnitude of the coupling is strongly attenuated by the intervening bonds. The one-bond carbon–phosphorus couplings are relatively small and are not simply

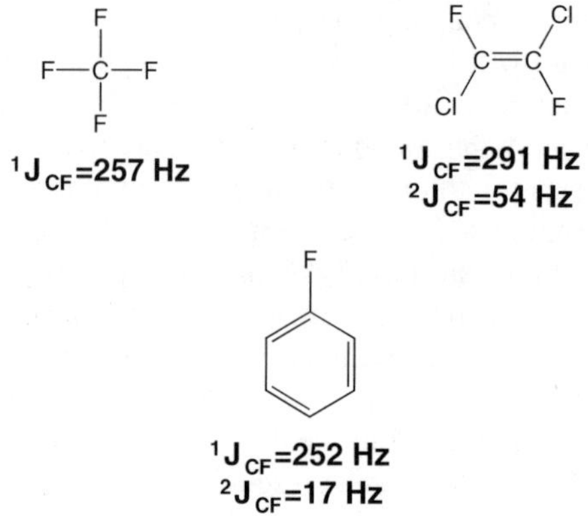

FIGURE 1.36 The carbon–fluorine scalar couplings in representative organic compounds.

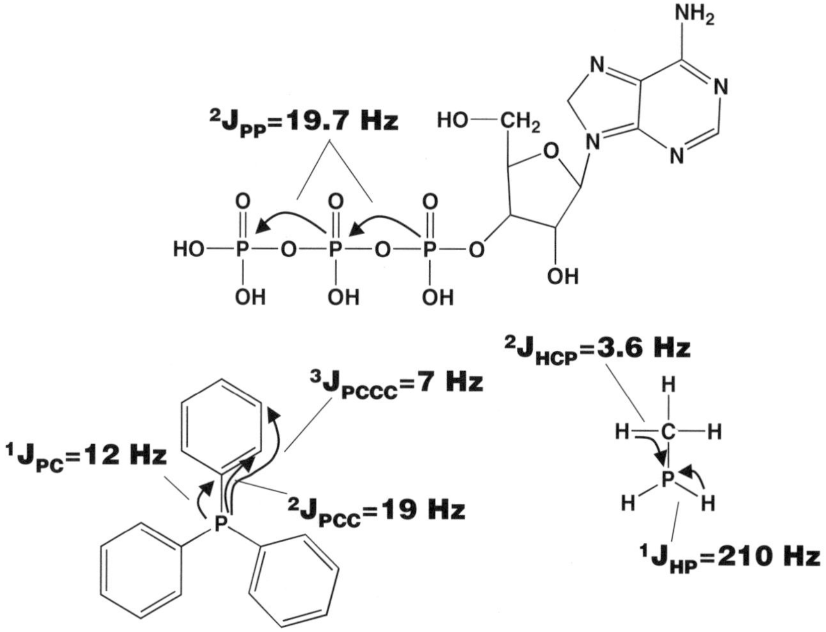

1.4 SPIN-SPIN COUPLING

FIGURE 1.37 The phosphorus–phosphorus, phosphorus–proton, and phosphorus–carbon scalar couplings in representative organic compounds.

scaled with the number of intervening bonds. The phosphorus–phosphorus couplings through oxygens are on the order of 20 Hz.

1.4.6.3 Silicon Couplings Silicon is a relatively common element in polymers that shows couplings to both carbons and protons. As shown in Figure 1.38, the one-bond silicon–proton couplings are on the order of 200 Hz and rapidly decrease with the number of intervening bonds. The carbon–silicon couplings constants are on the order of 50 Hz.

1.4.6.4 Nitrogen Couplings Nitrogen-15 has a low natural abundance and a low sensitivity. The scalar coupling of interest would be the one-bond proton–nitrogen coupling, which is on the order of 90 Hz (Figure 1.39). In favorable cases the nitrogen spectrum can be observed using proton-detected heteronuclear correlations where the magnetization is transferred via the one-bond proton–nitrogen coupling.

1.4.7 Homonuclear Couplings in Insensitive Nuclei

The scalar couplings between heteronuclei such as neighboring carbon atoms and between nitrogen and carbons are difficult to observe because of the low natural abundance of these nuclei. The carbon natural abundance is 1.1%, so the probability

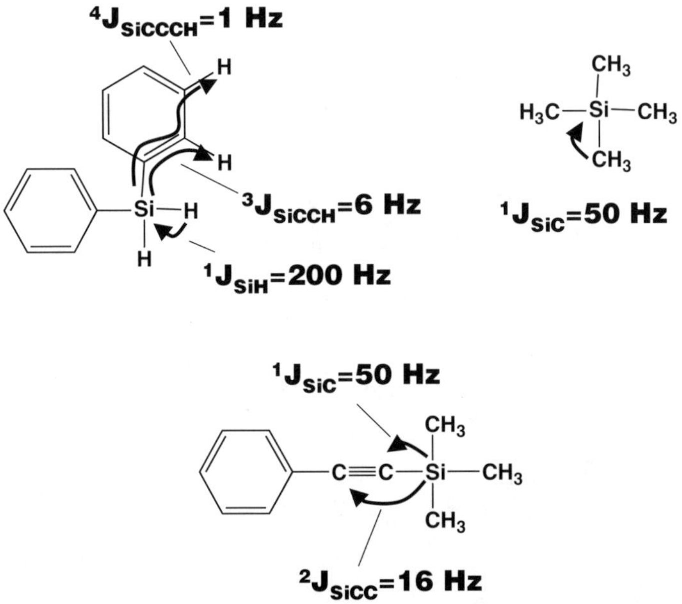

50 INTRODUCTION TO NMR

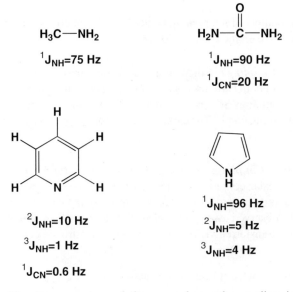

FIGURE 1.38 The silicon–silicon, silicon–proton, and silicon–carbon scalar couplings in representative organic compounds.

H_3C-NH_2

$^1J_{NH}=75$ Hz

$H_2N-\overset{\overset{O}{\|}}{C}-NH_2$

$^1J_{NH}=90$ Hz
$^1J_{CN}=20$ Hz

(pyridine structure)

$^2J_{NH}=10$ Hz
$^3J_{NH}=1$ Hz
$^1J_{CN}=0.6$ Hz

(pyrrole structure)

$^1J_{NH}=96$ Hz
$^2J_{NH}=5$ Hz
$^3J_{NH}=4$ Hz

FIGURE 1.39 The nitrogen–proton and nitrogen–carbon scalar couplings in representative organic compounds.

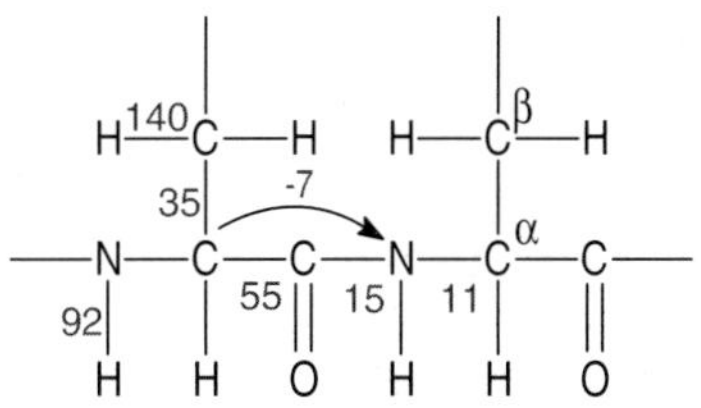

1.5 NMR RELAXATION

$$\begin{array}{c}
\text{H}\overset{140}{-}\text{C}-\text{H} \quad \text{H}-\overset{\beta}{\text{C}}-\text{H} \\
\underset{92|}{35|} \quad \overset{-7}{\curvearrowright} \quad |\alpha \\
-\text{N}-\text{C}-\text{C}-\text{N}-\text{C}^{\alpha}-\text{C}- \\
| \quad 55\| \quad 15| \quad 11| \quad \| \\
\text{H} \quad \text{H} \quad \text{O} \quad \text{H} \quad \text{H} \quad \text{O}
\end{array}$$

FIGURE 1.40 The carbon–carbon, carbon–nitrogen, and nitrogen–nitrogen scalar couplings in peptides.

of having two ^{13}C atoms next to each other is one in eight thousand. The natural abundance of ^{15}N is even lower (0.3%), so the probability of having an ^{15}N–^{13}C pair is about one in thirty thousand. This low natural abundance is also the reason that the spectra of protons bound to carbons or that nitrogens do not show splittings.

It is possible to observe heteronuclear couplings in samples isotopically enriched with ^{13}C or ^{15}N. This has been extensively used to study the structure and dynamics in biomolecules. Figure 1.40 shows the chemical structure of a peptide backbone along with the coupling constants expected in istopically enriched molecules. The carbon–carbon couplings are on the order of 35 Hz for aliphatic carbons and 55 Hz between aliphatic carbons and carbonyls. The carbon–nitrogen coupling is between 10 and 15 Hz, and in favorable cases the two-bond couplings (7 Hz) can also be observed.

1.5 NMR RELAXATION

1.5.1 Introduction

One of the effects of rf pulses is to cause the spin populations to deviate from their equilibrium populations. The rate at which the spin system returns to equilibrium contains important information about the molecular dynamics and the internuclear distances. There are several relaxation rates that can be measured, including the spin-lattice and spin-spin relaxation times and the nuclear Overhauser enhancement. Each of these relaxation rates has a different dependence on the molecular dynamics, and several relaxation times are usually measured to provide complementary information. For relaxation studies in solutions the relaxation times are very sensitive to molecular motions on the MHz time scale. Solid-state relaxation measurements are sensitive to the same time scale of motion, but also to dynamics on the kHz time scale. For both solutions and solids, nD exchange NMR can be used to probe the dynamics in the slow motion limit (millisecond to many seconds).

Nuclear relaxation is caused by molecular motion, since atomic fluctuations modulate the magnetic field experienced by a nearby atom. These fluctuations in the local magnetic fields can occur over a broad range of frequencies, and depend on the chemical structure and the molecular environment. Crystalline solids are a much more restrictive environment than a rubbery matrix, and this will be reflected in the molecular dynamics and relaxation times. To the extent that these fluctuations have components at the resonance frequency for the nuclei of interest, they will cause relaxation. The distribution of motional frequencies is called the *spectral density function* and is given by

$$J(\omega) = \frac{1}{2} \int_{-\infty}^{\infty} G(\tau) e^{-i\omega\tau} d\tau \qquad (1.55)$$

where $G(\tau)$ is the autocorrelation function for an internuclear vector—such as a ^{13}C–^1H vector—that is given by

$$G(t) = \langle F(t)F^*(t+\tau)\rangle \qquad (1.56)$$

The brackets denote the ensemble average over a collection of nuclei, and F represents a function, related to spherical harmonics, describing the position of the atoms. $J(\omega)$ may be thought of as expressing the power available at frequency ω to cause relaxation in the spins in question. The spectral densities and autocorrelation functions are Fourier inverses of each other in the time and frequency domains, respectively. If $G(\tau)$ decays to zero in a short time, this corresponds to a short correlation time τ_c, which means that molecular motion is rapid and that the molecules have only a short memory of their previous state of motion.

In order to model the relaxation of polymers and other molecules, we must adopt a dynamical model for $G(\tau)$. The simplest model is that of a rigid sphere immersed in a viscous continuum that is reoriented in small diffusive steps. The correlation time can be thought of as the interval between these alterations in the state of motion of the molecule. For such a molecule the loss of memory of the previous motional state is exponential with the time constant τ_c, and the correlation time is given by

$$G(t) = e^{-t/\tau_c} \qquad (1.57)$$

Fourier transformation of the correlation function gives the spectral density function as

$$J(\omega) = \frac{\tau_c}{1 + \omega^2 \tau_c^2} \qquad (1.58)$$

Figure 1.41 shows a plot of the spectral density function as a function of frequency for three values of τ_c. A long value of τ_c corresponds to the molecular motion of a large molecule, a stiff chain, or a small molecule in a viscous medium, and a short value of τ_c corresponds to the rapid motion of a small molecule or a very flexible polymer chain. Figure 1.41 also shows a plot for an intermediate case, when $\tau_c\omega_0 \cong 1$, where ω_0 is the resonant frequency of the observed nuclei. The areas under the curves are the same for all three correlation times, which means that the power available to cause relaxation is the same, and only the distribution varies with τ_c. For short correlation times the component at ω_0 is weak, and for the long τ_c the frequency spectrum is broad

1.5 NMR RELAXATION

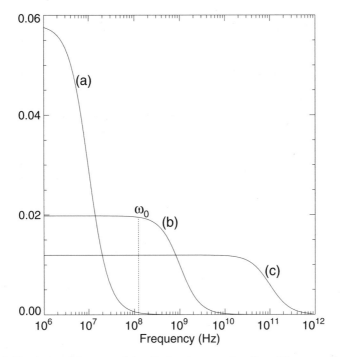

FIGURE 1.41 A plot of the spectral density functions as a function of frequency for (a) short, (b) intermediate, and (c) long correlation times.

so no one component, particularly that at ω_0, is very intense. At some intermediate value of τ_c the intensity at ω_0 will be at a maximum. Since the relaxation rate depends on the component at ω_0, it will be at a maximum for the intermediate value of τ_c, and the relaxation rates will be much slower for very fast or very slow motion. The exact dependence of the relaxation rate on the correlation time will depend on which experiment is performed and which relaxation rates are measured. A critical part of measuring the molecular dynamics of polymers is choosing the proper model for the correlation functions and spectral densities. Many studies have shown that a single exponential correlation function is not a good model to describe the molecular motion of an object as complex as a polymer chain (22).

Understanding the relaxation mechanism for the nuclei of interest is an important first step in using NMR to study the molecular dynamics of polymers. Depending on the nuclei and its chemical environment, there can be several mechanisms that contribution to the observed relaxation rate, including dipolar (DD) interactions, chemical shift anisotropy (CSA), and quadrupolar (QUAD) relaxation (23). The observed relaxation rate is the sum of all of these contributions, and for spin-lattice relaxation is given by

$$\frac{1}{T_1} = \frac{1}{T_1^{DD}} + \frac{1}{T_1^{CSA}} + \frac{1}{T_1^{QUAD}} \tag{1.59}$$

For a quantitative analysis of polymer relaxation it is important to understand the contribution that each of these mechanisms make to the observed relaxation rate.

1.5.2 Relaxation Mechanisms

Once a spin system has been perturbed from its equilibrium state, it returns to its equilibrium via one of several relaxation mechanisms. Any of the possible mechanisms may cause relaxation, but the relaxation mechanism is effectively determined by the mechanisms that is most efficient. Dipolar interactions, such as those between a ^{13}C and its directly bonded protons, are the most common relaxation mechanism. These carbons could also have a contribution from relaxation by chemical shift anisotropy, except for aliphatic carbons this pathway is not very effective and can be ignored. On the other hand, carbonyl carbons do not have directly bonded protons and their relaxation times are usually much longer and relaxation via the chemical shift anisotropy mechanism cannot be ignored. For nuclei such as deuterium, quadrupolar relaxation is the main pathway, and other sources of relaxation can effectively be ignored. Paramagnetic relaxation is an extremly efficient relaxation mechanism, but can be ignored if the polymer of interest does not have unpaired electrons. Intermolecular interactions with paramagnetic species (including oxygen) can also cause relaxation, so they must be removed from the polymer solutions before the relaxation times are measured.

1.5.2.1 Dipole–Dipole Interactions For most spin-1/2 nuclei in solution, dipole–dipole interactions are usually the most efficient relaxation mechanism. These dipole–dipole interactions arise as a consequence of direct through-space interactions between NMR-active nuclei. The energy of the dipolar coupling is given by

$$E_{DD} \propto \frac{\mu_1 \cdot \mu_2}{r^3} \tag{1.60}$$

where μ_1 and μ_2 are the dipole moments of the two nuclei. Note that the strength of the dipolar coupling depends both on nuclei involved and the distance. Those nuclei with larger moments (i.e., protons and fluorines) will be most effective at causing relaxation. The total dipolar coupling is the sum over all the nearby nuclei, so the dipolar coupling (for groups with the same correlation time) will scale with the number of attached protons. As we will see later (Section 1.5.3), the equations for the spin-lattice and spin-spin relaxation times depend on the inverse sixth power of the internuclear distance. For those groups without directly bonded protons (such as quaternary carbons and carbonyls) the dipolar couplings are mostly due to nearest-neighbor protons.

1.5.2.2 Quadrupolar Interactions Nuclei with a spin of $1/2$ have a spherical distribution of nuclear charge and are not affected by the electrical environment within the molecule. Nuclei with a spin of 1 or more have electric quadrupole moments and a nonspherical distribution of charge. The energies of the charges will depend on their orientation in the molecular electrical field gradient. In molecules where substantial electric field gradients are present that can interact with the nuclear quadrupole, molecular motion leads to large changes in spin states that cause relaxation. The effectiveness of this relaxation mechanism is related to the magnitude of the electric quadrupole moment eQ. Quadrupolar relaxation is the dominant mechanism for quadrupolar nuclei such as deuterium. The relaxation times depend on the quadrupole coupling constant $e^2qQ/\hbar$ and the asymmetry parameter η, as well as the rotational correlation time. The field gradient eq is actually the principal component of the field-gradient tensor and η is a measure of how much the electric field gradient deviates from axial symmetry.

1.5.2.3 Chemical Shift Anisotropy The relaxation due to chemical shift anisotropy is the only relaxation mechanism that depends on the strength of the magnetic field. The magnetic field experienced by a nucleus depends on the electronic shielding (the chemical shift). If the electronic distribution is nonspherical, the secondary magnetic field caused by electronic currents need not be parallel to the direction of the magnetic field and may have a perpendicular component. This perpendicular component fluctuates as the molecule rotates, leading to relaxation. The effectiveness of chemical shift anisotropy as a relaxation mechanism depends on $\Delta\sigma$, the difference in chemical shift for the molecule parallel and perpendicular to the magnetic field. Chemical shift anisotropy is not usually important for proton relaxation, but can contribute to the relaxation of nuclei with a large chemical shift range, such as carbon, fluorine, or phosphorus. For carbon nuclei, relaxation by chemical shift anisotropy is most important for anisotropic groups such as carbonyls and aromatic carbons that do not have directly attached protons. The chemical shift anisotropy contribution to the relaxation can be identified as the only mechanism that depends on the square of the magnetic field strength.

1.5.2.4 Paramagnetic Relaxation Paramagnetic relaxation is similar to dipolar relaxation, except it results from the dipolar interaction with an unpaired spin rather than another nucleus. The magnetic moment for an electron is 657 times that of a proton. This means that the relaxation by the unpaired electron is 431,649 (657^2) times more efficient at causing relaxation. Thus, paramagnetic relaxation is the dominant relaxation mechanism in any molecule containing an unpaired electron. Another consequence of efficient paramagnetic relaxation is that the spin-spin relaxation is very efficient and the lines become very broad. In many cases they become so broad that they cannot be observed above the baseline.

For high-resolution studies of polymers it is important to remove the paramagnetic species. This is not generally a problem, since most polymers do not contain unpaired electrons. However, contamination with a paramagnetic species can artificially shorten the relaxation times for polymers, leading to incorrect conclusions about the molecular structure and dynamics. This is most commonly encountered with paramagnetic oxygen, which should be removed from NMR samples by degassing or repeated freeze–thaw cycles prior to making relaxation measurements.

It is sometimes observed in polymers that the relaxation times can be extremely long for some nuclei such as silicon and carbon. This can lead to prohibitively long acquisition times, since it is important to wait at least five times the longest spin-lattice relaxation time between scans for quantitative spectra. In these cases a paramagnetic compound, such as chromium(III) actylacetonate, can be added to the solution to artificially shorten the relaxation times.

Chromium(III) Acetylacetonate

1.5.2.5 Other Relaxation Mechanisms In addition to the above mechanisms, there are a few other mechanism that are able to cause relaxation in small molecules, but they are usually not important for polymers. These include relaxation by spin-rotation, which results from direct interaction of the nuclear moment with the magnetic field. This is a very ineffective mechanism for polymers, and typically only contributes to the relaxation of gasses, where collisions can lead to energy transfer from nuclear spins to the molecular rotation. Scalar relaxation can be caused either by scalar coupling to a chemically exchangeable proton or by scalar coupling to a quadrupolar nuclei. These are rarely observed in polymers, except for carbons next to a ^{14}N where broadening is observed in the solid-state spectrum.

1.5.3 Spin-Lattice Relaxation

The spin-lattice relaxation rates are among the easiest to measure, and provide important information about the dynamics of polymers. These measurements are sensitive to both the molecular structure and the dynamics on a picosecond to nanosecond time scale. Knowledge of these relaxation times is also important for quantitative measurements, since the spin-lattice relaxation time determines the repetition rate when signal averaging multiple scans.

The spin-lattice relaxation times of insensitive nuclei (^{13}C, ^{29}Si, ^{31}P, etc.) are often used to study the polymer dynamics. To properly interpret the relaxation times it is necessary to understand the relaxation mechanism, which will depend on the polymer under study, the nuclei under observation and the local environment. The observed relaxation rate is the sum of all of the possible contributions and is given by

$$\frac{1}{T_1^{obs}} = \frac{1}{T_1^{DD}} + \frac{1}{T_1^{CSA}} + \frac{1}{T_1^{QUAD}} + \frac{1}{T_1^{other}} \qquad (1.61)$$

where T_1^{DD}, T_1^{CSA}, T_1^{QUAD}, and T_1^{other} are the relaxation times for dipole-dipole relaxation, chemical shift anisotropy, quadrupolar, and any other sources. In favorable

1.5 NMR RELAXATION

cases the relaxation is dominated by one of these mechanisms and the interpretation is relatively simple.

The relaxation times can be measured for any NMR-active nuclei, and depending on the nucleus and its environment, the relaxation will be caused by one of the mechanisms discussed in the preceding relaxation mechanisms. Proton relaxation times are the easiest to measure, but the interpretation of the proton relaxation rates can be complicated by magnetization exchange between protons in close proximity ($r < 5$Å). When the rate of molecular motion is less than the Larmour frequency (the slow motion limit, $\omega^2 \tau_c^2 \gg 1$), protons can efficiently relax by spin exchange with their neighbors. This process provides important information about the proximity of pairs of protons, but makes the spin-lattice relaxation times difficult to interpret. The rate of magnetization exchange, which can be measured by 2D exchange NMR, provides important information about the structure, since it depends on the inverse sixth power of the internuclear distance. Magnetization exchange is very efficient in solids and is known as *spin diffusion*, since the propagation of magnetization in solids follows the diffusion equation. Spin diffusion in solids can extend over hundreds of angstroms and provides important information about the length scale of phase separation in semicrystalline polymers, block copolymers, blends, and mixtures.

The carbon relaxation times are most commonly measured to obtain information about the polymer molecular dynamics. The relaxation for protonated carbons is usually due to dipole–dipole interactions, so the interpretation is relatively simple. Carbonyl and aromatic carbons can have contributions from chemical shift anisotropy, and the relaxation times must be measured as a function of magnetic field strength to measure the relative contributions from dipole-dipole interactions and chemical shift anisotropy (24). Deuterium NMR is also used to investigate the dynamics of polymers that have been labeled with deuterium (25). The interpretation of the deuterium relaxation times is relatively simple because the relaxation is dominated by quadrupolar interactions.

The longitudinal or spin-lattice relaxation measures the return to equilibrium of the magnetization along the z axis. We can understand this process by considering a two-spin system, with spins I and S that have equilibrium magnetization along the z axis (M_I^z and M_S^z). The rate of return to equilibrium following a perturbation of the spin system is given by

$$\frac{dM_I^z}{dt} = -\rho \left(M_I^z - M_I^0 \right) - \sigma \left(M_S^z - M_S^0 \right) \tag{1.62}$$

where M_I^0 and M_S^0 are the equilibrium magnetizations for the I and S spins. The terms σ and ρ contain numerical factors and spectral density function that depend on the relaxation mechanism (dipolar, quadrupolar, etc.) and will be discussed in more detail in Chapter 5. It is clear from Equation (1.62) that the rate of return to equilibrium depends on the spin populations of the spins at the beginning of and during the experiment.

There are two basic types of spin-lattice relaxation rates that can be measured: *selective* relaxation rates, where pulses are applied only to the I spin, and *nonselective* relaxation rates, where pulses are applied to both the I and S spins. In a selective experiment it is common to invert the magnetization of spin I while leaving the magnetization of the S spin at equilibrium. The starting conditions for this experiment are

$$M_I^z - M_I^0 = -2M_I^z \tag{1.63}$$

and

$$M_S^z - M_S^0 = 0 \tag{1.64}$$

and the relaxation is given by

$$\frac{dM_I^z}{dt} = -\rho \left(M_I^z - M_I^0 \right) \tag{1.65}$$

In the case of nonselective relaxation, both the I and S spins are inverted to start the experiment and we have

$$M_I^z - M_I^0 = -2M_I^z \tag{1.66}$$

and

$$M_S^z - M_S^0 = -2M_S^z \tag{1.67}$$

For homonuclear relaxation $M_I^z = M_S^z$ and the relaxation is given by

$$\frac{dM_I^z}{dt} = -(\sigma + \rho) \left(M_I^z - M_I^0 \right) \tag{1.68}$$

Selective relaxation is most commonly used to measure the relaxation of insensitive nuclei like carbon, silicon, phosphorus, and nitrogen. The main source of relaxation in most polymers is heteronuclear dipolar interactions with the nearby protons. The protons and these insensitive nuclei are widely separated in frequency, so it is easy to excite the carbons with pulses while not perturbing the protons, for example. For those nuclei present at a low natural abundance there is a very small probability that two NMR-active isotopes will be close enough to affect the spin-lattice relaxation. In carbon NMR, for example, the natural abundance is 1.1%, so the fraction of ^{13}C's that are near each other is $(0.01)^2$ or 0.0001. Therefore, the carbons can be considered as isolated spins that are relaxed by protons.

The situation is more complex for homonuclear proton relaxation. The protons cause the relaxation of their neighbors and the signals are not widely separated in frequency, so selective excitation can be difficult. It is more common in proton NMR to perform a relaxation rate measurement by inverting all of the protons and measuring the nonselective relaxation-rate. The selective relaxation rates are easier to interpret and can be measured using shaped pulses to excite only a limited frequency range or in 2D exchange NMR.

1.5.3.1 Heteronuclear Spin-Lattice Relaxation

The heteronuclear spin-lattice relaxation rates (usually carbon) are often used to study the molecular dynamics of polymers. For the simple case of a methylene carbon relaxed by the directly attached protons, the relaxation is predominantly due to dipolar interaction, and the relaxation rate is given by

$$\frac{1}{T_1^{DD}} = \frac{n}{10}\left[\frac{\mu_0}{4\pi}\right]\frac{\gamma_C^2 \gamma_H^2 \hbar^2}{r^6}\{J(\omega_H - \omega_C) + 3J(\omega_C) + 6J(\omega_H + \omega_C)\} \quad (1.69)$$

where n is the number of attached protons, μ_0 is the vacuum magnetic permeability, γ_H and γ_C are the magnetogyric ratios for protons and carbons, ω_H and ω_C are the resonant frequencies for protons and carbons, and r is the internuclear distance. The spectral density functions $J(\omega)$ depend on the rate and amplitude of molecular motion. For the simple case, isotropic rotational motion of the spectral density is given by

$$J(\omega) = \frac{\tau_c}{1 + \omega^2 \tau_c^2} \quad (1.70)$$

where τ_c is the rotational correlation time.

Equation (1.69) shows that the relaxation rates depend on the inverse sixth power of the internuclear distance, the carbon and proton frequencies, which of course depend on the magnetic field strength, and the correlation time. For protonated carbons the relaxation is dominated by the dipolar interactions with the directly bonded protons. The carbon–hydrogen distance is taken as 1.08 Å for methine and methylene carbons, and 1.09 Å for aromatic carbons. In such cases, the only unknown variable is the correlation time.

One of the limitations of using relaxation rates to probe the molecular dynamics of polymers is that it is possible for different models for molecular motion to give the same relaxation rates. For this reason it is often desirable to gather as much information as possible by measuring the relaxation rates as a function of temperature and magnetic field strength. For the purposes of this discussion we will use the isotropic-motion model for the correlation time. Many studies have shown that this model is inadequate to describe the complex molecular motions of polymers (22). This topic is discussed in more detail in Section 5.2.

Figure 1.42 shows a plot of the carbon spin-lattice relaxation rate as a function of correlation time at magnetic field strengths of 11.7 and 4.6 T, corresponding to proton frequencies of 200 and 500 MHz and carbon frequencies of 50 and 125 MHz. These plots have characteristic features that have important implications for using NMR to study the molecular dynamics of polymers. One important feature is the T_1 minimum observed at the inverse of the observation frequency ($\omega/2\pi = 1.2$ and 3.1 ns). The minimum in the relaxation rate is due to the fact that terms in Equation (1.69) contribute differently to the relaxation rate in the fast- and slow-motion limit. These limits are defined by the relative ratio of the correlation time and the observation frequency. The dynamics are said to be in the fast motion limit when the correlation

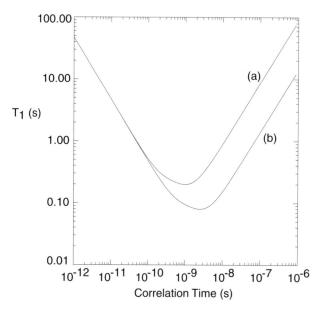

FIGURE 1.42 A plot of the carbon spin lattice relaxation time as a function of correlation time for carbons at (a) 125 MHz and (b) 50 MHz.

time is much less than the inverse of the resonance frequency ($\omega\tau_c < 1$), while the slow-motion limit is when the correlation time is much longer than the inverse of the resonance frequency ($\omega\tau_c \gg 1$). The denominator in Equation (1.70) is equal to 1 in the fast-motion limit, but much larger than 1 in the slow-motion limit.

We can conclude from the hyperbolic shape of the relaxation rate as a function of correlation time that, except for the T_1 minimum, it is not possible to calculate the correlation time from a single relaxation-rate measurement. For correlation times other than at the T_1 minimum, two possible correlation times are consistent with the measured relaxation time. It is often possible to determine whether the molecular motion is on the fast or slow side of the T_1 minimum by making relaxation-rate measurements at two temperatures. Since chain motion is a thermally activated process, a faster correlation time will be observed at the higher temperature. If the T_1 decreases with increasing temperature, then the molecular motion must be on the slow side of the T_1 minimum. Conversely, if the relaxation time increases with increasing temperature, then the molecular motion is fast relative to the T_1 minimum.

It is important to note that there is not enough information in the spin-lattice relaxation-rate measurements to completely characterize the molecular dynamics of polymers. A complete description of the chain dynamics requires the measurement of the relaxation times as a function temperature and magnetic field strength over as wide a range as possible. The molecular dynamics are thermally activated so it is possible to change the correlation times by changing the temperature. The shape of the relaxation time versus the correlation time plot, and the position and depth of the

1.5 NMR RELAXATION

T_1 minimum contain important information about the details of the molecular motion. The relaxation times are not proof of a particular motional model, but they show that the motional model is consistent with the NMR data.

1.5.3.2 Homonuclear Spin-Lattice Relaxation

The spin-lattice relaxation arising from homonuclear dipolar interactions among protons is one of the easiest relaxation rates to measure, but the results are often difficult to interpret in terms of the molecular dynamics. Selective relaxation rates are the easiest to interpret, but these experiments are difficult to perform because of signal overlap in the proton spectrum. In favorable cases it is possible to excite one type of protons using a selective rf pulse. In this case the relaxation is due both to the σ and ρ terms in Equation (1.68) and the relaxation rate from dipolar interactions is given by

$$\frac{1}{T_1^{DD}} = \frac{n}{20}\left(\frac{\mu_0}{4\pi}\right)^2 \frac{\gamma_H^4 \hbar}{r^{-6}} \{J(0) + 3J(\omega_H) + 6J(2\omega_H)\} \quad (1.71)$$

The return to equilibrium following selective excitation is often not a simple exponential process because of the cross relaxation term σ in Equation (1.68). The $M_S^z - M_S^0$ term changes during relaxation as magnetization is exchanged between the I and S spins. The spin I can relax by spin exchange (a mutual spin flip), so M_S^z builds up as relaxation proceeds. The relaxation can be modeled as a system of coupled differential equations, or the T_1 can be measured from the initial relaxation rate.

It is more common to measure the relaxation following excitation of the entire spin system. This is known as nonselective relaxation and the rate is given by

$$\frac{1}{T_1^{DD}} = \frac{n}{20}\left(\frac{\mu_0}{4\pi}\right)^2 \frac{\gamma_H^4 \hbar}{r^{-6}} \{3J(\omega_H) + 12J(2\omega_H)\} \quad (1.72)$$

Note that Equation (1.72) does not contain the $J(0)$ term of Equation (1.71). In the fast motion limit the $J(0)$ term is small and it is possible to interpret the relaxation times in terms of the molecular motions. In the slow motion limit ($\omega\tau_c > 1$) the $J(0)$ or spin diffusion term can be very large and dominates the relaxation of the spin system. The net effect is that spin diffusion averages the nonselective relaxation rate, so the same relaxation times are measured for all protons, regardless of differences in their chain dynamics or internuclear distances. Unfortunately the dynamics of many polymers fall in the slow-motion limit and extreme caution should be used for the interpretation of the nonselective proton spin-lattice relaxation rates. The relaxation times of protons in solids are dominated by spin diffusion, making it difficult to extract molecular-level information about the dynamics from these measurements.

1.5.4 Spin-Spin Relaxation

The transverse or spin-spin relaxation time, T_2, measures the loss of magnetization in the xy plane, and it is this process that leads to signal decay in the free-induction decay. In favorable cases, the T_2 can be measured from the linewidth as

$$T_2 = \frac{1}{\pi \delta \nu_{1/2}} \tag{1.73}$$

where $\delta \nu_{1/2}$ is the observed linewidth. While it is possible in principle to measure the spin-spin relaxation times from the linewidths, it is often difficult in practice because other factors may also contribute to the linewidth. Magnetic field inhomogeneities can also contribute to the observed linewidths, and the T_2 measured from the linewidth is often denoted T_2^* and is given by

$$\frac{1}{T_2^*} = \frac{1}{T_2} + \frac{\gamma \delta B}{2} \tag{1.74}$$

where $\gamma \delta B_0/2$ is the magnetic field inhomogeneity. In polymers there is the additional problem that local structures (stereochemistry, defects, etc.) can lead to peaks that are slightly shifted from each other. If the shifts are not large enough that separate signals are observed, then the peak will appear to be broadened. Broadening by this chemical shift dispersion is called inhomogeneous broadening and is not related to spin-spin relaxation. The spin-spin relaxation times can be measured using a spin-echo experiment, as discussed in Section 2.6.1.2. Modifications of the standard spin-echo that remove the contribution from inhomogeneous broadening from the relaxation can be used, so the "real" T_2 can be measured.

As with the spin-lattice relaxation, there are several possible mechanisms that can contribute to the spin-spin relaxation, including dipolar interactions, chemical shift anisotropy, quadrupolar interactions, and chemical exchange. The observed relaxation rate is given by the sum of all of these interactions

$$\frac{1}{T_2^{obs}} = \frac{1}{T_2^{DD}} + \frac{1}{T_2^{CSA}} + \frac{1}{T_2^{QUAD}} + \frac{1}{T_2^{other}} \tag{1.75}$$

The spin-spin relaxation of a carbon by a nearby proton from dipolar interactions is given by

$$\frac{1}{T_2^{DD}} = \frac{n}{20} \left(\frac{\mu_0}{4\pi}\right)^2 \frac{\gamma_C^2 \gamma_H^2 \hbar}{r^{-6}} \{4J(0) + J(\omega_H - \omega_C) + 3J(\omega_C) \\ + 6J(\omega_H) + 6J(\omega_H + \omega_C)\} \tag{1.76}$$

Figure 1.43 shows a plot of the T_2 as a function of correlation time at carbon frequencies of 50 and 125 MHz. Note that there is no minimum in the T_2, but rather the T_2 becomes shorter as the correlation time decreases. This leads to a decrease in the resolution from an increase in the linewidth, as given by Equation (1.73). It was initially expected that polymers would give very poorly resolved spectra because they are high molecular-weight materials and long correlation times are expected. Fortunately,

1.5 NMR RELAXATION

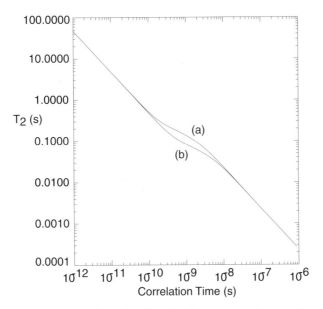

FIGURE 1.43 A plot of the carbon spin-spin relaxation time as a function of correlation time for carbons at (a) 125 MHz and (b) 50 MHz.

the relaxation is not dominated by the overall reorientation of the entire chain, but rather by more localized atomic motions such as librations and *gauche–trans* isomerizations. These segmental motions drastically shorten the effective correlation times so polymer linewidths on the order of 1–10 Hz are commonly observed.

1.5.5 The Nuclear Overhauser Effect

The nuclear Overhauser effect (NOE) is another relaxation parameter often used to characterize the chain dynamics of polymers in solution. The steady-state NOE is measured as the intensity change observed when a nearby spin is saturated with an rf field and is given by

$$\frac{M_I(S \text{ irradiated})}{M_I^0} = 1 + \frac{\sigma}{\rho} \cdot \frac{\gamma_S}{\gamma_I} \qquad (1.77)$$

Both hetero- and homonuclear NOEs can be observed. The magnitude of the NOE depends not only on the chain dynamics, but also on the magnetogyric ratio of the observed and irradiated nuclei. NOEs arise from through-space dipolar interactions, so relaxation by any mechanism other than dipolar interactions diminishes the NOE from its theoretical value. In most cases the signal intensity is increased from the heteronuclear NOE, so the SNR can be improved with proton irradiation. However, the NOEs can vary from atom to atom and must be suppressed if a quantitative interpretation of the spectra is desired.

1.5.5.1 Heteronuclear Nuclear Overhauser Effects The most commonly measured heteronuclear NOEs are for carbons with steady-state proton irradiation. In this case the NOE is given by

$$\frac{M_C(^1\text{H irradiation})}{M_C^0} = 1 + \frac{6J(\omega_C + \omega_H) - J(\omega_C - \omega_H)}{J(\omega_C - \omega_H) + 3J(\omega_C) + 6J(\omega_C + \omega_H)} \cdot \frac{\gamma_H}{\gamma_C} \quad (1.78)$$

or

$$\frac{M_C(^1\text{H irradiation})}{M_C^0} = 1 + \eta_{CH} \quad (1.79)$$

where η_{CH} is the $^{13}\text{C}\{^1\text{H}\}$ NOE enhancement factor. Figure 1.44 shows a plot of the NOE as a function of correlation time at carbon observation frequencies of 50 and 125 MHz. The NOE is insensitive to the correlation time for correlation times that are much faster or slower than the observation frequency, but the NOE changes rapidly at correlation times near the inverse of the observation frequency. In cases where $\omega_C \tau_c \cong 1$, the NOE will be very sensitive to the magnetic field strength. In all cases the NOE leads to a signal enhancement for carbons. However, since different parts of the chain may have different correlation times, as for main-chain and

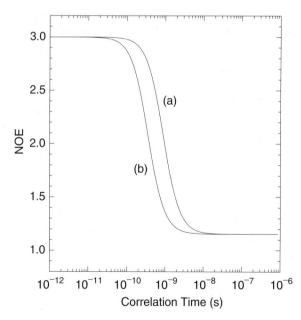

FIGURE 1.44 A plot of the carbon–proton NOE as a function of correlation time for carbons at (a) 125 MHz and (b) 50 MHz.

1.5 NMR RELAXATION

side-chain atoms, the NOE enhancement will be different. If the magnetogyric ratio is negative, as for nitrogen, then the spectrum will appear inverted.

1.5.5.2 Homonuclear Nuclear Overhauser Effects The proton–proton NOEs can be used to measure the chain structure and dynamics and to establish the resonance assignments. Since the magnetogyric ratios are the same for the irradiated and observed nuclei, the NOE is given by

$$\frac{M_H(^1H \text{ irradiation})}{M_H^0} = 1 + \frac{6J(2\omega_H) - J(0)}{J(0) + 3J(\omega_H) + 6J(2\omega_H)} \quad (1.80)$$

Figure 1.45 shows a plot of the NOE as a function of correlation time. The maximum enhancement is 0.5 and a NOE of -1 is observed in the slow motion limit ($\omega_H \tau_c \gg 1$). An important feature to note about Figure 1.45 is the point at which the NOE is zero ($\omega_H \tau_c \cong 1$). Therefore, even if two protons are near in space, it is possible that no NOE will be observed between them. These steady-state NOEs were used in the early days for resonance assignments and for structure determination, but it is now easier to make these measurements using 2D NMR methods (15).

A closely related measurement is the rotating-frame nuclear Overhauser effect (ROE). In this case the signal enhancement is measured using a spin-locking field,

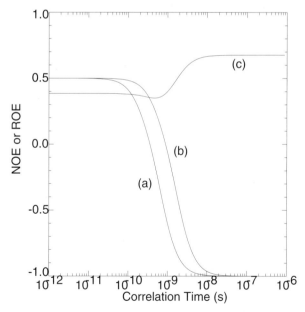

FIGURE 1.45 A plot of the proton–proton NOE as a function of correlation time for protons at (a) 200 MHz and (b) 500 MHz and (c) the ROE at 500 MHz.

and the cross relaxation rates depend on $T_{1\rho}$ relaxation. The ROE is given by

$$\frac{M_H(^1\text{H spin} - \text{lock})}{M_H^0} = 1 + \frac{3J(\omega_1) + J(0)}{J(0) + 3J(\omega_H) + 6J(2\omega_H)} \quad (1.81)$$

where ω_1 is the strength of the spin-lock field. Figure 1.45 also shows a plot of the ROE as a function of the correlation time. The enhancement in the slow-motion limit is 0.38 and increases to 0.68 in the slow-motion limit. The ROE has the advantage over the NOE that the value is always positive and does not have the zero crossing at $\omega_H \tau_c \cong 1$. As with the NOEs, the ROEs are most commonly measured using 2D NMR methods.

1.6 SOLID-STATE NMR

The NMR spectra of solids are fundamentally different from solutions because atomic motions are limited by the proximity of nearby chains. This means that the local interactions that cause line broadening, such as the chemical shift anisotropy and dipolar couplings, are not averaged by molecular motion, and some artificial means must be used to obtain a high-resolution spectrum. This makes the acquisition of solid-state NMR spectra more challenging than for solutions. However, because the information contained in solid-state NMR spectra can be so valuable for understanding structure–property relationships in polymers, many new methods have been developed and modern NMR spectrometers are often equipped with accessories that make the acquisition of solid-state NMR spectra routine.

1.6.1 Chemical Shift Anisotropy

In the discussion thus far we have been treating the chemical shift as a scalar quantity, but it is actually more complex. The chemical shift is *anisotropic* or directional, as it depends on the orientation of the molecule with respect to magnetic field direction. The chemical shift is expressed as a tensor (a mathematical quantity having both direction and magnitude) and is composed of three principal components, σ_{ii}

$$\sigma = \lambda_{11}^2 \sigma_{11} + \lambda_{22}^2 \sigma_{22} + \lambda_{33}^2 \sigma_{33} \quad (1.82)$$

where λ_{ii} are the direction cosines of the principal axes of the screening constant with respect to the magnetic field. The principal axis systems may lie along the bond direction, but this is not necessarily so. The orientation of the axis system cannot be predicted a priori, and must be experimentally determined (26). By convention, the lowest field resonance (largest chemical shift) is taken as σ_{33}. In some publications σ_{xx}, σ_{yy}, and σ_{zz} are used in place of σ_{11}, σ_{22}, and σ_{33}.

The chemical shift anisotropy lineshape depends on the nuclei under study and the dynamics. Nuclei with the largest chemical shift range tend to have the largest

1.6 SOLID-STATE NMR

chemical shift anisotropies. The anisotropic lineshapes depend on two factors: the electron distribution, which gives rise to differences in chemical shifts for the molecules aligned parallel and perpendicular to the field, and the orientational distribution of the molecule. Most samples are isotropic and have a uniform distribution relative to the magnetic field direction. The two parameters that describe the chemical shift anisotropy lineshapes are the shielding anisotropy $\Delta\delta$ and the asymmetry η, which are given by

$$\Delta\sigma = \sigma_{33} - \frac{1}{2}(\sigma_{11} + \sigma_{22}) \tag{1.83}$$

and

$$\eta = \frac{\sigma_{22} - \sigma_{11}}{\sigma_{33} - \sigma_{iso}} \tag{1.84}$$

where the isotropic chemical shift σ_{iso} is given by

$$\sigma_{iso} = \frac{1}{3}(\sigma_{11} + \sigma_{22} + \sigma_{33}) \tag{1.85}$$

For some applications the width of the chemical shift anisotropy lineshape, which is given by $|\sigma_{33} - \sigma_{11}|$, and the coupling constant δ, which is given by

$$\delta = \sigma_{33} - \sigma_{iso} \tag{1.86}$$

are also used. Rapid molecular motion (as for polymers in solution) averages the chemical shift anisotropy and a sharp line at the isotropic chemical shift is observed.

For a better understanding of the concept of chemical shift anisotropy and the lineshapes, it is instructive to consider the case of diphenyl carbonante, a model compound for polycarbonate, shown in Figure 1.46. Using a combination of NMR studies and quantum mechanical calculations the principal components of the chemical shift

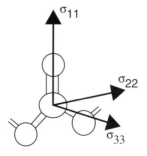

FIGURE 1.46 The chemical drawing and principal axis for the chemical shift anisotropy in diphenyl carbonate.

tensor have been determined (27). Three components are observed in the NMR spectrum of the carbonate at 88, 125, and 239 ppm. The highest field component σ_{11} lies along the axis defined by the carbonate carbon and the double-bonded oxygen. The lowest field component σ_{33} is perpendicular to σ_{11} and lies in the plane of the three oxygen atoms, and the σ_{22} component is perpendicular to the plane of the three oxygen (i.e., coming out of the drawing).

If we were to take the NMR spectrum of a single crystal of dipheny carbonate that had been site-specifically labeled ^{13}C at the carbonate carbon, we would observe a strong carbonate signal and very weak signals from all of the other carbons. If the single crystal was oriented such that the bond vector connecting the carbonate carbon and the double-bonded oxygen lies along the magnetic field direction, the peak frequency would be 88 ppm (corresponding to σ_{11}). If we rotate the crystal by 90° such that the σ_{33} direction lies along the magnetic field direction, we would observe a chemical shift of 239 ppm for the carbonate carbon. In a similar way, if the σ_{22} direction lies along the magnetic field direction, we would observe a chemical shift of 125 ppm.

It is clear from this discussion that the chemical shift anisotropy is a valuable tool for the study of oriented polymers. It is much more common, however, for the samples to be isotropically oriented, and we must consider the effect of having all possible orientations between the principal components of the chemical shift tensor and the magnetic field direction. The relationship between the frequency and orientation is given by

$$\omega = \omega_{\text{iso}} + \frac{\delta}{2}(3\cos^2\theta - 1 - \eta\sin^2\theta\cos\phi) \quad (1.87)$$

where θ and ϕ are the angles relating the principal axes of the molecule to magnetic field direction, and δ is the coupling constant.

Equation (1.87) shows that there is a direct relationship between the orientation of the principal axis of the chemical shift tensor and the frequency. However, from statistical considerations, not all orientations are equally likely. There is only one orientation directly along the field direction, for example, while there are many possible orientations that are perpendicular to the field. Since the probabilities of all orientations are not equally likely, not all frequencies are equally likely. This gives rise to a characteristic lineshape for the chemical shift anisotropy.

There are two characteristic lineshapes for the chemical shift anisotropy that depend on the molecular symmetry. The lineshapes for the general case ($\sigma_{11} \neq \sigma_{22} \neq \sigma_{33}$) and for the axially symmetric case are shown in Figure 1.47 ($\sigma_{11} = \sigma_{22} \neq \sigma_{33}$). If the spectra are well-resolved, the principal values of the chemical shifts can be determined directly from the spectra.

In this discussion we are considering only the chemical shift anisotropy lineshapes for polymers in the absence of molecular motion. Molecular motions lead to a change in the orientation for a particular atom, and (as expected from Equation (1.87)), a change in frequency. The effect that such mobility has on the chemical shift anisotropy lineshape depends on the amplitude and frequency of motion. This makes

1.6 SOLID-STATE NMR

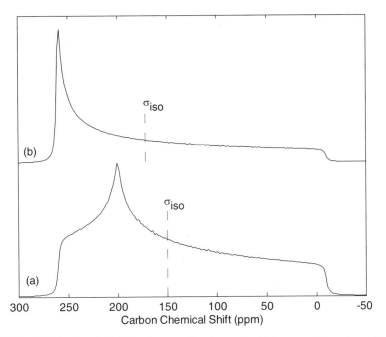

FIGURE 1.47 The chemical shift anisotropy lineshapes for (a) the general case ($\sigma_{11} \neq \sigma_{22} \neq \sigma_{33}$) and (b) the axial symmetric case ($\sigma_{11} = \sigma_{22} \neq \sigma_{33}$).

the anisotropic lineshapes a valuable tool for the study of polymer dynamics, a topic discussed in more detail in Section 5.3.2.

The chemical shift anisotropy is important for polymer NMR studies because it affects the resolution. In many cases the width of the chemical shift anisotropy lineshape is almost as large as the chemical shift range. Under these conditions all of the peaks overlap and a high-resolution spectrum cannot be obtained without some means to average the chemical shift anisotropy. Magic-angle sample spinning is the means most commonly used to obtain a high-resolution spectrum.

The magnitude of the chemical shift anisotropy depends on the nuclei under study and on the electron distribution, which can vary dramatically from molecule to molecule and group to group. As a general rule those nuclei with the largest chemical shift range have the largest chemical shift anisotropy. Thus the chemical shift anisotropy lineshapes are much greater for carbons (which have a chemical shift range of 200 ppm) compared to those of protons (which have a chemical shift range of 10 ppm).

Table 1.6 shows the principal values of the proton chemical shift tensor for several proton types. The carboxyl protons have the largest chemical shift anisotropy width of 21 ppm. The others are much smaller, and those protons in a more symmetric environment give a smaller anisotropy.

Table 1.7 shows the chemical shift anisotropy values for several carbon types. Note that very large values are observed for aromatic and carbonyl carbons, while

TABLE 1.6 Representative Values for the Proton Chemical Shift Anisotropy Lineshapes.

Structure	σ_{11} (ppm)	σ_{22} (ppm)	σ_{33} (ppm)	$|\sigma_{33}-\sigma_{11}|$ (ppm)
Carboxyl Aromatic Olefinic	−8	−4	13	21
Methylene Methyl	−2	−2	6	8

smaller anisotropies are observed for methylene and methyl carbons. Even though the anisotropies are smaller for the methyl and methylene carbons, the anisotropy would lead to extensive peak overlap in the absence of magic-angle sample spinning. As we will see later (Section 1.6.2), the width of the chemical shift anisotropy lineshape determines how fast we must spin to obtain a high-resolution spectrum without spinning sidebands.

The chemical shift anisotropy is also large for fluorine and phosphorus. In poly(tetrafluoro ethylene), for example, the chemical shift anisotropy width is 113 ppm (28), which corresponds to a linewidth of 53 kHz in a 11.7-T magnet. The anisotropies for phosphorus, silicon, and nitrogen can be as large as 400, 50, and 300 ppm, respectively.

1.6.2 Magic-Angle Sample Spinning

The chemical shift anisotropy lineshapes contain valuable information about orientational order in polymers, but the peaks tend to overlap in all but the simplest polymers. In order to observe a high-resolution spectrum, it is necessary to sacrifice this information and collapse the lineshape to a single sharp line. This can be accomplished by magic-angle sample spinning.

We know from Equation (1.87) that the chemical shift depends on the orientation of the chemical shift anisotropy axis relative to the magnetic field. Anything that changes the molecular orientation, including molecular motion, diffusion, or rotating the sample, leads to a change in the chemical shift. If we rapidly rotate the sample the orientations and chemical shifts become time dependent, and the time average under

TABLE 1.7 Representative Values for the Carbon Chemical Shift Anisotropy Lineshapes.

Structure	σ_{11} (ppm)	σ_{22} (ppm)	σ_{33} (ppm)	$|\sigma_{33}-\sigma_{11}|$ (ppm)
Nonprotonated Aromatic	30	100	240	210
Protonated Aromatic	30	130	280	250
Carbonyl	80	260	280	200
Methylene	10	40	50	40
Methyl	0	20	25	25

1.6 SOLID-STATE NMR

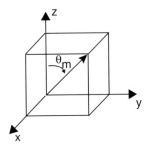

FIGURE 1.48 A diagram illustrating magic-angle sample spinning.

rapid rotation is given by

$$\sigma = \frac{1}{2} \sin^2 \beta (\sigma_{11} + \sigma_{22} + \sigma_{33}) + \frac{1}{2}(3\cos^2 \beta - 1) \qquad (1.88)$$

where β is the angle between the rotation axis and the magnetic field direction. When β is equal to the so-called magic angle (54.7°, or the body diagonal of a cube), $\sin^2 \beta$ is 2/3 and the first term becomes equal to one-third of the trace of the tensor (i.e., the isotropic chemical shift) and the $(3\cos^2 \beta - 1)$ term is equal to zero. Thus, under rapid magic-angle sample rotation, the chemical shift pattern collapses to the isotropic average, giving the high-resolution spectrum.

We can gain a further understanding of magic-angle sample spinning by considering the geometry of sample rotation. The rotation that most effectively averages the chemical shift anisotropy is one in which the x, y, and z components interchange. We can visualize this using a cube centered at the x, y, and z axis, with a vector connecting the axis to the [1,1,1] coordinate (Figure 1.48). If we rotate about this vector connecting the unit cube to the origin, each 120° rotation will interchange the axes (x to y, y to z, and z to x). The vector connecting the origin to the [1,1,1] coordinate of the cube makes an angle of 54.7° relative to the magnetic field.

Rotation about the body diagonal of the unit cube is referred to as the magic angle because it is the only angle that leads to a complete averaging of the chemical shift anisotropy. Rapid rotation about other angles leads to incomplete averaging leads to partial averaging, but not a complete collapse of the lineshape. This is illustrated in Figure 1.49, which shows a chemical shift anisotropy line with rapid sample spinning as a function of the spinning angle. The chemical shift anisotropy lineshape is scaled with fast rotation and the breadth of the pattern is decreased as the angle is increased from zero to 54.7°. As the angle is increased beyond the magic angle, the breadth is again increased and the features are reversed relative to the static pattern.

Thus far, we have discussed the situation when the rate of magic-angle spinning is fast compared to the width of the chemical shift anisotropy lineshape, and spinning at the magic angle leads to the complete collapse of the broadening from the chemical shift anisotropy. It is often the case that the breadth of the chemical shift anisotropy

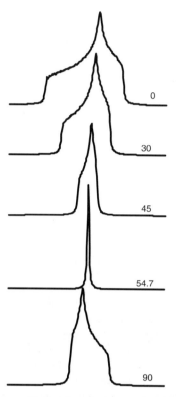

FIGURE 1.49 The effect or rapid sample spinning on the chemical shift anisotropy as a function of the spinning angle.

pattern is larger than the available spinning speed. As we will see in Section 2.3.1, the maximum achievable spinning speed depends on the probe design and the rotor diameter. For relatively insensitive nuclei like carbon, larger rotors that have a greater sample volume are required, while smaller rotors can be used for higher sensitivity nuclei like protons and fluorines. A typical 7-mm rotor used for carbon NMR may have a maximum spinning speed of 7–8 kHz. As we saw in Table 1.7, the breadth of the chemical shift anisotropy pattern for a protonated aromatic carbon may be on the order of 250 ppm. In an 11.7-T magnet where the carbon frequency is 125 MHz, this corresponds to a linewidth of greater than 30 kHz, which is clearly beyond the spinning frequency for the 7-mm probe. Under these conditions the lineshape is split into a series of sharp spinning sidebands that trace out the chemical shift anisotropy lineshape. The separation between the sidebands is the spinning speed. The sideband at the isotropic shift grows in intensity at the expense of the other sidebands as the spinning speed is increased.

The effect of spinning speed on the carbon spectrum of crystalline alanine is shown in Figure 1.50. The chemical shift anisotropy lineshapes for the carbonyl, methine, and methyl carbons for alanine can be recognized in the static spectrum acquired with proton decoupling to remove the dipolar broadening. As the spinning speed is

1.6 SOLID-STATE NMR

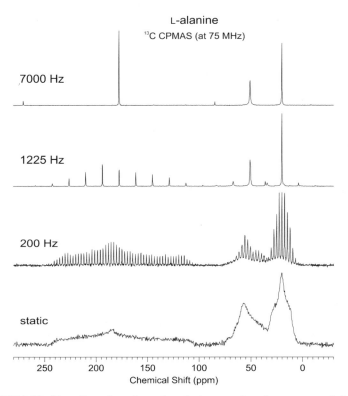

FIGURE 1.50 The effect of magic-angle spinning speed on the spectrum of alanine.

increased to 200 Hz, the chemical shift anisotropy patterns are traced out by a series of sharp sidebands, where each of the sidebands is separated from its neighbor by the spinning speed. Note that as the spinning speed increases, the isotropic peak increases in intensity, while all of the spinning sidebands become less intense.

The spinning sidebands can present problems for the high-resolution analysis of polymers because they can overlap with the peaks of interest. The simplest solution is spinning fast enough such that the sidebands have a negligible intensity, but this is not always possible. The spinning sidebands can be easily identified because they change their position with spinning frequency. In some cases it is necessary to change the spinning speed to ensure that the peaks of interest do not overlap with any of the sidebands. The sidebands can be eliminated using a solid-state spectral editing method where 180° pulses are applied at specific points during the sample rotation using the so-called total suppression of sidebands (TOSS) method (29).

1.6.3 Dipolar Broadening and Decoupling

The dipolar couplings are a major source of line broadening in solids, and high-resolution spectra can only be observed if we remove this broadening. The field experienced by a nucleus is the sum of the external magnetic field and the local fields from nearby nuclei. The separation between the energy levels due to dipolar couplings depends both on the distance r between the nuclei and the orientation of the vector connecting the nuclei θ and the magnetic field and is given by

$$\Delta E = 2\mu \left[B_0 + B_{\text{loc}}\right] = B_0 \pm \frac{3\mu}{r^3}\left(3\cos\theta - 1\right) \tag{1.89}$$

The $\pm$ is due to the fact that the local field may add or subtract from the local field depending on whether the nucleus is aligned with or against the magnetic field. If we consider the case of an isolated CH pair in a single crystal that has a fixed orientation relative to the external magnetic field, the dipolar coupling D (in Hz) is given by

$$D = \frac{\hbar \gamma_C \gamma_H}{2\pi r^3}\left(3\cos^2\theta - 1\right) \tag{1.90}$$

The spectrum for this isolated pair would appear as a doublet where the peak separation is given by Equation (1.90). The dipolar couplings are often larger than the chemical shift range, and they depend on the nuclei involved through the magnetogyric ratios. For carbon–proton pairs, the dipolar couplings are on the order of 30–50 kHz.

Most polymers of interest are not single crystals, but rather semicrystalline materials with randomly oriented crystallites or isotropic amorphous materials. Under these conditions we do not have isolated spin pairs with a fixed orientation relative to the external field, but rather each nucleus experiences a variety of dipolar interactions that differ in distance and orientation relative to the field. Both of these factors affect the dipolar lineshape.

To better understand the effect of orientation on the dipolar lineshape let us consider an isolated carbon–proton pair that is isotropically oriented relative to the field. In this case the carbon–proton distance is fixed and the dipolar coupling for each pair depends only on the orientation relative to the field. The lineshape can be calculated by summing the spectra for all possible orientations of the pair. The result is the so-called "Pake" powder pattern shown in Figure 1.51 (30). Note that the dipolar coupling contains the same dependence on orientation $(3\cos^2\theta - 1)$ as the chemical shift anisotropy (Equation (1.87)) and that the Pake powder pattern can be obtained by reversing and adding the lineshapes from axially symmetric chemical shift anisotropy lineshapes. If such dipolar lineshapes can be observed, they provide information about the magnitude of the dipolar couplings that can provide important information about internuclear distances.

In most polymer samples, the nuclei of interest experience dipolar couplings from several nearby NMR-active nuclei. For carbons, for example, there can be dipolar couplings to both directly bonded and nearest-neighbor protons. Under these conditions, the observed dipolar lineshape results from the sum of all of these interactions, which is often observed as a broad, unresolved lineshape (Figure 1.51c).

It is necessary to remove the dipolar couplings to observe a high-resolution spectrum. This is most commonly accomplished with dipolar decoupling in which high-power irradiation is applied to one of the spins (protons) to cause rapid transitions

1.6 SOLID-STATE NMR

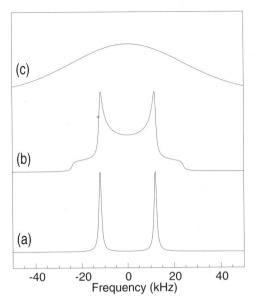

FIGURE 1.51 The effect of dipolar coupling on the lineshape for (a) an isolated single crystal with a fixed orientation relative to the field, (b) an isotropic power with a single carbon–proton distance, and (c) an isotropic powder with a variety of distances.

between the spin states. The magnitude of the rf field must be larger than the dipolar coupling for efficient line narrowing. Since the dipolar coupling has the same dependence on the orientation as the chemical shift anisotropy, the line broadening can also be removed by fast magic-angle sample spinning, provided that the spinning speed is on the same order of magnitude as the dipolar coupling. With recent advances in probe design it is now possible to spin small rotors faster than 30 kHz to remove the dipolar couplings (31). However, this method is mostly used for abundant nuclei such as protons and fluorines, because the sample volumes in the rotors that are able to spin at 30 kHz are very small. The dipolar couplings can also be averaged by molecular motions, such as those in polymers above their glass transition temperatures. It is also possible to use multiple-pulse irradiation to remove the dipolar couplings so that the proton spectrum can be directly observed in solids using combined rotation and multiple-pulse spectroscopy (CRAMPS) (28), as discussed in Section 2.5.3.

1.6.4 Cross Polarization

Solid-state NMR is an important method for polymer characterization, but many of the nuclei of interest, including carbon, nitrogen, and silicon, suffer from a low sensitivity. This results from the inherent low sensitivity of these nuclei (Table 1.1) and from the very long spin-lattice relaxation times in solids where the molecular motion is restricted by the local environment. This sensitivity problem can be partially overcome using *cross polarization*. Cross polarization works by forcing the insensitive nuclei and the protons to precess at the same frequency in the rotating frame even though they differ in frequency by many MHz in the laboratory frame. As with many other spectroscopies, efficient energy transfer occurs when the energy levels are matched. Under these conditions magnetization can equilibrate between the protons and the less sensitive nuclei. Since the protons have a higher magnetogyric ratio, this leads to sensitivity enhancement. The observed signal is derived from the proton spin bath, so it is the shorter proton relaxation time that determines the pulse repetition time for signal averaging. This leads to a better SNR, since the protons typically have shorter spin-lattice relaxation times.

Cross polarization was first demonstrated by Hartmann and Hahn in 1962 (33), when it was shown that energy transfer between nuclei with widely differing Larmor frequencies can be made to occur when the nuclei were simultaneously irradiated with matched rf fields. For carbons and protons this can occur when

$$\gamma_C B_{1C} = \gamma_H B_{1H} \tag{1.91}$$

the so-called *Hartmann–Hahn condition* is satisfied. Since γ_H is four times γ_C, the Hartmann–Hahn match occurs when the strength of the applied carbon field B_{1C} is four times the strength of the applied proton field, B_{1H}. Equilibration of the carbon and protons spins leads by rapid spin exchange to a fourfold increase in the carbon magnetization because

$$\frac{\gamma_H}{\gamma_C} = 4 \tag{1.92}$$

The signal intensity during cross polarization represents a compromise between several processes, including the buildup of intensity from carbon–proton cross polarization and the decay of proton and carbon magnetization in the spin-locking fields. The intensity at cross polarization contact time, t, is given by

$$M(t) = \frac{M_0 \left[1 - e^{-\left(1 + \frac{T_{CH}}{T_{1\rho}(H)} - \frac{T_{CH}}{T_{1\rho}(C)}\right)\frac{t}{T_{CH}}} \right] e^{-t/T_{1\rho}(H)}}{1 + \frac{T_{CH}}{T_{1\rho}(H)} - \frac{T_{CH}}{T_{1\rho}(C)}} \tag{1.93}$$

where T_{CH} is the carbon–proton cross-polarization time constant, and $T_{1\rho}(C)$ and $T_{1\rho}(H)$ are the carbon and proton spin-lattice relaxation times in the presence of the spin-locking fields. The decay times during the spin-locking period are called the rotating-frame spin-lattice relaxation time constants. In most cases, $T_{1\rho}(C) \gg T_{CH}$ and $T_{1\rho}(H) \gg T_{CH}$, so Equation (1.93) simplifies to

$$M(t) = M_0 \left[1 - e^{-t/T_{CH}} \right] e^{-t/T_{1\rho}(H)} \tag{1.94}$$

In most polymers the signal intensity builds up rapidly during cross polarization, and then slowly decays due to $T_{1\rho}(H)$ relaxation. The observed signal intensity depends on the relative values of these relaxation-rate constants, which in turn depend on

1.6 SOLID-STATE NMR

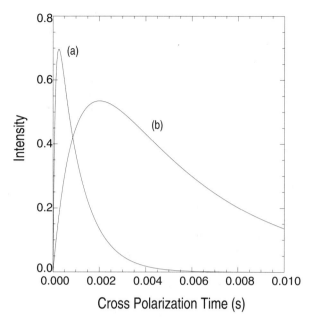

FIGURE 1.52 The time-dependent signal intensities for cross polarization for (a) a rigid polymer and (b) a mobile polymer.

the chain dynamics. This is illustrated in Figure 1.52, which shows the buildup and decay of magnetization during cross polarization. The maximum signal depends on the relative ratios of the T_{CH} and the proton $T_{1\rho}$ processes. The value of T_{CH} is very sensitive to the molecular dynamics, and is much shorter for crystalline polymers relative to amorphous polymers and rubbers. The $T_{1\rho}$ tends to be very different for crystalline polymers than for amorphous or rubbery polymers. This leads to a situation where the cross-polarization dynamics are very different for the different morphologies. For crystalline polymers the magnetization builds up quickly and slowly decays away, leading to a maximum in intensity for short cross-polarization times (<1 ms). Other morphologies build up more slowly, so the maximum intensity is observed at a longer cross-polarization time. If a sample contains a mixture of crystalline and amorphous or rubbery polymer, the cross-polarization contact time can be adjusted to enhance the signal from one of these phases. It is often difficult to quantitatively measure peak intensities using cross polarization because the relative intensities will depend on the molecular dynamics and the rate constants for the buildup and decay of magnetization.

1.6.5 Quadrupolar NMR

The NMR lineshapes and relaxation times of quadrupolar nuclei can be very different from spin-$\frac{1}{2}$ nuclei. Deuterium is the most commonly encountered quadrupolar nucleus in polymers. Deuterium is an insensitive nucleus, both because of its low

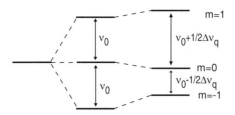

FIGURE 1.53 The energy levels for deuterium showing two transitions that are perturbed by the quadrupolar interaction.

frequency (76 MHz in an 11.7-T magnet) and its very low natural abundance (0.015%). Deuterium NMR studies of polymers are only possible with isotopically labeled polymers.

Deuterium has a spin of 1, so there are three quantized energy levels, $-1, 0$, and $+1$, as shown by Figure 1.53. The quadrupolar interaction perturbs the spacing of the energy levels so that the two transitions are observed between the -1 and 0 states, and between the 0 and $+1$ states (i.e., $\Delta m = 1$). The deuterium signal is observed at

$$\omega = \omega_0 \pm \frac{3e^2qQ}{8\hbar}(3\cos^2\theta - 1 - \eta \sin^2\theta \cos 2\phi) \quad (1.95)$$

where θ and ϕ are the angles relating the deuterium bond vector to the static magnetic field, η is the asymmetry parameter, and $e^2qQ/\hbar$ is the quadrupolar coupling constant. In most cases the deuterium is axially symmetric ($\eta = 0$), so the deuterium frequency is given by

$$\omega = \omega_0 \pm \frac{3e^2qQ}{8\hbar}(3\cos^2\theta - 1) \quad (1.96)$$

As with the dipolar coupling and the chemical shift anisotropy, the deuterium frequency depends on the orientation of the deuterium bond vector relative to the field via the $3\cos^2\theta - 1$ term. However, unlike the chemical shift anisotropy, there are two transitions in the deuterium spectrum. Thus a single deuterium in the field would give rise to a doublet, with the spacing between the peaks determined by the product of the quadrupolar coupling constant and the orientation factor. The maximum splitting would be observed for a deuterium oriented along the field direction ($\theta = 0$), and a single peak would be observed for a deuterium oriented at the magic angle when the orientation term is equal to zero.

The magnitude of the quadrupolar coupling (250 kHz) is much larger than the other interactions (dipolar, chemical shift anisotropy, scalar coupling) and dominates the spectra. Since the quadrupolar coupling is much larger than the range of deuterium chemical shifts, high-resolution spectra are not observed for deuterium-labeled polymers. The deuterium lineshape is very sensitive to the amplitude and frequency of molecular motions, so deuterium NMR is most frequently used to study the molecular

1.7 MULTIDIMENSIONAL NMR

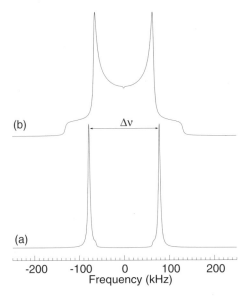

FIGURE 1.54 The static spectrum for (a) an single deuteron and (b) an isotropic distribution of deuterons showing the so-called "Pake" pattern.

dynamics of polymers. Since the splitting in the deuterium spectrum depends on the orientation of the CD bond vector relative to the field, deuterium NMR can be used to study the orientation of polymers. For unoriented polymers, the deuterium lineshape is the result of adding the contributions from all possible orientations of the deuterium bond vector, as was done for the dipolar and chemical shift anisotropy lineshapes. Figure 1.54 shows the "Pake-type" deuterium spectrum expected from an unoriented sample. Very different lineshapes are observed in the presence of molecular motion (Section 5.3.4). Because of the large quadrupolar coupling the lines are very broad and special experimental methods (the quadrupolar-echo) are required to record the spectra (Section 2.5.4).

1.7 MULTIDIMENSIONAL NMR

One factor that often limits polymer NMR studies is the signal overlap that results from different mircrostructures in the polymer chains. This makes it difficult to observe the features of interest, such as defect sites, that may not give signals that are well resolved from the normal (head-to-tail) monomers. One way to overcome these limitations is to expand the chemical shifts into two or more frequency dimensions using nD NMR. There are many kinds of nD NMR experiments that can be classified as *correlated* or *resolved* experiments. In correlated experiments the resonance frequency of one signal is related to those of its neighbors and molecular connectivities or distances between atoms can be determined. In resolved experiments the frequency axes show different interactions. In one kind of resolved experiment, for example, the carbon chemical shift may appear along one axis and the proton–carbon scalar couplings along the other. Extension of these same principles leads to 3D NMR experiments with three independent frequency axes. This has the potential of providing still greater resolution for those cases where the 2D spectrum exhibits extensive overlap.

Experiments with nD NMR are related to the more familiar experiments in that they consist of a series of pulses and delays. They differ from the more common experiments in that we allow the spin system to *evolve* instead of immediately transforming the free-induction decay after a pulse. The pulse sequence for nD NMR experiments can be divided into four periods: *preparation, evolution (t_1), mixing,* and *detection (t_2)*. During the preparation period, the spins are allowed to come to equilibrium, that is the populations of the spin states are allowed to equilibrate with their surroundings. This interval allows the establishment of reproducible starting conditions for the remainder of the experiment. During the evolution or t_1 period, the spins evolve under all the forces acting on the nuclei, including the chemical shifts, dipolar couplings and scalar couplings. One of the things that commonly occurs during the t_1 period is *frequency labeling*, where the spins are labeled (i.e., prepared in a nonequilibrium state) by their frequency in the NMR spectrum. The mixing period may consist of either pulses or delays and results in the transfer of magnetization between spins that have been frequency labeled in the t_1 period. The final period, t_2, is the signal acquisition that is common to all pulsed NMR experiments. The second frequency is introduced by acquiring spectra while systematically incrementing the length of the t_1 period.

Data collection in nD experiments consists of gathering many free-induction decays, each obtained with a different value of the incremented time variable. This is illustrated in Figure 1.55, which shows the pulse sequence diagram for 2D COSY, a homonuclear correlation experiment that allows us to identify nuclei that are connected to each other by scalar couplings. The pulse sequence for COSY consists of two 90° pulses separated by the incremented time variable t_1 and followed by the acquisition time t_2. The first spectrum in the COSY experiment is collected with a very short (a few μs) value for the incremented variable t_1. Each of the following spectra is acquired by systematically incrementing the value for t_1, as shown in Figure 1.56. The free induction decays are collected for each spectrum as a function of the t_1 variable and stored on the computer.

The free-induction decays are Fourier transformed with respect to t_2 to obtain a series of spectra in which the peak intensities or phases are modulated as a function of the t_1 delay. This is illustrated in Figure 1.57, which shows the peak intensity as a function of the delay time t_1. We can see that the intensity is periodically modulated as

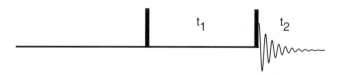

FIGURE 1.55 The pulse sequence diagram for 2D correlated (COSY) NMR spectroscopy.

1.7 MULTIDIMENSIONAL NMR

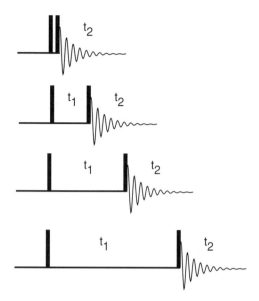

FIGURE 1.56 The pulse sequence diagram for 2D COSY NMR showing a systematic increase in the t_1 delay time.

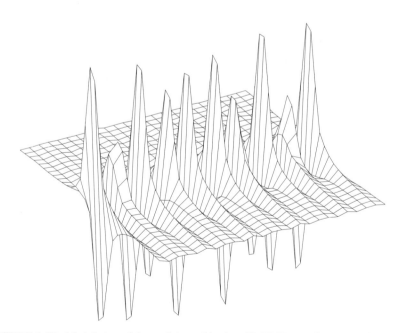

FIGURE 1.57 Modulation of the peak intensities in a 2D NMR experiment as a function of the t_1 delay time.

a function of t_1, and Fourier transformation with respect to t_1 converts these frequency modulations into peaks in the 2D spectrum, which is actually a surface in 3D space. The three coordinates in the 2D spectrum are the two NMR frequencies along the x and y axes and the intensity along the z axis.

The extension to experiments with a higher dimensionality (3D, 4D) simply involves adding additional evolution and mixing periods. The pulse sequence for a 3D experiment would be

preparation-evolution (t_1)–mixing evolution (t_2)–mixing detection (t_3)

For a 3D experiment we have three independent time periods during which the chemical shift evolves. The data are recorded in the usual manner during the detection period. The other two time periods are independently incremented. In a typical experiment we would begin data collection with very short values for t_1 and t_2. Keeping t_2 fixed, we would then collect a series of data by systematically incrementing t_1. We would then increment t_2 and again collect a series of spectra, again incrementing the value of t_1 for each spectrum. Thus, we are essentially collecting a full 2D data set for each value of t_2 in the 3D experiment. The data are processed by transforming the detected data, followed by Fourier transformation of the data with respect to the t_1 and t_2 variables. The result is a 4D plot, with three frequency dimensions and the intensity associated with each point.

There are several ways in which these plots can be displayed and analyzed. The two most popular formats are the contour and stacked plots shown in Figure 1.58. Contour plots are very good for visualizing the correlations between peaks, but are not as good at showing the relative peak intensities. Stacked plots, which are a plot of the intensity as a function of the two frequency dimensions, are better at showing the relative intensities of the peaks, but they make it more difficult to visualize the connectivities. In some cases it is preferable to show cross sections through the 2D spectra at a particular frequency, to show the relative intensities of the interactions.

Another way to analyze these plots is to use projections, or sums, of all the data along one of the frequency axes. In a J-resolved spectrum, for example, the sum of all of the data along the chemical shift dimension results in a pseudo one-dimensional spectrum in which all of the coupling constants have been removed.

Since the data representation for a 2D plot is a 3D object (two frequency dimensions plus intensity), it is difficult to plot nD NMR experiments with three or more frequency dimensions. In many cases the simplest way to visualize the data is to take cross sections through one of the frequency dimensions, as illustrated in Figure 1.59. The cross section through a 3D experiment would correspond to a 2D spectrum associated with one particular frequency in the third dimension. In a 3D spectrum correlating the carbon spectrum with two frequency dimensions, for example, we might take a cross section through the spectrum at a particular carbon frequency. The 2D cross section would show all the proton–proton correlations associated with that particular carbon frequency.

There are many nD NMR experiments that can be used to study the structure and dynamics of polymers. The experiments differ in the mechanism of magnetization

1.7 MULTIDIMENSIONAL NMR

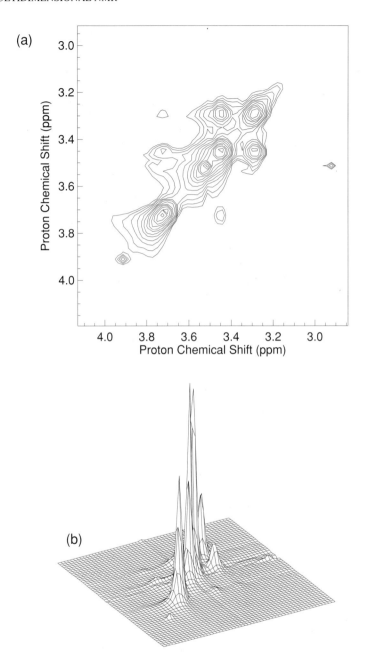

FIGURE 1.58 The (a) contour and (b) stacked plot representation of nD NMR data.

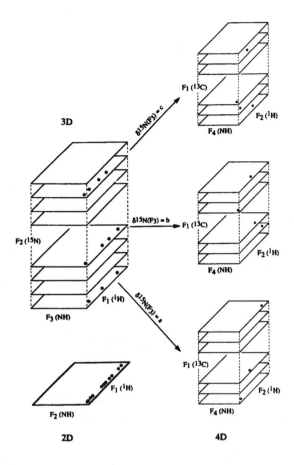

FIGURE 1.59 A schematic diagram showing cross sections through a 3D NMR data set.

transfer and which types of interactions are resolved from the chemical shifts. Among the most common homonuclear methods are COSY (34) and total corrleation spectroscospy (TOCSY) (35) that are used to correlate protons using two- and three-bond proton scalar couplings. Nuclear overhauser effect spectroscopy (NOESY) (15) or 2D exchange spectroscopy provides similar information about proton connectivities, but uses through-space dipolar interactions or chemical exchange to correlate pairs of protons. Heteronuclear multiple-quantum coherence (HMQC) (36) spectroscopy provides a correlation between different types of nuclei, such as carbon and protons via the one- or two-bond heteronuclear couplings. More complex experiments can be constructed by combining different magnetization transfer mechanisms. For example, the carbon–proton HMQC–NOESY combines magnetization transfer from the one-bond carbon–proton J couplings with the through-space dipolar interactions. In the 3D NMR, the magnetization transfer may be the same in all dimensions (e.g., NOESY–NOESY) or different (e.g., HMQC–TOCSY).

1.7 MULTIDIMENSIONAL NMR

1.7.1 Magnetization Transfer in nD NMR

Because it provides the means to assign the peaks and measure the structure and dynamics of polymers, nD NMR has emerged as an important research tool in polymer science. One of the most important uses of NMR is for microstructural characterization, because as new synthetic methods are developed, new analytical methods are required to characterize the new polymers. In many cases the changes in the bulk properties of polymers are the result of relatively small changes to the structure (stereochemistry, regioisomerism, etc.) and NMR provides the means to establish structure–property relationships. Multidimensional NMR has become an important method because it allows us to correlate neighboring sites on the polymer chain through magnetization exchange.

Magnetization exchange can occur as a consequence of either through-space dipolar interactions, as in NOESY NMR, or through-bond scalar couplings, as in TOCSY and HMQC NMR. The fundamental step in the through-space magnetization exchange results from dipolar interactions between nuclei that are close in space ($r < 5$ Å), and depends on the inverse sixth power of the distance. Correlations between nuclei separated by more than 5 Å can occur through a sequential series of transfers. The through-bond magnetization exchange depends on the magnitude of the scalar couplings. The couplings can be either homo- or heteronuclear, although heteronuclear magnetization transfer is much more efficient because the one-bond J couplings are usually much larger than the three-bond proton–proton couplings. With the proper choice of a mixing pulse sequence, multistep exchange through scalar couplings is possible.

1.7.1.1 Through-Bond Magnetization Transfer Thus far we have discussed the nuclear coupling of spins as something that affects the appearance of the spectrum. In reality, these couplings can have many effects, one of the most important being that they provide a mechanism for the transfer of magnetization between nuclei along the polymer chains. The pulse sequences for through-bond magnetization exchange are chosen to control this magnetization transfer. The sequences can restrict the exchange to a single exchange or promote multiple exchanges. The mixing times can be chosen to filter the spectra, such that only correlations for specific values of the scalar coupling constants are observed. In this way it is possible, for example, to observe two-bond heteronuclear correlations while suppressing magnetization transfer from the larger one-bond couplings.

It is difficult to understand all of the effects of spin coupling from the vector diagrams that have been used thus far to introduce pulse sequences. These processes are more easily understood using the POF introduced in Section 1.2.7. The POF provides a framework to understand the complete magnetization transfer and all of the spin states that are generated by a pulse sequence. Many of these spin states do not lead to observable magnetization, and some lead to unwanted peaks that can be removed through phase cycling. For the purpose of this discussion we will focus only on those processes that generate the desired diagonal and cross peaks, and readers are referred to the original literature for a more complete discussion of the POF and nD NMR (10).

We noted earlier that only the magnetization in the xy plane is detected in NMR experiments during the acquisition period. The key to understanding through-bond magnetization transfer is to understand how equilibrium magnetization of one type is converted to observable magnetization of another type. For a two-spin example (I and S), we would like to understand how to convert equilibrium magnetization from one spin (e.g., I_z) into observable magnetization for the other spin (e.g., S_y) and visa versa. The spins I and S can represent two of the same type of nuclei (i.e., two protons with different chemical shifts) or two different types of nuclei (i.e., carbons and protons) that are connected by a scalar coupling. Pulses to the spin system can either be nonselective or selective, depending on whether a homonuclear or heteronuclear spin system is being studied.

Let us first consider a heteronuclear spin system connected by a one-bond scalar coupling, J_{IS}. Since we are considering nuclei that differ widely in frequency, we can selectively apply pulses to the I and S spins. The pulse sequences typically begin with a 90° pulse to tip the magnetization into the xy plane, which in the POF is given by

$$I_z \xrightarrow{\frac{\pi}{2}I_y} I_x \tag{1.97}$$

The spin system is allowed to evolve under the influence of the chemical shift Ω, and the spin state is given by

$$I_x \xrightarrow{\Omega_I t I_z} I_x \cos(\Omega_I t) + I_y \sin(\Omega_I t) \tag{1.98}$$

where t is the evolution time. In a coupled system the spins also evolve under the influence of the scalar coupling. In the nomenclature of the POF, the spins evolve under the influence of the coupling operator $\pi J t 2 I_z S_z$ (10). If we first consider how I_x magnetization evolves under the influence of the coupling operator, the POF gives

$$I_x \xrightarrow{\pi J_{IS} 2 I_z S_z} I_x \cos(\pi J_{IS} t) + 2 I_y S_z \sin(\pi J_{IS} t) \tag{1.99}$$

Note that even though we started with I_z magnetization, there is now a mixed spin state that contains both I and S terms. Such terms are of critical importance because (with additional pulses and delays) they result in the transfer of magnetization from I to S spins. Note also that the term with mixed spin states is modulated by $\sin(\pi J_{IS} t)$. This modulation is related to the efficiency of magnetization transfer and is at a maximum when $t = 1/2J_{IS}$. It is for this reason that we often insert delays into correlation experiments with a delay time equal to $1/2J_{IS}$.

The full expression for the evolution of the spin system under the influence of the chemical shifts and scalar couplings following a pulse applied to the I spins is

$$\begin{aligned} I_z \xrightarrow{\frac{\pi}{2}I_y} \xrightarrow{\pi J_{IS} 2 I_z S_z} & I_x \cos(\Omega t)\cos(\pi J_{IS} t) + 2 I_y S_z \cos(\Omega t)\cos(\pi J_{IS} t) \\ & + I_y \cos(\Omega t)\cos(\pi J_{IS} t) - 2 I_x S_z \sin(\Omega t)\cos(\pi J_{IS} t) \end{aligned} \tag{1.100}$$

1.7 MULTIDIMENSIONAL NMR

The algebra for generating this equation using the POF is straightforward but tedious, as there are many terms. Not all of the terms in Equation (1.100) give rise to observable signals, but they must be carried through the pulse sequence. In the following sections we will limit our discussions to only those terms that give rise to observable magnetization, which appears as diagonal or cross peaks in the nD spectra.

To demonstrate the magnetization transfer, let us consider the fate of the $I_y S_z$ term. If we now apply an x pulse to both the I and S spins, we obtain

$$2I_y S_z \xrightarrow{\frac{\pi}{x} I_x} 2I_z S_z \xrightarrow{\frac{\pi}{x} S_x} -2I_z S_y \quad (1.101)$$

where the sine and cosine terms have been omitted for clarity. The evolution of the $I_z S_y$ spin state under the influence of the I and S chemical shifts gives

$$-2I_z S_y \xrightarrow{(2\pi \Omega t) I_z} -2I_z S_y \xrightarrow{(2\pi \Omega t) S_z} -2I_z S_y + 2I_z S_x \quad (1.102)$$

and evolution under the influence of the IS scalar coupling gives

$$-2I_z S_y + 2I_z S_x \xrightarrow{(\pi J_{IS} t) 2I_z S_z} -2I_z S_y + S_x + 2I_z S_x + S_y \quad (1.103)$$

where the terms S_x and S_y are observable magnetization. Thus, by following the POF through a simple pulse sequence of a 90° pulse and a delay followed by another 90° pulse and delay, the starting I_z magnetization has been converted into observable S magnetization. This result is difficult to visualize using the vector picture, but is given by a straightforward application of the POF. In addition, we will begin to recognize some of the terms in the POF as leading to diagonal and cross peaks in the nD spectra that have specific properties (*vida infra*).

At first glance, many of the nD pulse sequences seem impossibly complex. A closer inspection shows that the pulse sequences are constructed from specific building blocks, and we can often understand the function of a pulse sequence by recognizing these building blocks. In this section we will introduce some of the most common and useful building blocks for nD NMR pulse sequences, including those for homo- and heteronuclear magnetization transfer via the scalar couplings and from magnetization transfer via the through-space dipolar interactions.

Heteronuclear nD NMR experiments often use polarization transfer from protons to a less sensitive nucleus, such as carbon or nitrogen, at the beginning of the experiment. Polarization transfer pulse sequences take advantage of the higher sensitivity of protons and were originally used for sensitivity enhancement in 1D NMR (37). Since protons have the highest sensitivity, it is always better to detect the protons in an nD NMR experiment. For this reason, many heteronuclear nD NMR experiments end with a reverse polarization step to transfer polarization back to the protons for the detection period. In a proton-detected heteronuclear carbon–proton correlation experiment, for example, the experiment begins with polarization transfer from the protons to carbons followed by an evolution period, during which the signals evolve under

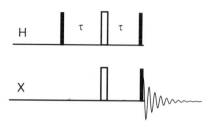

FIGURE 1.60 The pulse-sequence diagram for the INEPT pulse sequence.

the influence of the carbon chemical shifts (the t_1 dimension). A reverse-polarization transfer is used to transfer the magnetization back to the protons for detection (the t_2 dimension). This experiment has a high sensitivity because the experiment both begins and ends with proton magnetization.

One of the most commonly used polarization transfer sequences is the so-called insensitive nuclei enhanced by polarization transfer (INEPT) (37), which is shown in Figure 1.60. INEPT begins with a proton pulse and a delay period $\tau/2$ followed by 180° pulses to both nuclei that refocus the chemical shift, but not the heteronuclear couplings. After another $\tau/2$ delay, 90° pulses are applied to both the protons and the other nuclei. The phase of the pulse for the insensitive nuclei is x, while the phase of the 90° proton pulse is alternated between $+y$ and $-y$ on every other scan and the phase of the receiver is cycled between $+x$ and $-x$ on every other scan, effectively leading to subtraction of the FIDs from every other scan.

Following the initial 90° pulse to the protons, the IS spin system evolves under the influence of the chemical shifts and the coupling constants during the first delay period. The 180° pulse to the protons refocuses the proton chemical shifts and the 180° pulse to the S spins inverts the S spin labels prior to the second delay period. The spin state σ after the final 90° S pulse for the experiment with a $90°_{+y}$ pulse to the protons is

$$\sigma_+ = I_y \cos(\pi J_{IS}\tau) + 2I_z S_y \sin(\pi J_{IS}\tau) + \frac{\gamma_s}{\gamma_I} \qquad (1.104)$$

and the spin state following the $90°_{-y}$ pulse to the protons is

$$\sigma_- = I_y \cos(\pi J_{IS}\tau) - 2I_z S_y \sin(\pi J_{IS}\tau) + \frac{\gamma_s}{\gamma_I} \qquad (1.105)$$

Taking the difference between these experiments gives

$$\sigma = -2I_z S_y \sin(\pi J_{IS}\tau) \qquad (1.106)$$

As we saw earlier, terms containing the $I_z S_y$ product operator can evolve into observable magnetization during the detection period. The signal intensity is at a maximum for $\tau = 1/2J_{IS}$ and the sensitivity is enhanced by the ratio of γ_I/γ_S, which is a factor

1.7 MULTIDIMENSIONAL NMR

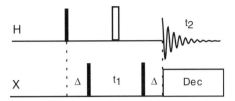

FIGURE 1.61 The pulse-sequence diagram for heteronuclear multiple-quantum coherence (HMQC).

of 4 for carbons and protons. It is important to note that the magnetization generated from the evolution of the $I_z S_y$ terms under the influence of the scalar couplings is *antiphase* magnetization. For a two-spin system this would be a doublet where one half of the double is phased up and the other half is phased down. Such antiphase peaks are often observed in nD NMR experiments that utilize scalar couplings for correlation.

Another common pulse sequence element in heteronuclear nD NMR is the one for generating HMQC (36,38). This simple pulse sequence element, illustrated in Figure 1.61, consists of a pulse to the protons, a delay time Δ, followed by a pulse to the S spins. The 90° proton pulse creates I magnetization that evolves under the influence of the IS scalar coupling to give

$$-I_y \xrightarrow{(\pi J_{IS}\tau)2I_z S_z} -I_y \cos(\pi J_{IS}\tau) + 2I_x S_z \sin(\pi J_{IS}\tau) \qquad (1.107)$$

and application of the 90° pulse to the S spins after the delay time gives

$$-I_y \xrightarrow{(\pi J_{IS}\tau)2I_z S_z} -I_y \cos(\pi J_{IS}\tau) + 2I_x S_y \sin(\pi J_{IS}\tau) \qquad (1.108)$$

If we choose $\tau = 1/2J_{IS}$, the cosine term is zero and we have created a *multiple quantum coherence* that is represented by the $I_x S_y$ operator. These multiple quantum coherences are not observable magnetization, but they can evolve under the influence of the carbon chemical shift during the t_1 period. If we then transfer the magnetization back to the protons using a reverse HMQC block, we would observe that the proton signal is modulated by the carbon chemical shift. HMQC is most frequently used to show the correlation between carbons and directly bonded protons, but if the delay time is lengthened, it can be used to correlate via the two-bond carbon proton couplings. This experiment is called *heteronuclear multiple-bond correlation* (HMBC) (39).

A related pulse sequence, the *heteronuclear single quantum coherence* (HSQC) is also frequently used as a pulse sequence element in heteronuclear nD NMR experiments. The HSQC pulse sequence (Figure 1.62) is very similar to the INEPT pulse sequence described earlier. The difference between the HMQC and HSQC is that the single quantum coherence ($2I_z S_y$), rather than double quantum coherence ($2I_x S_y$), is generated. The HSQC pulse sequence elements have more pulses and the possibility of more magnetization loss from imperfect pulses. HSQC is sometimes superior to

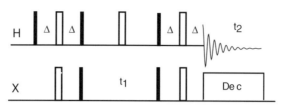

FIGURE 1.62 The pulse-sequence diagram for heteronuclear single-quantum coherence (HSQC).

HMQC for very crowded carbon spectra in which the lines are not broadened in the proton dimension by homonuclear scalar couplings.

COSY experiments use proton–proton couplings for magnetization transfer. The pulse sequence element for a single-step magnetization transfer is $90°-t_1-90°$ sequence, shown in Figure 1.55. In such homonuclear experiments the I and S spins are relatively close in frequency and are affected equally by the $90°$ pulses. The $90°-t_1-90°$ pulse sequence generates $I_y S_z$ and $-I_z S_y$ magnetization that gives rise to cross peaks between I and S. The cross-peak intensities depend on the same $\sin(\pi J_{IS} t)$ terms as discussed for the INEPT experiment. However, because the three-bond scalar couplings of most interest in polymers are much smaller than the one-bond couplings in the heteronuclear correlation experiments, longer delay times are required and the cross peaks are generally smaller. These delays are typically long enough that COSY-type cross peaks are often reduced in intensity from spin-spin relaxation during the long delays. This makes it difficult to observe cross peaks between pairs of protons that are coupled with a coupling constant less than 1 Hz. Multiple COSY-type building blocks can be included in a sequence to observe multiple-step magnetization transfer.

The through-bond magnetization transfer typically involves delay times on the order of $1/2J$ for maximum transfer efficiency. This is a potential problem in polymers, because the signals can also decay from T_2 relaxation during these delay times. The one-bond carbon–proton and nitrogen–proton coupling constants are on the order of 140 and 95 Hz, leading to maximum transfer efficiency for delay times of 3.5 and 5.2 ms. Fortunately, most polymers in solution have relaxation times that are significantly longer than a few ms, and the spectra can be detected with a high sensitivity. On the other hand, the proton–proton couplings are on the order of 7 Hz, leading to a maximum transfer efficiency with a delay time of 71 ms. This time can be long relative to the T_2, so it is sometimes necessary to choose a shorter value for the delay time as a compromise between the maximum transfer efficiency and the T_2 relaxation.

In COSY-type experiments, magnetization is transferred between directly coupled spins with a mixing pulse in a stepwise fashion after the multiplets have evolved into an antiphase state. For multiple transfers, multiple evolution times are required. This can be prohibitive in polymers, where the spin-spin relaxation times are on the order of $1/2J$. An alternative is to use TOCSY-type mixing sequences in the nD NMR sequence (35). The idea behind TOCSY-type sequences (Figure 1.63) is that

1.7 MULTIDIMENSIONAL NMR

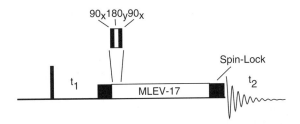

FIGURE 1.63 The pulse-sequence diagram for TOCSY magnetization transfer.

the mixing occurs during a spin-lock pulse sequence. This is also known as isotropic mixing, because all the spins have the same chemical shifts in the spin-lock field. Under these conditions the spin system behaves as a very strongly coupled system and rapidly distributes the magnetization among all of the coupled spins. Using TOCSY it is possible to observe correlations between spins that are not directly coupled to each other. TOCSY correlations can be observed between pairs of protons separated by five or more bonds, as long as there is a chain of coupled protons in between. Because antiphase magnetization is not required for exchange, the TOCSY-type cross peaks are in-phase. This makes it less likely that the overlap of different parts of a multiplet with different phases will lead to cross-peak cancellation.

A number of composite-pulse decoupling schemes have been introduced to provide efficient magnetization exchange in TOCSY-type-pulse sequence elements. The most common method, MLEV-17 (Figure 1.63) is a modification of the MLEV-16 decoupling sequence. Such composite-pulse decoupling sequences are constructed from trains of pulses that differ in phase and length and result in a larger bandwidth. By systematically varying the phase of pulse-sequence elements (such as the 90_x-180_y-90_x element of MLEV-16), the performance of the composite-pulse decoupling sequences can be improved. In most TOCSY sequences the MLEV-17 sequence is sandwiched between short (1–2 ms) spin-lock pulses to select only for I_y magnetization. The power level during the spin-lock period is quite high and care must be taken to avoid damaging the probe by leaving the high power on for too long a time (35).

1.7.1.2 Through-Space Magnetization Transfer Though-space magnetization transfer can be used to establish the resonance assignments in polymers and for structure determination, since the peak intensities depend on the inverse sixth power of the internuclear distance. The most useful applications are homonuclear (proton or fluorine), since most heteronuclear correlations can be established more efficiently using through-bond scalar couplings.

The two most useful means of through-space magnetization transfer are the NOE or the ROE. In both cases the magnetization exchange is a consequence of the through-space dipolar interactions that give rise to spin-lattice relaxation. The differences between the two methods are that in the ROE the magnetization exchange occurs in a spin-locking field and has a different dependence on the rotational correlations times when compared to the NOE (Section 1.5.5.2).

The rate of proton magnetization transfer in NOE exchange experiments is given by

$$\sigma_{IS} = \frac{n}{20}\left(\frac{\mu_0}{4\pi}\right)^2 \frac{\gamma_H^4 \hbar}{r^6}\{6J(2\omega_H) - J(0)\} \quad (1.109)$$

where n is the number of protons, μ_0 is the vacuum magnetic permeability, and r is the internuclear separation. Since the rate depends on the inverse sixth power of the separation, and the NOE falls off quickly with increasing internuclear distance. This makes it difficult to observe cross peaks between pairs of protons separated by more than 5 Å. The cross relaxation rate also depends on the difference between the $J(0)$ and $J(2\omega)$ spectral density terms, so the rate of magnetization exchange is very sensitive to the rotational correlation times. Figure 1.45 showed a plot of the dependence of the NOE on the rotational correlation time. In the fast motion limit, a value of 0.5 is observed, and the NOE is insensitive to the correlation time. As the correlation time approaches the inverse of the spectrometer frequency, the NOE becomes very sensitive to the correlation time as the $J(0)$ and $J(2\omega)$ spectral density terms become comparable. A negative NOE is observed in the slow-motion limit. Note that the NOE goes to zero for correlation times near the observation frequency ($\omega_H \tau_c \cong 1$) and magnetization exchange cannot be observed.

In a similar way, the magnetization exchange from the ROE is given by

$$\sigma_{IS} = \frac{n}{10}\left(\frac{\mu_0}{4\pi}\right)^2 \frac{\gamma_H^4 \hbar}{r^6}\{3J(\omega_1) + J(0)\} \quad (1.110)$$

where ω_1 is the strength of the spin-locking field. The two important features to note about Equation (1.110) is that the ROE rate is twice as fast as the NOE and that the rate does not depend on the difference between the spectral densities, but rather on the sum. Therefore, the ROE is always positive (Figure 1.45). This makes the ROE very valuable for studying polymers with correlation times near the proton observation frequency. Since the magnetization exchange takes place in a weak spin-lock field (1–3 kHz), it can also be a valuable probe of the low-frequency dynamics of polymers.

1.7.2 Solution 2D NMR Experiments

Solution-state nD NMR experiments are extensively used for materials characterization in polymers. Most polymers are readily soluble in organic solvents and the linewidths are sufficiently narrow that it is possible to observe well-resolved nD NMR spectra. For many polymers there is a chain of J-coupled nuclei along the polymer main chain and side chains that can be used to establish the resonance assignments. The carbon spectra usually has much better resolution than the proton spectra, so carbon–proton correlation spectra are frequently used for resonance assignments.

1.7.2.1 COSY Figure 1.64 shows the pulse-sequence diagram for COSY spectroscopy and the 2D spectrum of polycaprolactone. As discussed earlier, the

1.7 MULTIDIMENSIONAL NMR

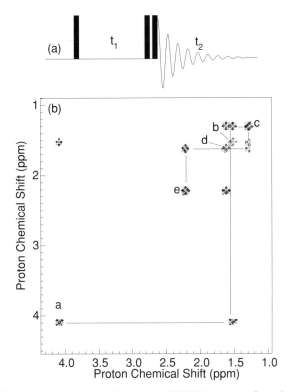

FIGURE 1.64 The pulse-sequence diagram and COSY spectrum for polycaprolactone.

magnetization transfer is accomplished with the $90°-t_1-90°$ part of the pulse sequence. This particular variant of COSY uses a double-quantum filter (40) to suppress the intense diagonal peaks, so that the cross peaks between protons that are close in frequency can be more easily observed. The correlations in the COSY spectra are between pairs of protons that have scalar couplings between 2 and 15 Hz. Among the couplings that can be observed in polymers are the two- and three-bond scalar couplings, such as those between the methylene protons in polycaprolactone. It is possible to follow the magnetization transfer along the chain to establish all of the assignments for polycaprolactone.

$$\begin{array}{cccccc} a & b & c & d & e & \\ -CH_2-CH_2-CH_2-CH_2-CH_2-\overset{O}{\underset{f}{C}}-O- \end{array}$$

Polycaprolactone

1.7.2.2 TOCSY TOCSY also utilizes the through-bond scalar couplings to correlate nearby protons, but a spin-locking pulse is used for magnetization transfer (35). Under these conditions, cross peaks are observed not only to directly coupled pairs, but also between indirectly coupled pairs of protons such as methylenes *a* and *c*

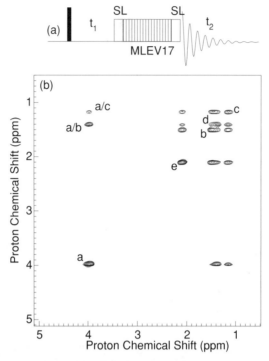

FIGURE 1.65 The pulse-sequence diagram and TOCSY spectrum for polycaprolactone.

in polycaprolactone (35). The pulse-sequence diagram and the TOCSY spectrum for polycaprolactone are shown in Figure 1.65. The mixing sequence in this TOCSY variant is MLEV-17, a composite-pulse decoupling sequence. The spin-locking sequence is sandwiched between two 2-ms spin-locking pulses so that everything except I_y magnetization is dephased by the start of the spin-locking period. The proton power during the spin-locking (or mixing) period is 26 kHz and the length of the mixing period is 35 ms. In addition to all of the peaks between directly coupled protons observed in the COSY spectrum, the TOCSY spectrum shows peaks between pairs of protons (such as *a* and *c* or *b* and *d*) that are not directly coupled. The TOCSY spectrum is extremely useful for identifying groups of protons that are near each other, such as those in polymer side chains.

1.7.2.3 Heteronuclear Multiple Quantum Coherences HMQCs are extremely useful for correlating the proton spectrum with the spectrum of a less sensitive nucleus (38,41). This is a high-sensitivity experiment, since it starts with proton magnetization and the protons are detected during the t_2 period. The heteronuclear magnetization transfer is very efficient because it takes advantage of the large one-bond heteronuclear coupling constants ($J_{CH} = 140$ Hz, $J_{NH} = 95$ Hz). Because of its high sensitivity, HMQC has virtually replaced the older heteronuclear correlation (HETCOR) experiments that detect the insensitive nuclei.

1.7 MULTIDIMENSIONAL NMR

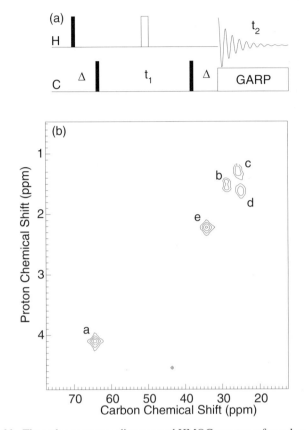

FIGURE 1.66 The pulse-sequence diagram and HMQC spectrum for polycaprolactone.

Figure 1.66 shows the pulse sequence diagram and the HMQC spectrum for polycaprolactone. Multiple-quantum coherences are created by the first proton pulse, a delay of $1/(2J_{CH})$ and the following carbon pulse. These coherences evolve during the t_1 period and the 180° proton pulse refocuses the proton couplings in the t_1 dimension. The carbon pulse and delay transfers the magnetization back to the protons where the spectrum is recorded with carbon decoupling. The end result is the correlation of the carbon and proton chemical shifts. If some of the peaks in the carbon spectrum have been assigned, then we can establish the proton assignments from the correlation.

1.7.2.4 2D Exchange NMR 2D NOESY or exchange spectroscopy provides information about the structure and dynamics of polymers (42,43). Magnetization exchange takes place during the mixing time (τ_m) either via through-space dipolar interactions or chemical exchange (15). The pulse-sequence diagram and 2D exchange spectrum for poly(styrene-*co*-methyl methacrylate) are shown in Figure 1.67. The off-diagonal peaks (cross peaks) arise from magnetization exchange between nearby

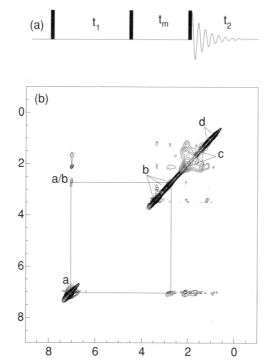

FIGURE 1.67 The pulse-sequence diagram and NOESY spectrum for poly(styrene-*co*-methyl methacrylate).

protons during the mixing time. The cross peaks can be both from direct and relayed correlations. The 2D exchange experiments are useful for making the assignments in polymers where the scalar couplings along the chain are interrupted, as in methyl methacrylates, where the scalar coupling along the main chain is interrupted by the carbonyl group. The through-space correlations allow us to make assignments for nearest-neighbor monomers. Figure 1.67 shows, for example, the correlations between the aromatic protons of styrene and the neighboring methoxyl protons of the methyl methacrylate. 2D exchange NMR can also be used to monitor intermolecular interactions between polymer chains (44).

2D exchange NMR spectroscopy can be used to quantitatively measure the structure and dynamics of polymers in solution. The equation describing the time dependence of the diagonal and cross peaks in the NOESY spectrum as a function of τ_m is given by (15)

$$A(\tau_m) = A_0 e^{-R\tau_m} \quad (1.111)$$

where $A(\tau_m)$ is the matrix of peak volumes obtained at mixing time τ_m, A_0 is the matrix of peak volumes at $\tau_m = 0$ (i.e., the diagonal spectrum), and R is the relaxation-rate

1.7 MULTIDIMENSIONAL NMR

matrix given by

$$R = \begin{bmatrix} \rho_{11} & \sigma_{12} & \sigma_{13} & \cdot \\ \sigma_{21} & \rho_{22} & \sigma_{23} & \cdot \\ \sigma_{31} & \sigma_{32} & \rho_{33} & \cdot \\ \cdot & \cdot & \cdot & \cdot \end{bmatrix} \quad (1.112)$$

where the ρ_{IS} and σ_{IS} are the terms for selective spin-lattice relaxation and cross relaxation given by Equations (1.65) and (1.109). If the spectrum is well enough resolved that all of the diagonal and cross peaks can be measured or estimated, the peak volumes can be used to solve for the relaxation rates, from which the distances can be measured and the rates of molecular motion determined. Equation (1.111) can be rewritten as

$$\frac{-\ln(A(\tau_m)/A_0)}{\tau_m} = R \quad (1.113)$$

and solved as

$$\frac{-T \ln[D] T^{-1}}{\tau_m} = R \quad (1.114)$$

where D is the diagonal matrix of eigenvalues of the normalized peak volume matrix, and T and T^{-1} are its eigenvector matrix and its inverse. Thus, at a single mixing time it is possible to measure all of the diagonal and cross relaxation rates, although gathering spectra at several mixing times is usually more desirable (45).

In the event that the peaks are not well enough resolved for this analysis, the cross relaxation rates can be measured from the initial buildup of cross-peak intensity (15). The rate of cross-peak buildup is approximately given by

$$\sigma_{IS} = \sigma_{IS} \cdot \tau_m + \frac{1}{2}\sigma_{IS}^2 \tau_m^2 + \cdots \quad (1.115)$$

and for short values of τ_m, the cross relaxation rate can be obtained from the initial buildup rate.

Once the relaxation rates are obtained from the 2D exchange data, these rates must be interpreted in terms of the molecular dynamic and the internuclear distances. In favorable cases it is possible to calculate the correlation times from an analysis of these and other relaxation rates. In other cases, the correlation times can be calculated from the cross relaxation rates between protons with fixed internuclear distances using Equation (1.109). Among the fixed distances found in polymers are those between geminal protons (1.78 Å) and neighboring protons on an aromatic ring (2.45 Å). The initial rate of magnetization exchange between pairs of protons with a known separation also provides an easy way to estimate other distances, since the ratio of

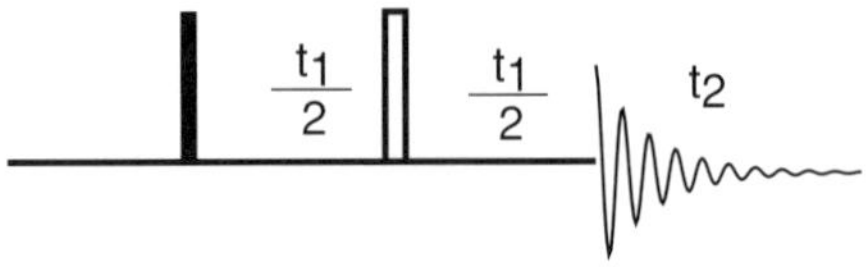

FIGURE 1.68 The pulse-sequence diagram and J-resolved proton 2D spectrum of polycaprolactone.

distances is related to the cross relaxation rates by

$$\frac{r_{ab}}{r_{cd}} = \left[\frac{\sigma_{cd}}{\sigma_{ab}}\right]^{1/6} \tag{1.116}$$

where r_{ab} and r_{cd} are the known and unknown distances.

1.7.2.5 J-Resolved NMR J-resolved nD NMR methods are an example of a separation rather than a correlation type of nD NMR experiment. The purpose of this experiment is to separate the chemical shift information from the J-coupling information, such that the chemical shifts appear along one axis and the J couplings appear along the other. Figure 1.68 shows the pulse-sequence diagram and the J-resolved spectrum for polycaprolactone. The 180° pulse in the middle of the evolution period refocuses the proton chemical shifts, but not the scalar couplings, so that the normal spectrum appears at zero frequency in the 2D spectrum and the couplings appear in F_1 dimension. The number of coupled protons and the magnitude of the coupling constant can be easily determined by inspection of these data. The peaks in the F_1 dimension are not exactly perpendicular to the chemical shift axis, but appear at the chemical shift $\pm J$. This offset is small and is typically corrected by a computer routine. J-Resolved spectroscopy can also be used to measure the heteronuclear couplings.

1.7.3 Solid-State 2D NMR Experiments

Solid-state nD NMR has emerged more recently than solution-state nD NMR as a method for polymer characterization. The experiments in solids differ from their solution counterparts in that the line-narrowing methods routinely used in solid-state NMR must be incorporated into the pulse sequence. The lines are broader in solids, making it difficult to utilize the scalar couplings that are so useful in solution-state nD NMR experiments. Magnetization exchange in the solid-state experiments rely primarily on dipolar interaction and chemical exchange. For high-resolution experiments, magic-angle sample spinning and high-power proton decoupling are required. Cross polarization is often used for sensitivity enhancement and because the spin-lattice relaxation times can be very long for carbons and other rare nuclei. Some studies of the molecular dynamics of polymers utilize wideline NMR to correlate different parts of the wideline spectrum.

1.7 MULTIDIMENSIONAL NMR

1.7.3.1 2D Exchange NMR Solid-state 2D exchange NMR can be used to study the molecular dynamics of polymers, and provides a valuable tool to study the ultraslow motions in polymers. In most cases the cross peaks are due to chemical exchange during the mixing time. In the studies with magic-angle spinning, cross peaks may arise from carbons that have changed environments during the mixing time, such as carbons that have diffused from a crystalline to an amorphous environment in a semicrystalline polymer. In the experiments without magic-angle spinning for line narrowing, the correlations are observed between different parts of the chemical shift anisotropy lineshape. As noted in Section 1.6.1, the anisotropic carbon lineshapes arise from a distribution of orientations relative to the external magnetic field. If a carbon undergoes motion during the mixing time, it can change its orientation relative to the magnetic field, and hence its frequency. Well-resolved cross peaks are not observed in these wideline experiments, but the off-diagonal intensity appears as ridges, and the geometry of the ridges relative to the diagonal are related to the reorientation angle. Experiments that are similar in concept, but different in detail, can be used to study the very slow reorientation in polymers that have been site-specifically labeled with deuterium.

Figure 1.69 shows the pulse-sequence diagram for 2D exchange NMR and the 2D exchange spectrum for semicrystalline polyethylene (46). The major difference between the solution and the solid-state spectrum is that the carbon magnetization is generated by cross polarization, and proton decoupling is applied during the t_1 and t_2 periods. Two peaks are observed in the solid-state carbon spectrum of polyethylene that can be assigned to the crystalline phase at 33 ppm and amorphous phase at 31 ppm. At high temperatures and for long mixing times, the rate of chain diffusion

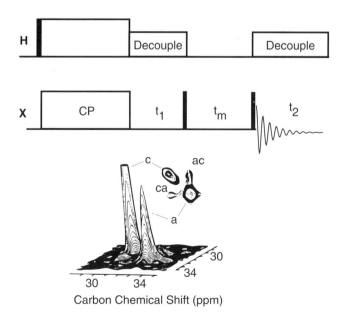

FIGURE 1.69 The pulse-sequence diagram and 2D exchange spectrum for semicrystalline polyethylene.

is fast enough that chains can diffuse between the crystalline and amorphous regions, and this gives rise to the cross peaks observed in Figure 1.69. 2D exchange NMR has also been used to study polymers with solid-state proton NMR and fast magic-angle sample spinning. If the spinning is fast compared to the proton–proton dipolar couplings, then a high-resolution proton spectrum can be observed (47). Under these conditions, the solution-state 2D exchange pulse sequence can be used to monitor magnetization exchange between the resolved proton resonances.

1.7.3.2 Wideline Separation Spectrsocopy The proton linewidths contain information about the chain dynamics of polymers and can be measured using wideline separation (WISE) 2D NMR (48). Figure 1.70 shows the pulse-sequence diagram for 2D WISE NMR and the WISE spectrum for a blend of polystyrene and poly(vinyl methyl ether). The experiment begins with a 90° pulse to the protons and the t_1 evolution

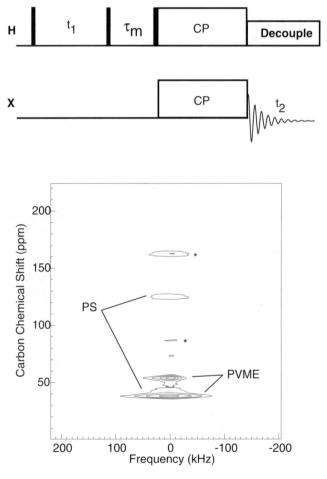

FIGURE 1.70 The pulse-sequence diagram and WISE spectrum for the blend of polystyrene and poly(vinyl methyl ether).

1.7 MULTIDIMENSIONAL NMR

period, during which the protons evolve under the influence of their chemical shifts and the dipolar couplings. After an optional mixing time, the signal is detected with magic-angle spinning and cross polarization. The final result is a correlation of the high-resolution carbon spectrum with the wideline proton spectrum. The dipolar couplings (50 kHz) are typically much larger than the range of chemical shifts (5 kHz), so wideline spectra are observed in the indirectly detected dimension. However, if there is large-amplitude chain motion that is fast on the time scale of the dipolar couplings, then the proton lines will be narrowed. Figure 1.70 shows the 2D WISE spectrum for a blend of polystyrene and poly(vinyl methyl ether). This blend is miscible, but different linewidths are observed for the two types of polymer chains, showing that the poly(vinyl methyl ether) chains have a much greater mobility than the polystyrene chains in the blend. The WISE pulse sequence also contains a mixing time to allow for magnetization exchange. If the length scale of mixing is small, magnetization exchange during the mixing time leads to equilibration between the two chains, and a single averaged lineshape is observed for both polymers.

1.7.3.3 Heteronuclear Correlation Heteronuclear correlation is a valuable tool for the study of polymers in the solid state. Figure 1.71 shows the pulse-sequence

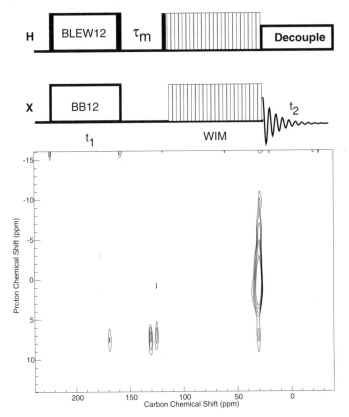

FIGURE 1.71 The pulse-sequence diagram and 2D HETCOR spectrum for polycarbonate.

diagram for HETCOR and the spectrum for polycarbonate (49,50). This version of HETCOR uses through-space dipolar couplings to transfer magnetization between the carbons and the protons. The proton linewidths are typically broad in solids, so multiple-pulse decoupling is used for line narrowing in the t_1 dimension. Multiple-pulse NMR is also used for cross polarization using the windowless isotropic mixing (WIM) pulse sequence that allows for cross polarization while proton magnetization exchange (which can occur during cross polarization) is suppressed. The resolution in the proton dimension is typically low (1 ppm), but this is often sufficient for solids. The pulse sequence also includes a mixing time to allow for magnetization exchange before the WIM sequence.

REFERENCES

1. Alpert, N. L. *Phys. Rev.*, 1947, *72*, 637.
2. Purcell, E. M.; Torrey, H. C.; Pound, R. V. *Phys. Rev.*, 1946, *69*, 37.
3. Bloch, F.; Hansen, W. W.; Packard, M. E. *Phys. Rev.*, 1946, *69*, 127.
4. Bovey, F. A. *High Resolution NMR of Macromolecules*, Academic Press, New York, 1972.
5. Tonelli, A. *NMR Spectroscopy and Polymer Microstructure: The Conformational Connection*, VCH Publishers, New York, 1989.
6. Schmidt-Rohr, K.; Speiss, H. W. *Multidimensional Solid-State NMR and Polymers*, Academic Press, New York, 1994.
7. Bloch, F. *Phys. Rev.*, 1946, *70*, 460.
8. Croasmun, W. R.; Carlson, M. K., eds. *Two-Dimensional NMR: Applications for Chemists and Biochemists*, VCH Publishers, New York, 1994.
9. Cooley, J. W.; Tukey, J. W. *Math. Comput.*, 1965, *19*, 297.
10. Sorensen, O. W.; Eich, G. W.; Levitt, M. H.; Bodenhausen, G.; Ernst, R. R. *Prog. NMR Spectrosc.*, 1983, *16*, 163.
11. Arnold, J. T.; Packard, M. E. *J. Chem. Phys.*, 1951, *19*, 1608.
12. Bovey, F. A. *Nuclear Magnetic Resonance Spectroscopy*, Academic Press, San Diego, 1988.
13. Waugh, J. S.; Fessenden, R. W. *J. Am. Chem. Soc.*, 1957, *79*, 846.
14. McConnell, H. M. *J. Chem. Phys.*, 1957, *27*, 226.
15. Jeneer, J.; Meier, B.; Bachmann, P.; Ernst, R. *J. Chem. Phys.*, 1979, *71*, 4546.
16. Stothers, J. B. *Carbon-13 NMR Spectroscopy*, Academic Press, New York, 1972.
17. Ditchfield, R.; Ellis, P. D. *Topics in Carbon-13 NMR Spectroscopy*, Vol. 1, Academic Press, New York, 1974.
18. Pople, J. A.; Gordon, M. S. *J. Am. Chem. Soc.*, 1967, *89*, 4253.
19. Levy, G. C.; Lichter, R. L.; Nelson, G. L. *Carbon-13 Nuclear Magnetic Resonance Spectroscopy*, 2nd ed.; Wiley-Interscience: New York, 1980.
20. Breitmaier, E.; Voelter, W. *Carbon-13 NMR Spectroscopy*, 3rd ed.; VCH: Weinheim, 1987.
21. Bovey, F. *Nuclear Magnetic Resonance Spectroscopy*, 2nd ed, Academic Press, New York, 1988.
22. Heatley, F. *Prog. NMR Spectroscopy*, 1979, *13*, 47.

REFERENCES

23. Bovey, F. A.; Mirau, P. A. *NMR of Polymers*, Academic Press, New York, 1996.
24. Glowinkowski, S.; Gisser, D. J.; Ediger, M. D. *Macromolecules*, 1990, *23*, 3520.
25. Zhu, W.; Ediger, M. D. *Macromolecules*, 1995, *28*, 7549.
26. Mehring, M. *High Resolution NMR in Solids*, Springer-Verlag, Berlin, 1983.
27. Robyr, P.; Utz, M.; Gan, Z.; Scheurer, C.; Tomaselli, M.; Suter, U. W.; ERnst, R. R. *Macromolecles*, 1998, *31*, 5818.
28. Brandolini, A. J.; Alvey, M. D.; Dybowski, C. *J. Polym. Sci., Polym. Phys. Ed.*, 1983, *21*, 2511.
29. Dixon, W. T.; Schaefer, J.; Sefcik, M.; Stejskal, E. O.; McKay, R. A. *J. Magn. Reson.*, 1982, *49*, 341.
30. Pake, G. E. *J. Chem. Phys.*, 1948, *16*, 327.
31. Brown, S. P.; Schnell, I.; Spiess, H. W. *J. Am. Chem. Soc.*, 1999, *121*, 6712.
32. Taylor, R.; Pembleton, R.; Ryan, L.; Gerstein, B. *J. Chem. Phys.*, 1979, *71*, 4541.
33. Hartmann, S. R.; Hahn, E. L. *Phys. Rev.*, 1962, *128*, 2042.
34. Aue, W.; Bartnoldi, E.; Ernst, R. *J. Chem. Phys.*, 1976, *64*, 2229.
35. Bax, A.; Davis, D. *J. Magn. Reson.*, 1985, *63*, 207.
36. Muller, L. *J. Am. Chem. Soc.*, 1979, *101*, 4481.
37. Morris, G. A.; Freeman, R. *J. Am. Chem. Soc.*, 1979, *101*, 760.
38. Bax, A.; Griffey, R. H.; Hawkins, B. L. *J. Am. Chem. Soc.*, 1983, *105*, 7188.
39. Bax, A.; Summers, M. F. *J. Am. Chem. Soc.*, 1986, *108*, 2093.
40. Rance, M.; Sorensen, O. W.; Bodenhausen, G.; Wagner, G.; Ernst, R. R.; Wuthrich, K. *Biochem. Biophys. Res. Commun.*, 1983, *117*, 479.
41. Mueller, K. T.; Jarvie, T. P.; Aurenta, D. J.; Roberts, B. W. *Chem. Phys. Lett.*, 1995, *242*, 535.
42. Heffner, S.; Bovey, F.; Verge, L.; Mirau, P.; Tonneli, A. *Macromolecules*, 1986, *19*, 1628.
43. Kogler, G.; Mirau, P. *Macromolecules*, 1992, *25*, 598.
44. Mirau, P.; Tanaka, H.; Bovey, F. *Macromolecules*, 1988, *21*, 2929.
45. Mirau, P. *J. Magn. Reson.*, 1988, *80*, 439.
46. Schmidt-Rohr, K.; Spiess, H. *Macromolecules*, 1991, *24*, 5288.
47. Mirau, P. A.; Heffner, S. A. *Macromolecules*, 1999, *32*, 4912.
48. Schmidt-Rohr, K.; Clauss, J.; Spiess, H. *Macromolecules*, 1992, *25*, 3273.
49. Caravatti, P.; Bodenhausen, G.; Ernst, R. R. *Chem. Phys. Lett.*, 1982, *89*, 363.
50. Burum, D. P.; Bielecki, A. *J. Magn. Reson.*, 1991, *94*, 645.

2

EXPERIMENTAL METHODS

2.1 INTRODUCTION

Solution and solid-state NMR are extensively used for polymer characterization, as they provide information about the polymer microstructure, morphology, and dynamics, and permit a unique insight into polymer structure–property relationships. The quality of the data depends on the sample preparation methods, the spectrometer, the NMR experiment of choice, the NMR parameters used for data acquisition, and the methods of data processing. NMR is an important analytical method in polymer science, and modern NMR spectrometers are able to perform a wide array of experiments. The solution NMR experiments discussed in this book can be routinely executed on any liquids spectrometer, but more specialized hardware (high-power amplifiers, magic-angle spinning probes, etc.) are required for solid-state NMR experiments.

The collection of high-quality data for understanding the structure and dynamics of polymers requires a mastery of a small number of fundamental principles, including the basic functions of an NMR spectrometer. To obtain high-quality data it is important to understand how sample preparation, NMR relaxation, and data processing affects the spectra. This is particularly important for the quantitative analysis of polymer NMR spectra.

Sample preparation is critical for both solution and solid-state NMR. For solution analysis, high-purity solvents and samples are important to avoid reaching erroneous conclusions about the polymer under study. Prior to data acquisition, it is important to properly set up and calibrate the spectrometer. After the data are acquired, data processing is required before the final analysis, and improper weighting and phasing of the data can affect the results. Experiments are available to edit the spectra using specific pulse sequences so that only subsets of the data can be viewed.

A Practical Guide to Understanding the NMR of Polymers, by Peter A. Mirau
ISBN 0-471-37123-8 Copyright © 2005 John Wiley & Sons, Inc.

Solid-state NMR is more demanding than solution NMR, since it requires high-powered amplifiers and specialized probes for magic-angle sample spinning. Because of the high powers involved, careful attention to the power settings is required, as excess power can destroy probes and other valuable pieces of equipment. The samples for solid-state NMR must be packed into rotors and spun at several kHz.

The NMR relaxation rates can provides information about the molecular dynamics of polymers. Relaxation data are relatively easy to acquire, but care must be taken in the analysis, since more than one kinetic model may fit equally well to the decay curves. To avoid these potential pitfalls, it is necessary to acquire the data over as wide a range of temperatures and NMR frequencies as possible. Understanding the relaxation times is important for quantitative analysis, since the peak intensities may depend on both the relaxation times and the NOE's.

Many of the recent NMR advances in polymer science are a consequence of the introduction of nD NMR. This technique makes it possible to analyze the structure and dynamics of polymers in much greater detail by spreading the information contained in the NMR spectrum into two or more frequency dimensions. It also makes it possible to assign many of the peaks to specific microstructures in the polymer chain. While the nD NMR experiments may appear complicated, they are easily performed and analyzed on any modern spectrometer. Much of what happens on an NMR spectrometer is computer controlled, so it is possible to write macros to acquire and process nD NMR experiments without understanding all there is to know about the spectrometer. Once an experiment has been performed on one sample, it is possible to recall the parameters and run the same experiment on another sample. Since the acquisition times are longer and the data sets are larger in nD NMR experiments, additional care must be taken in setting up the NMR spectrometer and in data processing.

2.2 THE NMR SPECTROMETER

A modern NMR spectrometer is a complex instrument to build, but a relatively simple instrument to operate. The main components are the superconducting magnet, an rf console, probes, and a computer system. Many parts of the spectrometer are under computer control and do not require adjustment. Some adjustment of the magnetic field homogeneity and tuning of the probe is required for the highest resolution and sensitivity. The most common pulse sequences are stored on the computer, and some kind of scripting language can be used to create new pulse sequences. Each experiment has many possible parameters, and these are typically stored as parameter files that can be recalled to set the spectrometer up for a new experiment. Typically only minor adjustment is required after recalling a parameter set.

2.2.1 The Magnet

A high-field superconducting magnet is one of the principal components of an NMR spectrometer. Since very high fields are required for NMR, special superconducting metals are used to make the magnet coils. Superconducting materials are able to

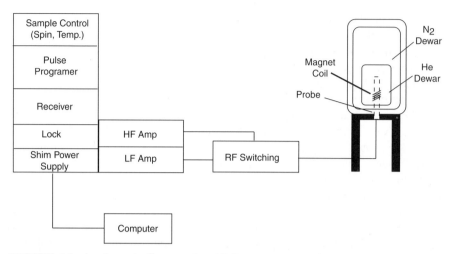

FIGURE 2.1 A schematic diagram of an NMR spectrometer showing the magnet, probe, rf console, and computer.

conduct electricity without generating resistance, making it possible to have high-current magnets that generate very little heat. Once a superconducting magnet is energized, it is said to be persistent if no additional electricity is required to maintain the magnetic field. The metals used to make the coils are only superconducting at very low temperatures, so the magnet coil is immersed in a bath of liquid helium at 4 K. The helium dewar is surrounded by a second dewar containing liquid nitrogen at 77 K for further thermal isolation. An important part of maintaining the NMR spectrometer is regularly filling the inner and outer dewars with liquid helium and nitrogen. Figure 2.1 shows a schematic diagram of a magnet and illustrates how the coil is immersed in the liquid helium dewar and surrounded by the liquid nitrogen dewar. The magnet has a bore tube down the middle so that the probe and sample can be placed in the magnetic field.

2.2.2 Shim Coils

To obtain the highest resolution it is necessary to have the magnetic field as homogeneous as possible over the sample volume. Under the best conditions, we would like the field to be homogeneous to one part in a billion. It is possible to construct coils that are relatively homogeneous, but not to this extent. To make the field as homogeneous as possible a second set of coils is added to the bore of the magnet. These so-called shim coils contain many (20 or more) independent coils that can cause minor changes in the field to make it more homogeneous. Each of these coils has a different geometry and a different effect on the field. Some of the shim coils, the so-called nonspinning shims, are used to maximize the homogeneity in the plane of the coil, while another set, the so-called spinning shims, is used to adjust the homogeneity along the direction of the bore tube. For the highest resolution it is necessary to adjust

2.2 THE NMR SPECTROMETER

the field homogeneity each time the sample is changed. This is done automatically on most modern spectrometers.

2.2.3 RF Console

The rf console is the main electronic component of the NMR spectrometer and consists of amplifiers, pulse programmers, switches, gates, and mixers. The rf console is under computer control and the user does not have to be concerned with the switches, gates, and so forth. In a typical experiment the computer sends a signal to the rf console to give a pulse and the rf console generates the proper frequency, turns the rf transmitter on for the pulse length, amplifies the pulse, and delivers it to the probe. The signal is detected by the reciever and stored on the computer. Most of the activity is automatic and requires only the input of parameters, such as the pulse length, the number of data points to be acquired, and the sweep width. The rf console also controls the homogeneity by controlling the current applied to the shim coils.

2.2.4 NMR Probes

The probe is the device inserted into the magnet to hold the sample. It contains an rf coil and is usually connected to a box of electronics near the magnet leg. Pulses from the rf console are sent to the probe and the coil. After the pulses are transmitted, the same coil detects the free induction decay. Most probes also contain other electronic elements to measure the spinning rate and the temperature. Probes also contain heaters that warm flowing air (or nitrogen) to regulate the sample temperature.

Probes for solution NMR fit into the bottom of the magnet, and the samples are inserted from the top of the magnet by air pressure. Another gas stream spins the sample (15–30 Hz) for the best homogeneity. Solution probes also contain another coil tuned to the deuterium frequency. When a deuterated solvent is used, the deuterium signal from the solvent is used as a field frequency *lock* to maintain the field homogeneity during signal acquisition. Probes are built to accommodate a specific sample tube. Proton probes are typically tuned to a single frequency and accept 5 mm NMR tubes. Broadband probes are tunable over a wide frequency range (typically nitrogen through phosphorus) and usually require larger (10-mm) NMR tubes. The sample volumes for the 5- and 10-mm probes are on the order of 0.5 and 4 mL, respectively. These broadband probes also have a second rf input for proton decoupling.

Probes for solid-state NMR are also inserted from the bottom of the magnet, but the probe is usually removed from the magnet each time the sample is changed. Wideline NMR probes (i.e., deuterium) are often single-frequency probes that accept special 5-mm sample tubes. Probes for magic-angle sample spinning have a housing with nitrogen inputs for rapidly spinning the samples. The samples are packed in rotors with fluted end caps. It is these end caps that cause the rotors to spin under the high-pressure nitrogen stream. The rotors are typically ceramic and vary in size from 2.5 to 14 mm. It is much easier to rapidly spin the smaller rotors, but they do not have the large sample volume required for low-sensitivity nuclei like carbon. Most carbon experiments use rotors with a diameter of 5 to 7 mm.

2.2.5 Computer

Computers are an integral part of modern spectrometers. They are usually specialized to control the high-speed functions, such as applying ns-long rf pulses and controlling the gates to send pulses to the probe and route the NMR signals back to the computer for storage. The computer also controls the spectral parameters, such as the number of data points and the number of scans through some graphical user interface. The computer also contains programs for data processing and plotting.

2.3 TUNING THE NMR SPECTROMETER

NMR spectrometers are complex instruments that are relatively simple to operate because most functions are under computer control and transparent to the user. The users do not have to be concerned with how pulses are gated and amplified, but they do have to be concerned with adjusting the homogeneity, properly setting the receiver gain, and measuring the pulse widths. The homogeneity and gain have to be set for every sample, but the pulse widths are not usually changed between similar samples.

2.3.1 Adjusting the Homogeneity

The resolution (linewidth) in NMR experiments is often of critical importance in NMR studies of polymers because the highest resolution may be required to distinguish subtle features in the polymer chains. The resolution is adjusted by changing the currents in the shim coils to make the field as homogeneous as possible. In solution probes, the lock circuit measures the intensity of the deuterium signal from the solvent and applies a feedback current to maintain the field homogeneity. As the field becomes more homogeneous, the deuterium linewidth decreases and the intensity of the lock signal is increased. The process of adjusting the homogeneity is known as shimming for historical reasons. The shimming is done iteratively. First the homogeneity is maximized, as the sample spins between 15 and 30 Hz using the coils along the z axis of the magnet. The spinning is stopped and the field in the plane of the coil is maximized. The resolution after shimming should be less than 0.5 Hz. Shimming can be a tedious process and is automated on many modern spectrometers

2.3.2 Adjusting the Gain

The receiver on the NMR spectrometer is built for high sensitivity, so adjustment may be require for concentrated samples. The receiver gain is adjusted to adequately digitize the noise while not saturating the analog-to-digital converter in the receiver. Saturating the receiver leads to clipping of the most intense parts of the FID and distortions in the spectra. Figure 2.2 shows two FIDs and their Fourier transforms. The first FID has been properly digitized and gives a good spectrum with a flat, undistorted baseline. The initial part of FID in the second case has been clipped by receiver saturation, leading to distortions in the spectrum. Receiver saturation can be

2.3 TUNING THE NMR SPECTROMETER

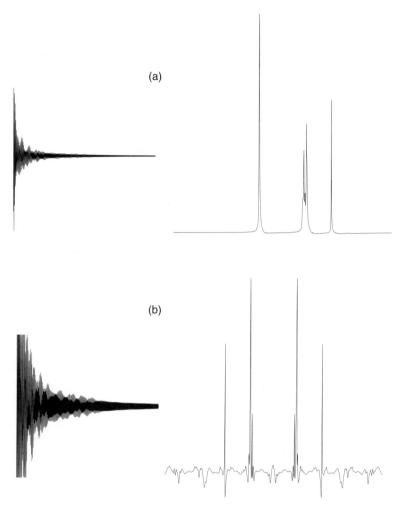

FIGURE 2.2 The effect of receiver gain settings on the FID and spectra. The FIDs and spectra are shown for (a) proper and (b) improper setting of the receiver gain.

identified by examining the FID after a single scan. Setting the receiver gain from proton samples is most important, because the proton signals are so strong.

2.3.3 Tuning the Probe

To get the highest SNR it is important to have the probe tuned to the proper frequency. The high-field probes are very sensitive and the tuning can depend on the dielectric constant of the material under study. Most probes have two adjustable capacitors, one for tuning the frequency and one for matching. The tuning procedure depends on the spectrometer and the probe, and the tuning is done iteratively, as the two capacitors are mutually interacting. Tuning is required when changing solvent, but not necessarily

110 EXPERIMENTAL METHODS

between samples with the same solvent. For heteronuclear experiments it is necessary to tune the circuits for both the observation and decoupling channels.

2.3.4 Adjusting the Pulse Widths

Adjusting the pulse width is important for quantitative NMR and critical for nD NMR experiments. The pulse width can be calibrated by observing the spectrum as the pulse length is changed. This is illustrated in Figure 2.3, which shows a plot of the spectrum for a water sample as a function of the pulse width. The peak intensity follows a sinusoidal pattern with a maximum when the pulse length is equal to the 90° pulse length. It is difficult to find the point at which the peak is a maximum, so it is usually easier to identify the null point at which the pulse length is equal to the 180° pulse. The 90° pulse is half of the 180° pulse length.

It is important to choose the proper sample for pulse-width calibration. The sample should have a relatively short T_1, since it is necessary to weigh the T_1 five times between scans, and the sample should have a good SNR after only a single scan. The tune-up sample should also have a dielectric constant similar to the sample of interest, as large differences in the dielectric constant may change the probe tuning and the pulse widths.

The preceding tune-up procedure is adequate for most studies, since high accuracy in the pulse widths are not usually required. The problem with this approach is that the null point in Figure 2.3 used to determine the pulse width has the lowest SNR. For a more accurate determination of the pulse width, the data in Figure 2.3 may be fitted to a decaying sinusoidal function given by

$$f(t) = \sin(a_1 t)\, e^{-t a_2} \qquad (2.1)$$

where a_1 and a_2 are the fitted parameters. The 90° pulse width is given by $0.5/a_1$.

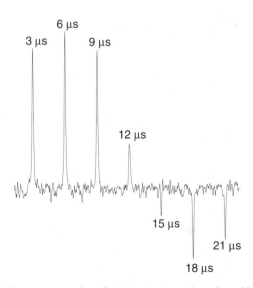

FIGURE 2.3 The spectrum of a reference sample as the pulse width is increased.

2.4 SOLUTION NMR METHODS

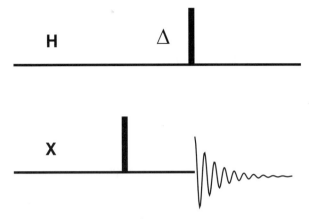

FIGURE 2.4 The diagram for the pulse sequence used to measure the decoupler field strength through observation of the carbon spectra.

For double resonance experiments is necessary to tune both the X nucleus (carbon, silicon, etc.) and decoupling (proton) channels. The 90° pulse for the X nucleus is determined from the preceding procedure. For many spectrometers it is important to determine the proton power while observing the X channel to ensure that the same rf path is followed in the tune-up sequence and the actual pulse sequence. The 90° pulse for the proton channel can be determined through the X channel signals using the pulse sequence shown in Figure 2.4 (1) on a sample, such as chloroform dissolved in deuterated acetone, that gives a doublet in the carbon spectrum from the one-bond carbon–proton coupling. If the delay time between the carbon and proton pulse is set to $1/2J$, the doublets in the spectrum will be out of phase if the proton pulse is absent. The POF treatment shows that the first pulse and delay of $1/2J$ generates a $2I_x S_z$ spin state that evolves into observable magnetization during the detection period. The 90° proton pulse converts the $2I_x S_z$ spin state into a $2I_x S_y$ double-quantum spin state that does not generate observable magnetization. Thus, the null point is observed when the pulse length is equal to the 90° pulse length. An accurate value of the proton pulse length is needed both for nD NMR experiments and composite pulse decoupling (Section 2.4.3). The relationship between the pulse length and the field strength γ_{B_1} is given by

$$\gamma B_1 = \frac{1}{4\tau_{90}} \qquad (2.2)$$

where τ_{90} is the length of the 90° pulse.

2.4 SOLUTION NMR METHODS

Solution NMR is one of the most useful analytical methods for polymer characterization because the spectra are sensitive to the polymer microstructure. High-resolution spectra were not initially expected for high molecular-weight polymers, because the broad lines are usually expected for large molecules. Fortunately, the dynamics in most polymers are such that there is sufficient segmental motion to average the dipolar interactions (2). In many polymers the solution linewidths are on the order of 1–10 Hz. The peaks from defects are often resolved from the main-chain resonances, and it is possible to characterize very low levels of defects, as well as chain architecture and stereochemistry. The information content of the spectra can be maximized with careful sample preparation, data acquisition, and data processing.

2.4.1 Sample Preparation

High-resolution spectra can often be observed for polymers in solution if the samples are properly prepared. The proton spectrum can be observed with a high sensitivity, but the chemical shift dispersion is often better in the carbon spectrum. High concentrations are required for low-sensitivity nuclei like carbon and nitrogen, but this is not a serious limitation, since most polymers are soluble in common organic solvents.

The solution spectra of polymers are acquired by dissolving the polymer of interest in a suitable solvent that contains deuterons in place of protons. Deuterated solvents are used for two reasons. The most important reason is that the solvent signal would overwhelm the proton signals of the polymer if we used a protonated solvent. In addition, the deuterated solvent signals can be used for frequency locking to ensure magnetic field stability. The field stability is controlled by a feedback circuit that adjusts and maintains the field stability for long periods of time so that the SNR can be improved by averaging many transients.

A large number of deuterated solvents are available for polymer NMR studies, and Table 2.1 lists some of the common deuterated solvents along with their densities, melting points, boiling points, and the frequencies of the residual solvent in the proton and carbon spectra. Many of the solvents can be obtained with a very high level of deuteration ($> 99.9\%$), but the sensitivity of modern NMR spectrometers is such that 1% signals are readily observed in the proton spectra. The solvent signals are always observed in the carbon spectra, since the solvent typically contains carbon atoms and is present in high concentration. The carbon signals of deuterated solvents with directly attached to protons will appear as multiplets in the carbon spectra, since the carbon–deuterium coupling is not removed by proton decoupling. The multiplicity is given by $2nI + 1$, where n is the number of attached deuterons. Since deuterium has a spin of 1, solvents such as chloroform-d or benzend-d_6 will give rise to a triplet, while solvents with two attached deuterons per carbon, such as dioxane, will give rise to a pentet in the carbon spectrum.

The choice of solvent depends on a number of factors, the most important being the polymer solubility. It is also important to consider the overlap of the residual solvent peaks with the polymer peaks of interst. For polymers without aromatic groups, benzene-d_6 would be an excellent choice because the polymer peaks would have no overlap with the residual solvent peaks. The solvent boiling point and melting points are also important factors to consider. The resolution can sometimes be increased by acquiring the spectra at a higher temperature, where increased chain motion leads to shorter T_2 and a decrease in the linewidths.

2.4 SOLUTION NMR METHODS

TABLE 2.1 The Solution Properties of Deuterated Solvents Commonly Used for Polymer NMR.

Solvent	Density (g/l)	Melting Point (°C)	Boiling Point (°C)	δ_H (ppm)	δ_C (ppm)
Acetone-d_6	0.87	−94	57	2.04 (br)	206 (13)
					29.8 (7)
Acetonitrile-d_3	0.84	−45	82	1.93 (br)	118.2 (br)
					1.3 (7)
Benzene-d_6	0.95	5	80	7.15 (br)	128.0 (3)
Chloroform-d	1.50	−64	62	7.24 (1)	77.0 (3)
Deuterium oxide-d_2	1.11	3.8	101.4	4.63	
Dimethyl formamide-d_7	1.04	−61	153	8.01 (br)	162.7 (3)
				2.91 (5) (3)	
				2.74 (br)	35.2 (7)
					30.1 (7)
Dimethyl sulfoxide-d_6	1.18	18	189	2.49 (5)	39.5 (7)
p-Dioxane-d_8	1.13	12	101	3.53 (br)	66.5 (5)
Methylene chloride-d_2	1.35	−95	40	5.32 (3)	53.8 (5)
Pyridine-d_5	1.05	−42	116	8.71 (br)	149.9 (3)
				7.55 (br)	
				7.19 (br)	135.5 (3)
					123.5 (3)
Tetrahydrofuran-d_8	0.99	−109	66	3.58 (br)	67.4 (3)
				1.73 (br)	25.3 (3)
Toluene-d_8	0.94	−95	111	7.09 (br)	137.5 (1)
				7.00 (br)	
				2.09 (5)	128.9 (3)
					128.0 (3)
					125.2 (3)
					20.4 (7)
Trifluoroacetic acid-d	1.50	−15	72	11.50 (br)	164.2 (4)
					116.6 (4)

For relaxation studies it is important to remove oxygen from the samples by sealing the sample tubes under an inert atmosphere, such as nitrogen or argon. Relaxation by dissolved oxygen can compete with the dipolar interactions or chemical shift anisotropy in samples with long relaxation times. It is also important to exclude dust or other particulate matter from the samples, since they can affect the field homogeneity, and the resolution.

Protons have a high sensitivity, so the samples are typically dissolved in the appropriate deuterated solvent and placed in 5-mm NMR sample tubes. The small sample tubes make it easier to adjust the field homogeneity. For carbon and other low-sensitivity nuclei it is better to use a larger tube (10 mm), so that the active volume of the NMR coil contains more of the nuclei of interest. High concentrations

(10–30%) are often required for the low-sensitivity nuclei. These solutions can be quite viscous, but they still give high-resolution spectra because the local viscosity is lower than the macroscopic viscosity.

2.4.2 Data Acquisition

The NMR data collection is relatively simple for 1D NMR experiments. Among the factors to be considered are the receiver gain, the number of data points used to collect the spectra, the sweep width, the acquisition time, and the delay time between scans. The parameters tend to be similar for similar types of spectra, so once these parameters have been set they can be used for data collection on many polymer samples.

The detected signal is a voltage from the coil in the probe in the magnet. For undistorted spectra it is important to set the spectrometer gain at a high enough level that the noise is adequately digitized, but that the most intense signals do not overload the detector. If the gain is set to high, the baseline appears distorted and the spectrum will be difficult to phase (Figure 2.2).

It is also important to choose an acquisition time that is sufficiently long that all of the signals are adequately digitized. The acquisition time (at) is the product of the dwell time (Δ, the time required to digitize each point) and the number of points (np), and is given by

$$at = \Delta \cdot np \qquad (2.3)$$

The dwell time is inversely related to the sweep width, the range of frequencies observed in the NMR spectrum. The sweep width is chosen so that all of the possible peaks are included in the spectrum. For carbons and protons on a 11.7-T magnet (500 MHz for protons and 125 MHz for carbons), typical sweep widths are 25 kHz (200 ppm) and 5 kHz (10 ppm). The number of data points is chosen such that the spectrum is well digitized, which means that there must be more than enough data points to digitize the sharpest feature in the spectrum. In a proton spectrum where 1-Hz lines might be observed, a *digital resolution* of 0.2 Hz per point or less would be required. Thus, more than 25,000 points would be required. Since 2^n data points are required for the FFT algorithm, some spectrometers only allow certain values for the number of data points. Thus it may be necessary to acquire 32,678 data points.

If broader lines are expected, then a much smaller number of data points may be acquired. It also may occur that the signal has completely decayed with long acquisition times and that only noise is digitized for much of the FID. In such cases we can collect a smaller number of data points and fill in the FID with zeros (*zero filling*) before Fourier transformation to obtain the desired digital resolution.

For nD NMR experiments it is not necessary to have such a high digital resolution, since the spectral features are spread out into two or more frequency dimensions. Such a high digital resolution would lead to extremely large nD NMR data sets and very long acquisition times. In many cases, only 1024 or 2048 data points are required.

The SNR in a spectrum is also of critical importance. For very concentrated samples a good proton spectrum can be acquired in a few (typically eight) scans. Many more

2.4 SOLUTION NMR METHODS

scans are required for insensitive nuclei. It is not uncommon for dilute samples to require an overnight accumulation. As noted earlier (Section 1.2.5) the SNR increases with the square root of the number of scans.

For a quantitative measure of the peak intensities, it is necessary to allow the spin system to return to equilibrium between acquisitions. This requires waiting at least five times longer than the longest polymer T_1. This is not a problem for protons where the T_1 are short, but long delay times can make the acquisition of carbon spectra difficult because some of the peaks, such as the carbonyls, can have very long T_1. In some cases it is possible to shorten the relaxation times by adding *relaxation reagents*, such as $Cr(acac)_3$ so that the spectra can be acquired more rapidly.

2.4.3 Decoupling

It is often desirable to simplify the spectra of insensitive nuclei (e.g., carbons) by removing the splittings from through-bond scalar couplings with heteronuclear decoupling. This can be accomplished by irradiating the protons while observing the insensitive nucleus. The basic idea behind decoupling is to cause rapid transitions in the proton spin states. If the transitions are fast compared to the scalar couplings (5–200 Hz), then the system will be in the fast exchange limit and a single (decoupled) peak will be observed at the average frequency. It is possible to remove the scalar couplings with single-frequency proton irradiation, but a relatively high power may be required to decouple those protons that are a few kHz off resonance. High-powered decoupling can lead to an uneven heating of the sample, which can affect the field homogeneity and the resolution.

To avoid sample heating and wear on the probe, it is desirable to use low-powered decoupling, which can be accomplished using *composite-pulse decoupling* (CPD) methods. A CPD pulse sequence is a concatenation of a series of rectangular pulses in which the pulse lengths, phases, frequencies, or amplitudes are varied in a regular way. The goal of the CPD is often to apply pulses with the widest possible bandwidth with the lowest power. Consider, for example, a very simple decoupling scheme consisting of a series of 180° pulses applied to the protons to cause rapid transitions. One approach to a CPD scheme would be to replace each of the 180° pulses with a sandwich of pulses to correct for off-resonance effects and pulse errors. Lower power could be used to decouple those protons that are further off resonance.

CPD has been extensively investigated, and many pulse sequences have been proposed (3). These composite-pulse schemes differ in their complexity, ease of implementation, and power required to achieve good decoupling. Low power and a wide frequency band of coverage are usually the motivating factors behind the design of CPD sequences. Many of the CPD designs are optimized using computer algorithms. The efficiency of the sequence can be determined by measuring the spectrum of an insensitive nucleus that is coupled to protons as the decoupler is moved further off resonance.

For proton-decoupled heteronuclear spectra, the CPD method of choice is WALTZ-16 decoupling (4). The basic pulse cycle for WALTZ decoupling is a series of (90_x)–(180_{-x})–(270_x) pulses applied to the protons while the heteronuclear signals are observed. The range of proton chemical shifts that must be decoupled is on the order of 10 ppm (or 5000 Hz in an 11.7-T magnet, where the proton frequency is 500 MHz). The decoupling bandwidth is given by $\Delta f/\gamma B_2$ where Δf is the frequency range decoupled with the application of pulses of strength B_2. The bandwidth for WALTZ-16 is 2, so a proton-decoupling field strength of 2500 Hz is required for proton decoupling across the range of proton frequencies. The decoupling efficiency can be increased with additional cycling of the pulse phases, such as inverting the phases of all the pulses for two of every four basic pulse cycles. To effectively set up WALTZ-16 decoupling it is necessary to accurately measure the proton 90° pulse.

The highest sensitivity nD NMR experiments are the so-called inverse experiments in which the proton signals are detected. In a typical experiment the magnetization from the protons is transferred to a heteronucleus where it evolves under the influence of the heteronuclear chemical shift before it is transferred back to the protons for detection. In such an experiment, only those protons coupled to the heteronucleus are detected. In a carbon–proton experiment this corresponds only to those protons bonded to a ^{13}C nucleus. The detected proton signals show splittings from the one-bond carbon–proton scalar coupling. In inverse experiments it is necessary to remove the carbon–proton couplings by decoupling the carbon spectra. This is a more demanding task, since the carbon signals are spread over a larger chemical shift range. The range of carbon chemical shifts for carbons with directly bonded protons is about 160 ppm. For an 11.7-T magnet (125 MHz for carbons) this corresponds to a frequency range of 20,000 Hz. It is possible to decouple the carbons using WALTZ-16 decoupling, but most probes would not be able to take the high power (10,000 Hz) required for good decoupling across the entire bandwidth.

Decoupling in the inverse detected experiments can be accomplished using composite-pulse decoupling sequences designed for larger bandwidths. One of the most popular is the so-called GARP sequence that has a decoupling bandwidth of 5 (5). This means that the carbon spectrum could be decoupled with a carbon field strength of 4000 Hz. The GARP pulse sequence consists of a series of 25 pulses that alternate in phase and tip angle.

The other important sequence used for composite-pulse decoupling is for magnetization transfer using the scalar couplings in TOCSY-type nD NMR experiments. One popular sequence is MLEV-17 (6). The basic pulse cycle for MLEV is (90_{-y})–(180_x)–(90_{-y}). The MLEV-16 pulse cycle consists of applying this pulse cycle 16 times as ABBA BBAA BAAB AABB, where A corresponds to the basic pulse cycle and B is the basic pulse cycle with all of the phases inverted. MLEV-17 consists of this pulse cycle followed by a 180_x pulse to compensate for pulse errors that accumulate during MLEV-16. The MLEV-17 pulse sequence is preceded and followed by a short (2 ms) spin-locking pulse to defocus magnetization that is not parallel to the x axis at the start of the mixing sequence. The proton power typically used for these experiments is on the order of 4000 to 7000 Hz, and the sequence is executed many times such that the total time for magnetization exchange is on the order of 30 ms. The DIPSI composite-pulse decoupling sequence (7) is also used for TOCSY-type magnetization exchange (8). This sequence has the advantage that cross relaxation from through-space magnetization transfer is suppressed.

2.4.4 Data Processing

It is important to properly set the acquisition parameters to gather high-quality NMR spectra. This includes setting the receiver gain, the sweep width, the carrier frequency, and the number of data points. It is equally important to properly process the data. Among the data processing steps that must be considered are offset corrections to the FID, zero-filling, the application of window functions, and phasing.

2.4.4.1 Baseline Corrections The data for a complex FT is collected in quadrature in most modern spectrometers. This is typically accomplished by collecting two FIDs that differ in phase by 90° and storing them separately. These are sometimes referred to as the real and imaginary parts of the FID, or the x and y components. It is the complex FT of these data that give rise to the normal spectrum. If the offsets for the two FIDs are not identical, Fourier transformation gives rise to a spike at the center of the spectrum. This can be removed by adjusting the real and the imaginary parts of the FID to have the same offset.

In addition to the offsets, other types of errors in setting up the experiment can give rise to baselines that are not flat. One common reason for distorted baselines is errors in accurately recording the first few points in the FID. This can happen if the waiting time between the last pulse and the beginning of data acquisition is too short. In favorable cases it is possible to fix the first few points in the FID by linear prediction (Section 2.7.2.4), which involves a backwards extrapolation of the FID to correct the first few points. If the baseline distortions are small, the baseline can be fit to a polynomial that is subtracted from the data to give a flat baseline.

2.4.4.2 Digital Resolution and Zero-Filling The digital resolution (points/Hz) required for a given experiment depends on the smallest frequency separation between two spectral features of interest. If the spectrum contains very sharp lines, then the digital resolution must be high enough to properly digitize such a sharp feature. For such spectra it is also important that the acquisition time be long enough that the signal has decayed to zero at the end of the FID. If the number of acquired data points is not enough to provide sufficient digital resolution, then the FID can be extended by zero-filling. Zero-filling consists of adding data points to the end of the FID with a zero amplitude. If, for example, the data are collected with 1024 data points, it can be zero-filled to 2048 data points before Fourier transformation to double the digital resolution.

2.4.4.3 Window Functions The quality of the NMR spectrum can often be improved by multiplying the FID by a window (or apodization) function prior to Fourier transformation. Depending on the type of window function, this can lead to an improvement in the SNR, the resolution, or the removal of the artifacts associated with truncation of the FID. The principle of matched filtering states that the maximum SNR is obtained when the window function matches the natural form of the decay of the FID.

The most common use for apodization is to increase the SNR. The most popular window function is exponential multiplication, where the FID is multiplied by

$$w(t) = e^{-\pi LBt} \tag{2.4}$$

where LB is the line broadening in Hz. Multiplication of the FID by this function will increase the SNR at the expense increasing the linewidth. When the line broadening is equal to the natural line width ($1/\pi T_2$), this leads to a significant increase in the SNR while only slightly increasing the linewidths. This is illustrated in Figure 2.5, which shows a carbon NMR spectrum with and without line broadening. In this case the line broadening of 30 Hz is on the order of the width of the sharpest line in the spectrum, and the SNR has been greatly improved.

The second most common use of window functions is to improve the resolution at the expense of the SNR. Broader lines decay more quickly and are emphasized in the initial part of the FID, while sharper lines have a longer decay time and make a significant contribution to the tail of the FID. Resolution enhancement can be accomplished by multiplying the FID by a function that enhances the middle and end of the FID relative to the initial part.

A number of resolution-enhancing window functions have been developed for high-resolution NMR. One of the most popular is the Lorentz–Gauss (LG) window function that is given by

$$w(t) = e^{\pi LBt} e^{-GBt^2} \tag{2.5}$$

where GB is the Gaussian broadening. The window function is a product of a growing exponential and decaying Gaussian function that enhances the center part of the FID

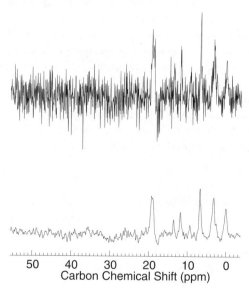

FIGURE 2.5 The carbon spectra of a polymer without (top) and with (bottom) exponential multiplication.

2.4 SOLUTION NMR METHODS

while suppressing the initial part of the FID that contains the rapidly decaying signals from the broad lines. The window function can be adjusted to give the best results by adjusting the parameters *LB* and *GB*. Strong resolution enhancement leads to undesirable negative lobes on the peaks. A number of other resolution enhancement window functions also have been reported that may have advantages under certain circumstances. It is important to note that resolution enhancement can affect the integrated intensity of peaks with different linewidths and should not be used for quantitative comparisons.

2.4.4.4 Phasing

Thus far we have assumed that the observable signal is exactly aligned along the *y* axis for observation. In reality we are detecting the signal relative to a reference signal from the NMR spectrometer. In the event that these signals are exactly in phase, a signal with a $\cos(\omega t)$ time dependence would be detected. There is no guarantee that this will happen, and it is more likely that the phase of the reference signal will be slightly offset from the desired signal. When the FID is Fourier transformed, this leads to a spectrum in which the peaks are not properly phased. This can be corrected by applying a *frequency-independent* phase correction.

A second problem is that signal detection does not begin exactly at zero time; there is some small delay (a few μs) after the pulse before signal detection begins. The chemical shifts can evolve during this delay, and signals with a larger offset from the reference frequency will evolve at a faster rate. Thus, the spectrum may also have *frequency-dependent* phase offsets.

These two types of phase offsets are easily corrected with an algorithm that mixes the real and imaginary parts of the spectra. Although the implementation differs from spectrometer to spectrometer, the procedure usually involves adjusting the phase of a peak at one end of the spectrum to correct the frequency-independent phase correction. A peak at the other end of the spectrum is then adjusted to correct for the frequency-dependent phase correction.

2.4.4.5 Quadrature Detection

It is usually desirable to acquire a signal with the reference frequency in the center of the spectrum to minimize off-resonance effects. However, to do this we need some way to distinguish positive and negative frequencies relative to the carrier. This can be accomplished by detecting the signals along two perpendicular directions, which is known as *quadrature detection*.

The signals in the NMR spectrometer are detected relative to some reference signal, which can be modulated by $\cos(\omega t)$. We could detect the perpendicular signal relative to another reference signal at $\sin(\omega t)$. These two signals are separately measured and used to construct a complex FID which, when subjected to a complex FT, allows for frequency sign discrimination. Quadrature detection is also important for nD NMR, but must be implemented in a different way.

2.4.4.6 Referencing

The chemical shifts are most useful if they are reported relative to some reference compound. The chemical shifts are reported in parts per million (ppm), as given by Equation (1.39). These units are particularly useful because, while the frequency difference between two peaks will change with magnetic field strength, the chemical shift in ppm will not.

TABLE 2.2 The reference compounds used for various nuclei in solution NMR studies

Compound	Nuclei	δ (ppm)
$(CH_3)_4Si$ Tetramethylsilane (TMS)	^1H	0.0
	^{29}Si	0.0
	^{13}C	0.0
$(CH_3)_3Si(CH_2)_5SO_3Na$ 4,4-dimethyl-4-silapentyl sodium sulfonate (DSS)	^1H	0.0015
CS_2 Carbon Disulfide	^{13}C	192.8
CH_3NO_2 Nitromethane	^{15}N	0.0
85% PO_4H_3 Phosphoric acid	^{31}P	0.0
$CFCl_3$ Fluoroform	^{19}F	0.0

The chemical shifts in solution NMR are typically determined by adding some reference compound to the solution and Table 2.2 lists some of the commonly used reference compounds. In organic solvents, the most common reference compound is tetramethylsilane (TMS), which has a chemical shift of zero in the proton, carbon, and silicon spectrum. TMS is a particularly useful reference compound because virtually all of the carbon and proton signals appear at a higher frequency than TMS, so any overlap between the signals of interest and the reference compound is minimized. Hexamethyldisilane is sometimes used, because it has a higher boiling point and can be used at higher temperature. 4,4-Dimethyl-4-silapentane sodium sulfonate (DSS) is water soluble and used as a reference compound for polymers in aqueous solution.

In the absence of a reference compound, the residual solvent peaks can be used as a secondary reference. Deuterated solvents typically have a small percentage of proton-containing solvent molecules, and if they are resolved from the polymer peaks, they can be used as a secondary internal reference. The chemical shifts for these residual solvent peaks are listed in Table 2.1.

Referencing in solid-state NMR is typically accomplished with an external reference. This involves measuring the chemical shift of some reference compound in one experiment and using the same chemical shift scale for the compound of interest. Solid-state carbon NMR spectra are often reference to the signals from hexamethyl benzene or adamantane, which are commonly used as tune-up samples. The solid-state reference compounds are listed in Table 2.3.

2.4.5 Quantitative NMR

One of the major uses of NMR in polymer science is for quantitative analysis, since under optimal conditions the signal intensity is directly proportional to the number of nuclei in the NMR sample. If the signals of interest, such as those from the different

2.4 SOLUTION NMR METHODS

TABLE 2.3 The Reference Compounds Used in Solid-State NMR.

Compound	δ_C (ppm)
Hexamethyl benzene	132.18
	17.36
Adamantane	29.50
	38.56
Poly(dimethyl siloxane)	1.50
Polyethylene	33.63

monomers in a copolymer, can be resolved, then the relative fractions of monomers can be directly calculated from the integrated signal intensities. Extreme care must be taken in the acquisition and processing of data for a quantitative analysis.

The collection of quantitative proton spectra is relatively straightforward, since the protons T_1 are typically short (a few seconds) and the SNR is high. As with all quantitative spectra, it is important to ensure that the spin system returns to equilibrium between acquisitions. The time required for the signals to relax depends on the pulse tip angle and the spin-lattice relaxation time. If the T_1's are relatively short and the SNR is high, we can wait five times the T_1 between scans to ensure that the spin system has returned to equilibrium between scans. If the peaks of interest are well-resolved from each other, then the integrated intensities can be directly compared with each other. If the peaks are overlapped, then it may be necessary to use a line-fitting program to extract the signal intensities.

Carbons and other insensitive nuclei often have very long T_1, making the acquisition of quantitative data very time-consuming. For these samples we can decrease the pulse tip angle and wait for a shorter period of time between scans to obtain quantitative spectra. The optimum tip angle α for acquiring spectra by waiting the time τ between scans is given by

$$\cos(\alpha) = e^{\tau/T_1} \tag{2.6}$$

The quantitative analysis of heteronuclear NMR spectra is slightly more complex than for protons since proton decoupling can lead to changes in the relative signal intensities through the NOE. As noted earlier, the NOE depends on the chain dynamics that can be very different for different parts of the polymer, such as the side-chain and main-chain resonances. For this reason it is necessary to collect the spectra using proton decoupling only during acquisition, as shown in the pulse-sequence diagram in Figure 2.5. The acquisition time is generally short enough that the NOE effects do not build up to a significant amount during signal acquisition, and the signal intensities can be directly related to the number of nuclei in the sample.

The acquisition time and data processing can also affect the quantitative analysis of the data. It is important that the acquisition time be long enough for the FIDs to decay to zero during this period. This is particularly important when comparing the intensities for two signals that have different linewidths. The intensities are only comparable if the acquisition time is long enough that complete signals are recorded for both the

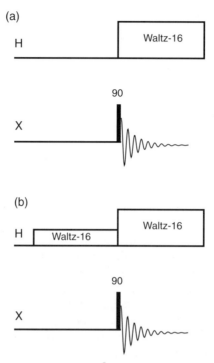

FIGURE 2.6 The pulse sequence diagrams for heteronuclear NMR (a) without and (b) with NOE enhancement.

long and short T_2 components. It is also important to avoid apodization functions that differentially affect the signals with different linewidths. Resolution-enhancement functions, for example, minimize the initial part of the FID and decrease the contributions from the broad components.

2.4.6 Sensitivity Enhancement

The sensitivity can be enhanced in heteronuclear NMR spectra by a variety of means, including the NOE and polarization transfer pulse sequences. These approaches lead to an increase in the SNR, but it is no longer possible to analyze the spectra quantitatively.

The simplest method for increasing the SNR is to use the NOE by irradiating the protons during the entire experiment, as shown in Figure 2.6. If the relaxation of the heteronuclei is dominated by dipolar interactions with the protons, then the signal intensity will change in a way that depends on the correlation time, as given by Equation (1.79). For carbons, the maximum NOE of three is observed for correlation times in the fast motion limit. The signal enhancement is diminished as the correlation time increases, as was shown by Figure 1.44. The obvious problem for polymers is that if the side-chain and main-chain atoms have different correlation times, then they

2.4 SOLUTION NMR METHODS

may experience different NOE enhancements and different intensities. Nitrogen has a negative magnetogyric ratio, so the NOE-enhanced signals will be inverted relative to the normal spectra.

Polarization-transfer pulse sequences, such as INEPT, can also be used for sensitivity enhancement. The INEPT pulse sequence was introduced in Section 1.7.1.1 and the pulse sequence was shown in Figure 1.60. The maximum sensitivity enhancement from INEPT is the ratio of the magnetogyric ratios, which for carbon is a factor of 4. This can be compared with a maximum enhancement factor of 3 for the NOE. The enhancement may be less than the maximum for polymers due to T_2 relaxation during the delay times in the pulse sequence that are set to $1/2J_{CH}$. Quantitative data cannot be obtained using INEPT because the pulse sequence is optimized for a particular value of J_{CH}, so different coupling constants, such as those between aromatic and aliphatic carbons, can give rise to different polarization-transfer efficiencies. Also, differences in the T_2 relaxation times can lead to differences in signal intensities for the different signals.

2.4.7 Spectral Editing

Spectral editing make it possible to distinguish between different types of nuclei based on the local interactions. Most spectral editing methods use polarization transfer based on the number of directly bonded protons to distinguish between different types of carbons or nitrogens. This information is often critical for establishing the assignments for complex polymers where, for example, the signals from the methine and methylene carbons overlap.

A number of pulse sequences have been developed for spectral editing. Among the most useful is distortionless enhancement by polarization transfer (DEPT) that is used to distinguish between carbons and nitrogens with different numbers of directly attached protons (9). The DEPT pulse sequences is shown in Figure 2.7. The sequence is similar to the refocused INEPT pulse sequence in that the $90°-\tau-180°-\tau$ sequence, where τ is equal to $1/2J_{CH}$, is for polarization transfer. In DEPT the signal acquisition is delayed to allow the multiplets to refocus so that the signal can be recorded with proton decoupling. The experiment differs from INEPT in that a series of spectra

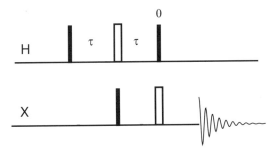

FIGURE 2.7 The pulse sequence diagram for distortionless enhancement by polarization transfer (DEPT).

(typically three) are recorded with different values for the final proton pulse width (θ). A simple product operator analysis, similar to that shown for INEPT in Section 1.7.1.1, shows that the intensity for a carbon with one attached proton is given by

$$M = \frac{\gamma_H}{\gamma_X} \sin \theta \tag{2.7}$$

An analysis for carbons with two and three attached protons gives

$$M = \frac{\gamma_H}{\gamma_X} \sin 2\theta \tag{2.8}$$

and

$$M = \frac{3}{4} \cdot \frac{\gamma_H}{\gamma_X} (\sin \theta + \sin 3\theta) \tag{2.9}$$

where θ is the tip angle for the final proton pulse. Figure 2.8 shows a plot of the intensity for the three types of carbons as a function of tip angle. The idea behind DEPT is to acquire a series of spectra with different values for θ and use linear combinations of these spectra to obtain subspectra showing the three types of carbons.

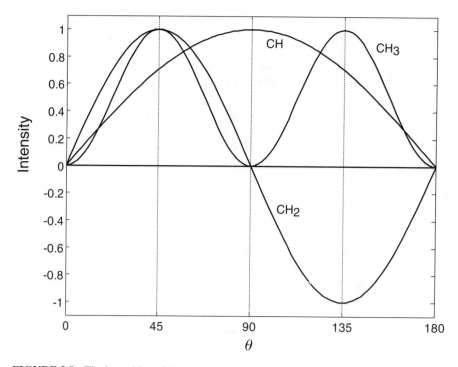

FIGURE 2.8 The intensities of the methyl, methylene, and methine signals as a function of the tip angle for the final proton pulse in the DEPT pulse sequence.

2.4 SOLUTION NMR METHODS

The spectra are acquired for θ values of 45°, 90°, and 135°. The spectra with $\theta = 45°$ contains all three types of signals, while the spectra with $\theta = 90°$ contains only the methine carbons, and the spectra with $\theta = 135°$ contains positive methine and methyl signals and negative phase methylene signals. The subspectra containing only the methylene signals is obtained by subtracting the suitably scaled spectra with $\theta = 90°$ and $\theta = 135°$ from the spectra with $\theta = 45°$. The subspectra containing only the methyl signals is obtained by subtracting the suitably scaled methylene and methine subspectra from the spectra acquired with $\theta = 45°$. The DEPT spectra for poly(propylene glycol) is shown in Figure 2.9.

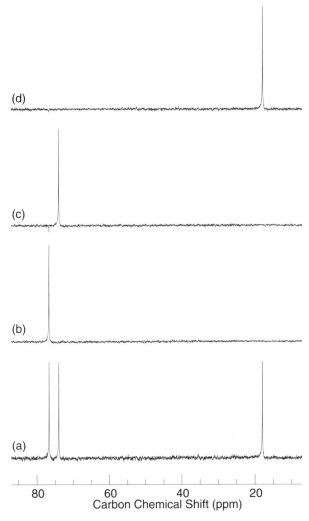

FIGURE 2.9 The DEPT spectra of poly(propylene glycol) showing (a) the 1-pulse spectrum and the DEPT subspectra for the (b) methine, (c) methylene and (d) methyl carbons.

2.5 SOLID-STATE NMR METHODS

Solid-state NMR is fundamentally different than solution NMR because the lines are broadened by the combination of dipolar couplings and chemical shift anisotropy, and possibly quadrupolar interactions. The factors that lead to line broadening contain important information about polymer structure and dynamics, and one approach to solid-state NMR is to directly measure the wideline spectra. Since the lines are so broad, however, the peaks will overlap and it is often not possible to associate the broad lines with specific molecular features. Molecular motions lead to averaging of the lineshapes, and the observation of the lineshape provides valuable information about the chain dynamics (Section 5.3.4). Deuterium lineshapes are often the most informative measure of the chain dynamics and special methods are required to observe these spectra because the lines are extremely broad (hundreds of kHz) and the T_2 relaxation times are extremely short (μs).

In many situations, it is desirable to obtain a high-resolution spectrum. Since chain motion does not average the dipolar interactions or the chemical shift anisotropy in solids, some artificial means, such as proton decoupling and magic-angle sample spinning must be used for line narrowing (10). Cross polarization is used both for sensitivity enhancement and for avoiding the long relaxation times for carbons in the solid state. Special methods are required to observe the solid-state proton spectrum in rigid solids, since it is difficult to observe the proton spectrum in the presence of high-power decoupling.

2.5.1 Magic-Angle Spinning

The chemical shift in solids is anisotropic and depends on the orientation of a particular molecule to the external magnetic field. The chemical shift anisotropy for carbons and nitrogen is within the range that can be averaged by rapid sample spinning at the magic angle, as expected from Equation (1.88). If the rate of sample rotation is fast compared to the width of the anisotropy pattern, then a single peak will be observed at the isotropic chemical shift. If the spinning speed is slower, then the anisotropy pattern will be broken up into a number of sidebands that are separated from the isotropic shift by the spinning speed. The isotropic peak frequency does not depend on the spinning speed, and can be identified by comparing the spectra acquired at several spinning speeds.

Special probes are required for magic-angle sample spinning. A number of probes are available that differ in spinning speed, sample size, and volume, and Figure 2.10 shows a schematic diagram of a magic-angle spinning apparatus. The sample is packed into the *rotor* and placed in the *stator*. The sample is suspended in the stator with a bearing gas flow and spun with drive gas that impinges on fluted end caps. Smaller rotors spin faster and 30 kHz and higher spinning speeds have been reported for 2.5 mm rotors (11). This spinning rate is sufficient to partially average the dipolar couplings for rigid solids, and proton spectra can be directly observed. The sample volume is very small in the 2.5-mm rotors and the SNR may be low for insensitive nuclei such as carbon and silicon. In addition, it is difficult to cross polarize the signals

2.5 SOLID-STATE NMR METHODS

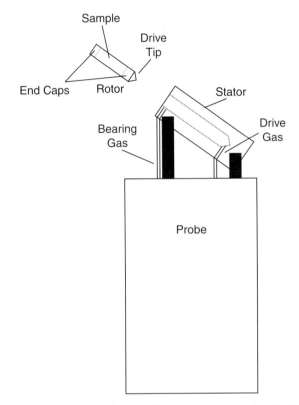

FIGURE 2.10 A schematic diagram of a magic-angle sample spinning probe.

with such fast magic-angle spinning. Four-mm rotors are available that are able to spin up at 20 kHz. These may be a better compromise between sample volume and fast spinning. Most routine experiments are performed on larger rotors (5–7 mm) that can accommodate 0.2 g of sample and provide a good sensitivity for insensitive nuclei. These rotors spin in the range of 8 kHz, which is fast enough to average the chemical shift anisotropy for most carbons of interest. Larger rotors (14 mm) are available for the study of very insensitive samples. These rotors accommodate 0.6 g and can spin up to 5 kHz.

The effect of magic-angle sample spinning on the carbon spectra of polycarbonate is shown in Figure 2.11. The anisotropies for the aromatic carbons are on the order of several kHz, so the spinning must be faster than this to obtain the spectrum without spinning sidebands. Figure 2.10a shows that a spinning speed of 3.5 kHz is fast enough to average the anisotropies for the methyl and quaternary carbons, but many sidebands are observed for the aromatic and carbonyl carbons. At this spinning speed it is difficult to separate the peaks and the sidebands. As the spinning speed is increased to 7.3 kHz, the sidebands are shifted from the isotropic peaks and are reduced in intensity and they are almost completely absent from the spectrum acquired with a spinning speed of 9.5 kHz.

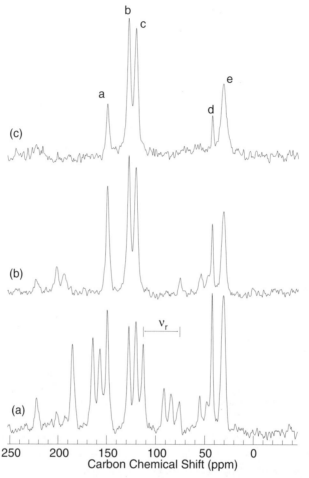

FIGURE 2.11 The effect of spinning speed on the carbon spectra of polycarbonate. The data are shown for spinning speeds of (a) 3.5 kHz, (b) 7.3 kHz and (c) 9.5 kHz.

2.5.2 Cross Polarization

The spectra of solids often require very long acquisition times, because the spin-lattice relaxation times can be very long. This limitation can be partially overcome using cross polarization. As noted in Section 1.6.4, this pulse sequence forces the protons and another nuclei to precess at the same frequency in the rotating frame, and results in effective magnetization transfer. Magnetization transfer via cross polarization is a consequence of strong dipolar interactions and is only useful for nuclei in close proximity (<5 Å) to a proton. Cross polarization is often useful because it is the proton spin-lattice relaxation times that determine waiting time between acquisitions, and these relaxation times are usually much shorter (<5 s) than the carbon, nitrogen, or silicon relaxation times. The other important advantage of cross polarization is

2.5 SOLID-STATE NMR METHODS

that it equilibrates the magnetization between protons and the less sensitive nuclei, thereby increasing the sensitivity. One disadvantage of cross polarization is that it is not quantitative, and the intensities of peaks cannot be directly compared without a full understanding of the cross-polarization dynamics.

One important feature of cross polarization is that the rate of magnetization buildup from cross polarization (T_{CH}) and the decay from rotating-frame spin-lattice relaxation ($T_{1\rho}$) depends on the chain dynamics. This was illustrated in Figure 1.52 for a hypothetical sample containing a rigid and a mobile phase. The rigid (or crystalline) phase cross polarizes most quickly and decays most rapidly. Therefore, the crystalline component can be enhanced in a spectrum acquired with a short cross-polarization time. With a very long cross-polarization time, the signals arise primarily from the mobile phase. Cross polarization provides an important method to distinguish between the phases in complex materials. Figure 2.12 shows the effect of cross-polarization contact time on the signal intensities for the matrix carbons and a mobile polymer in an organic/inorganic nanocomposite. The data in Figure 2.12 show that the spectrum with a short cross-polarization contact time contains mostly inorganic signals, while the spectrum with a longer contact time increases the contribution from the mobile polymer.

The efficiency of cross polarization depends on a good match of the Hartman–Hahn condition. This depends both on the magnitude of the dipolar couplings and

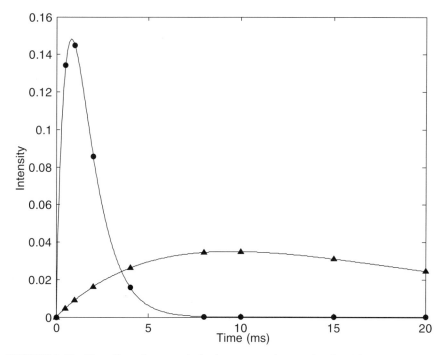

FIGURE 2.12 The effect of cross polarization contact time on the signal intensities for the (•) matrix and (▲) mobile polymer signals in an organic/inorganic composite.

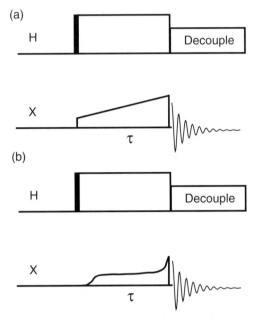

FIGURE 2.13 The pulse sequence diagram for cross polarization using (a) ramped cross polarization and (b) adiabatic cross polarization.

the matching rf fields. If the spinning speed is slow compared to the magnitude of the dipolar couplings then there will be a good Hartman–Hahn match over the entire spectrum. This is typically the case for crystalline polymers and glassy polymers below T_g. However, this may not be true for carbons without directly bonded protons (carbonyls, etc.) or for polymers above T_g. If the spinning speed is of the same magnitude as the dipolar couplings, then the Hartman–Hahn profile is broken down into a series of sidebands, and a good match is not obtained over the entire spectrum. Under these conditions, some peaks may not cross polarize as well as others, and in multicomponent mixtures the signals from some chains may be missing entirely. This problem becomes more severe as the spinning speed is increased to decrease the spinning sidebands from the chemical shift anisotropy.

This problem can be partially alleviated by using a shaped pulse for the cross polarization, as shown in Figure 2.13. A variety of shaped pulses have been proposed, including variable-amplitude (12), adiabatic-passage (13), or amplitude-ramped cross polarization (14). The ramped cross polarization is the easiest to implement and involves changing the rf power on one of the channels linearly during the cross polarization. With ramped cross polarization the power on one of the channels is ramped linearly from a starting to a final value, perhaps from 25% of the matched value to 100% of the matched rf level on the other channel. In the adiabatic passage Hartman–Hahn match, the waveform for one of the channels is based on the relative value of the magnitude of the dipolar coupling (d_{IS}) and the spin-locking field strength

2.5 SOLID-STATE NMR METHODS

(ω_{1I}). The value of α for the adiabatic passage Hartman–Hahn match is given by

$$\alpha = \frac{2}{\tau} \arctan \left(\frac{\omega_{1I}}{d_{IS}} \right) \qquad (2.10)$$

and the rf amplitude A is set to

$$A = -A_0 \tan \left(\alpha \left[\frac{\tau}{2} - t \right] \right) \qquad (2.11)$$

where A_0 is the amplitude for normal Hartman–Hahn matching and t is the time. The feature to note is that the cross-polarization waveform is adjusted for the ratio of ω_{1I}/d_{IS}. A good way to implement these sequences is to store the calculated amplitudes as a waveform and try several different waveforms to determine which has the best performance for the sample of interest. As with all cross-polarization experiments, the relative intensities are not quantitative.

2.5.3 Decoupling

High-power proton decoupling is used in solid-state NMR to average the dipolar couplings so that high-resolution spectra can be observed. The most common method is to use single-frequency irradiation with a power level greater than the magnitude of the dipolar couplings. A power level of 50 kHz is often sufficient for amorphous polymers where there is usually some molecular motion (methyl group rotation, etc.) to partially average the dipolar interactions. Higher powers are required for rigid crystalline polymers. The effect of decoupler power can be evaluated by measuring the effect of different power levels on the linewidth for polyethylene or some other crystalline material. As a general rule, a probe with smaller rotor diameters can accommodate higher decoupler powers, and 100-kHz decoupling is possible with 4-mm rotors. The upper limit on decoupling power is usually the amount of power that the probe is able to accommodate.

Single-frequency decoupling provides good decoupling for most cases and is the easiest to implement, but the decoupling efficiency can be increased using decoupler phase modulation. The so-called two-phase pulse modulation (TPPM) accomplishes efficient decoupling by applying a series of 180° pulses in which the phases are modulated by 10°–15° (15). The advantage of TPPM is that lower power can be used for more efficient decoupling.

2.5.4 Wideline NMR

The NMR lineshapes, particularly those due to quadrupolar coupling in deuterated polymers or the chemical shift anisotropy in the carbon, silicon, phosphorus, or nitrogen spectra, are a rich source of information about the molecular dynamics of polymers. The lineshapes have a characteristic appearance in the absence of molecular motion, and the changes in lineshape due to the molecular dynamics depend on the amplitude and frequency of molecular motion. The lineshapes for polymers experiencing large-amplitude, nearly isotropic motion (i.e., polymers above T_g) are averaged and solution-like spectra are observed.

Deuterium NMR is often used to study polymer dynamics because the relaxation is dominated by the quadrupolar coupling (16). Deuterium has a low sensitivity, so polymers with site-specific labels are required. This is a disadvantage in that new polymers must be synthesized, but an advantage in that the labels are incorporated into the polymer at well-defined sites, such as the main-chain or the side-chain aromatic rings. In the absence of molecular motion, the frequency of a given deuteron is given by

$$\omega = \omega_0 \pm \delta \left(3 \cos^2 \theta - 1 - \eta \sin^2 \theta \cos 2\phi\right) \quad (2.12)$$

where ω_0 is the resonance frequency, $\delta = 3e^2qQ/8\hbar$, $e^2qQ/\hbar$ is the quadrupolar coupling constant, η is the asymmetry parameter, and the orientation of the magnetic field in the principal axis of the electric field-gradient tensor is specified by the angles θ and ϕ. For C—D bonds in rigid solids $\delta/2\pi = 62.5$ kHz and $\eta \approx 0$. Thus, two lines are observed corresponding to the transitions for each deuteron. In isotropic samples, averaging over all possible orientations gives rise to the well-known "Pake" spectrum (17) shown earlier in Figure 1.54. Deuterium NMR is often used to study the molecular dynamics, because the lineshape is averaged by molecular motion, and the lineshape is very sensitive to the both the amplitude and frequency of molecular motion.

Deuterium NMR spectra are difficult to observe because the lines are extremely broad and the relaxation times are very short. To record such broad lines it is necessary to have a digitizer that runs at 1 MHz or faster. The spectra cannot be directly recorded because there is a "dead time" after the application of an intense rf pulse during which data cannot be recorded. This is not a problem for most nuclei, since the "dead time" is on the order of a few μs and significant relaxation does not occur during this period. However, since the T_2 relaxation times are so short for deuterium (μs), the signals would be distorted if they were recorded with the standard 1-pulse sequence. Because of this rapid relaxation it is necessary to record deuterium spectra using the quadrupolar echo pulse sequence shown in Figure 2.14. Since the spectrum is recorded as an echo, the delay time after the second pulse is longer than the dead time and undistorted spectra can be recorded. The use of such an echo pulse sequence assumes that molecular motion does not occur during the echo delay times, which are typically on the order of 10 μs. Molecular motion leads to a change in orientation and the frequency, and the signals from deuterons that undergo molecular motions are not refocused by the second pulse. It is necessary to record the spectra for several values of τ to ensure that signal is not being lost due to molecular motion.

Because of the broad lines in the deuterium spectrum, special probes, high-power amplifiers, and fast digitizers are required. As noted in Section 1.2.5, the spectral range

FIGURE 2.14 The pulse sequence diagram for the quadrupolar echo.

2.5 SOLID-STATE NMR METHODS

covered by a pulse is inversely proportional to the pulse length, and to give a pulse of equal tip angle across the entire spectrum, the spectral width should be much less than the inverse of the pulse length. This is difficult because the deuterium lineshape may cover a range of 200 kHz, and to give an equal tip-angle pulse across the entire deuterium spectrum requires a 90° pulse width of less than 3 μs. This requires a very high power, and thus a special probe. The sweep width for a deuterium spectrum is typically on the order of 1–2 MHz, so very fast signal digitization is required.

2.5.5 Solid-State Proton NMR

The solid-state proton spectrum of polymers is broadened by homonuclear dipole–dipole interactions that must be removed to observe a high-resolution spectrum. We noted in Section 1.6.3 that the dipolar coupling has the same $(3\cos^2\theta - 1)$ dependence on the orientation of the internuclear vector to the magnetic field as does the chemical shift anisotropy. This suggests that if we can perform magic-angle averaging, either by applying a series of pulses or by spinning fast enough, then we would be able to obtain a high-resolution proton spectrum. Protons are common to almost all polymers, and they can be detected with a high sensitivity.

High-resolution proton spectra can be acquired using either multiple-pulse NMR or very rapid magic-angle sample spinning to average the dipolar couplings. Multiple-pulse NMR works by applying a series of pulses such that on average the spins are at the magic-angle and the broadening disappears. Even if the multiple-pulse decoupling were perfect, the lines would still be broadened by the proton chemical shift anisotropy. The chemical shift anisotropy lineshape is usually less than a kHz for protons, so the anisotropy can be removed with relatively slow magic-angle sample spinning. The combination of multiple-pulse decoupling and magic-angle spinning is known as CRAMPS (18).

CRAMPS is one of the more difficult solid-state NMR methods because it requires high-power decoupling of the protons during observation, and this experiment must be done in such a way that the high-powered decoupling must not be applied while the signals are being acquired. This can be accomplished with "windowed" multiple-pulse decoupling, pulse sequences to average the dipolar couplings that contain small delays during which the transmitter is gated off and the receiver is gated on.

In an effort to understand how multiple-pulse decoupling works, let us consider one of the original pulse sequences, the so-called WAHUHA sequence (named for its inventors) shown in Figure 2.15 (19). Multiple-pulse decoupling is usually applied in cycles. The basic repeating element in the WAHUHA sequence is a four-pulse sequence in which the pulses are separated by short delay times of either one or two τ. Recall from the discussion of magic-angle spinning (Section 1.6.2) that averaging of the $(3\cos^2\theta - 1)$ term is possible by having the magnetization rapidly reorient between the x, y, and z axes such that the vector average lies along the body diagonal of a cube in the rotating reference frame. The angle of the averaged vector relative to the magnetic field is 54.7°, the so-called magic angle for which the $(3\cos^2\theta - 1)$ term goes to zero. The goal of multiple-pulse decoupling is to apply pulses such that the magnetization spends equal amounts of time along the x, y, and z axes so that on average vector is at the magic angle.

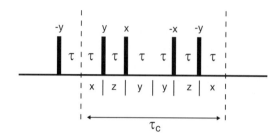

FIGURE 2.15 The pulse-sequence diagram for the WAHUHA multiple-pulse decoupling sequence.

We can understand how this averaging works by considering the magnetization in a hypothetical WAHUHA sequence in which the magnetization is rotated by instantaneous 90° pulses and the τ delay is too short to allow for any significant chemical shift evolution or relaxation. The pulse sequence begins with a 90_{-y} pulse to tip the magnetization along the x axis. The WAHUHA cycle begins after the first τ delay where the magnetization resides along the x direction. The 90_y pulse tips the magnetization along z and the 90_x tips the magnetization along the y axis where it resides for 2τ. The 90_{-x} pulse tips the magnetization back along the z axis for the delay τ and the final 90_{-y} tips the magnetization along the x direction for the final delay τ. The net result is that the magnetization spends equal time along the three axes and the average is a vector lying along the body diagonal of a cube (i.e., the magic angle) and the dipolar coupling goes to zero.

In practice, these multiple-pulse decoupling sequences can be difficult to implement. One problem is that any errors in the pulse length or phase greatly reduce the efficiency of the multiple-pulse decoupling. In addition, the cycle time must be short compared to the rotor cycle time in the magic-angle sample spinning. The cycle time τ_{cycle} is given by

$$\tau_{\text{cycle}} = 6\tau + 4\tau_{90} \qquad (2.13)$$

where τ_{90} is the 90° pulse width. To keep the cycle time short, extremely short 90° pulses ($<2\ \mu$s) are required. Such multiple-pulse decoupling often involves specialized probes and amplifiers. Since the chemical shift anisotropy is small for protons, CRAMPS spectra are usually acquired with magic-angle spinning speeds near 2 kHz.

While the WAHUHA sequence is useful for understanding the idea behind multiple-pulse decoupling, other pulse sequences give better averaging of the dipolar couplings. One of the most useful of the these is the MREV-8 pulse sequence (20–23) shown in Figure 2.16. In practice, one data point is acquired during one of the 2τ delays in each of the 128 to 512 multiple-pulse cycles to create the FID. An extensive tune-up procedure (24) is required for multiple-pulse proton NMR because even small errors in the pulse widths and phase angles cannot be tolerated.

The sweep width in the multiple-pulse spectrum is also different from conventional NMR. In most NMR experiments the sweep width is given by the inverse of the dwell time, the time allotted to acquire a single data point. In the multiple-pulse experiments

2.5 SOLID-STATE NMR METHODS

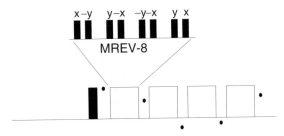

FIGURE 2.16 The pulse-sequence diagram for MREV proton decoupling.

the dwell time is given by the cycle time. In addition, the magnetization is not freely precessing in the xy plane of the rotating frame, but rather about the magic angle. The chemical shift is given by the projection of the vector at the magic angle onto the xy axis, and the chemical shift is effectively scaled by some factor. The scaling factor can be calculated for a particular sequence, but it is always a good idea to experimentally measure the chemical shift scaling by acquiring a proton spectrum for a compound with a known chemical shift separation between the peaks. Figure 2.17 shows the CRAMPS spectrum of of poly(vinyl phenol). Because of the difficulty in completely averaging the dipolar interactions, the resolution in the CRAMPS spectrum is on the order of 1 ppm.

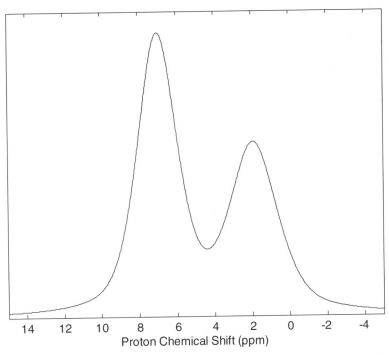

FIGURE 2.17 The solid-state proton CRAMPS spectrum of poly(vinyl phenol).

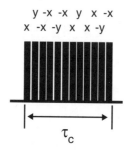

FIGURE 2.18 The pulse-sequence diagram for BLEW-12 multiple-pulse decoupling.

Special probes and amplifiers are required to directly observe the proton spectrum using CRAMPS. An alternative is to indirectly observe the proton spectrum in an nD NMR experiment such as HETCOR, which gives a correlation of the carbon and proton chemical shifts. If we are not trying to directly observe the proton spectrum we can use the more efficient "windowless" decoupling sequences such as BLEW-12 (Figure 2.18). The rf transmitter is always gate-on during a windowless pulse sequence, and the cycle time for BLEW-12 is given by

$$\tau_{\text{cycle}} = 12 \cdot \tau_{90} \tag{2.14}$$

In a HETCOR-type pulse sequence, the increment (or dwell time) in the indirectly detected dimension is usually equal to τ_{cycle}. The chemical shift scaling factors for BLEW-12 can be directly calculated, but like the CRAMPS experiments it is usually best to measure the chemical shift scaling from the chemical shifts in a known compound. The nD experiments where the proton chemical shifts are indirectly detected usually do not require special probes, amplifiers, or a difficult tune-up. The 1D solid-state proton spectrum can be obtained from the projection of the heteronuclear correlation spectrum along the proton dimension.

Another way to effectively remove the proton–proton dipolar couplings in heteronuclear correlation experiments is off-resonance decoupling, as pioneered by Lee and Goldberg (25). In the discussion thus far we have considered the effect of pulses on the magnetization using strong rf pulses on resonance. For off-resonance decoupling we must consider the effect of weaker fields further removed from the signals. If the pulse is applied off-resonance, the magnetization rotates around the effective field determined by the distance off resonance (Δv, in Hz) and the field strength (γB_1). The angle between the z axis and the effective precession angle is given by

$$\tan(\theta) = \frac{\Delta v}{\gamma B_1} \tag{2.15}$$

The trick of the Lee–Goldberg sequence is to choose the off-resonance setting and the field strength such that the effective precession vector is at the magic angle

$$\tan^{-1}\left(\frac{\Delta v}{\gamma B_1}\right) = 54.7° \tag{2.16}$$

2.5 SOLID-STATE NMR METHODS

Under these conditions the $(3\cos^2\theta - 1)$ term in the dipolar coupling is averaged to zero.

More recently it has been reported that better performance can be achieved using the so-called frequency-switched Lee–Goldberg (FSLG) pulse sequence (26,27). In this modification of the Lee–Goldberg sequence, the off-resonance decoupling is switched between $+\Delta\omega$ and $-\Delta\omega$ as the phase is inverted. The pulse length τ_{LG} that the offset and phase are maintained is given by

$$\tau_{LG} = \frac{\sqrt{(2/3)}}{\gamma B_1} \qquad (2.17)$$

This pulse sequence is often easier to implement and has a large scale factor ($\chi = 0.577$). Since the cycle time with the FSLG decoupling is shorter, this sequence can be used with fast magic-angle sample spinning.

As with the chemical shift anisotropy, fast magic-angle sample spinning can be used to average the proton–proton dipolar couplings. The only limitation is that the spinning speed must be large relative to the magnitude of the dipolar couplings, which can be on the order of 30–50 kHz for crystalline polymers and amorphous polymers below T_g. Probes are currently available that spin as fast as 35 kHz, which is fast enough to provide substantial narrowing for many rigid polymers (11). If there is some molecular motion, as for polymers above T_g, the slower spinning speeds can give rise to high-resolution spectra (28), since the dipolar couplings are partially averaged by chain motion. This is illustrated in Figure 2.19, which shows the solid-state proton spectrum for a poly(styrene-*co*-butadiene) copolymer at ambient temperature with 15-kHz

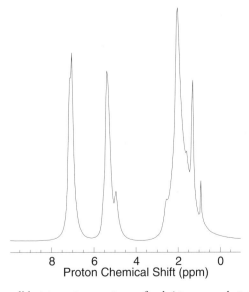

FIGURE 2.19 The solid-state proton spectrum of poly(styrene-*co*-butadiene) acquired with fast (15-kHz) magic-angle sample spinning.

magic-angle sample spinning. Once a high-resolution spectrum is obtained with fast magic-angle spinning, it is possible to use the common solution NMR methods, such as 2D exchange NMR, to measure the structure and local interactions (28).

2.6 NMR RELAXATION

The NMR relaxation parameters, including T_1, T_2, provide important information about the molecular dynamics of polymers, both in solutions and in the solid state, since the relaxation times can be related to the molecular correlation times. Molecular motion in polymers is often complex, and it is difficult to determine the correlation times from a single experiment. As a general rule, it is better to measure several relaxation times as a function of temperature and magnetic field strength and fit all of the data to a single model. It is also important to know the relaxation times for quantitative NMR experiments, since it is necessary to wait five time the longest spin-lattice relaxation time between acquisitions for the spin system to completely relax.

2.6.1 NMR Relaxation in Solution

Chain motion for polymers in solution leads to an effective averaging of the dipolar interactions and chemical shift anisotropy that cause relaxation. This motion can be characterized by measuring the T_1, T_2, and NOE relaxation because these measurements are sensitive to molecular motions on the same time scale as those that occur in polymers.

2.6.1.1 Spin-Lattice Relaxation The spin system will recover toward equilibrium along the z axis following some perturbation, such as an rf pulse, and the rate of return to equilibrium is the spin-lattice relaxation rate. The recovery is given by

$$\frac{dM(t)}{dt} = \frac{M_0 - M(t)}{T_1} \tag{2.18}$$

where M_0 is the equilibrium magnetization and $M(t)$ is the measured magnetization as a function of time after the rf pulse. Integration of Equation (2.18) gives

$$M_0 - M(t) = [M_0 - M(0)] e^{-\tau/T_1} \tag{2.19}$$

If the magnetization is inverted by an initial pulse, as in inversion-recovery experiments, the initial magnetization is $-M_0$ and the relaxation is given by

$$M_0 - M(t) = 2M_0 e^{-\tau/T_1} \tag{2.20}$$

2.6 NMR RELAXATION

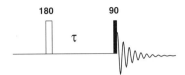

FIGURE 2.20 The pulse-sequence diagram for the inversion-recovery sequence used to measure the spin-lattice relaxation.

The relaxation data can be plotted as

$$\frac{M_0 - M(t)}{2M_0} = e^{-\tau/T_1} \tag{2.21}$$

and the T_1 can be directly determined from the slope of the semilog plot.

Inversion recovery is the most common method used to measure spin-lattice relaxation times, and the pulse sequence is shown in Figure 2.20. The magnetization is inverted with a 180° pulse and sampled with a 90° after a relaxation delay. The data appear inverted at short delay times and return to equilibrium via the spin-lattice relaxation. Figure 2.21 shows a typical example of spin-lattice relaxation data for a polymer sample, and Figure 2.22 shows a semilog plot of the fitted data for the peak at 3.5 ppm.

The potential problem with Equation (2.21) is that the equilibrium magnetization M_0 must be accurately determined, and errors in estimating M_0 may lead to errors in the relaxation time. An accurate measure of M_0 involves acquiring spectra with long

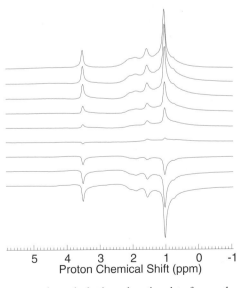

FIGURE 2.21 Representative spin-lattice relaxation data for a polymer nanocomposite.

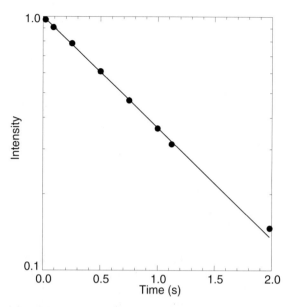

FIGURE 2.22 A semilog plot of the relaxation data of Figure 2.21.

delay times, which can increase the experiment time. A better alternative is to fit the data ($M(t)$ vs. time) as

$$y = M_0(1 - a_0)e^{-t/T_1} \tag{2.22}$$

where the three fitted parameters are the equilibrium intensity (M_0), the degree of inversion (a_0), and the T_1. A semilog plot of the data is still useful for identifying nonexponential relaxation.

One disadvantage of using inversion recovery to measure T_1 is that it is necessary to wait five times the longest T_1 for an accurate measure of the relaxation times. This is a serious disadvantage for insensitive nuclei with long relaxation times. The relaxation times can be measured with a higher SNR using the saturation-recovery pulse sequence shown in Figure 2.23. This pulse sequence begins with a train of pulses to completely saturate the spin system, and the return to equilibrium is monitored after the relaxation delay. In this case, $M(0) = 0$ in Equation (2.19) and an exponential recovery of magnetization is observed. Since the signal is saturated at the start of the experiment, it is not necessary to wait five times T_1 between acquisitions.

The relaxation-rate measurements for insensitive nuclei can be performed with nonselective pulses because the relaxation of nearby nuclei has only a small effect

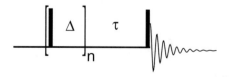

FIGURE 2.23 The pulse-sequence diagram for saturation-recovery NMR.

2.6 NMR RELAXATION

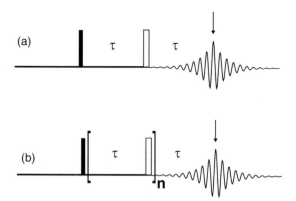

FIGURE 2.24 The pulse-sequence diagrams for (a) the Carr–Purcell spin echo and (b) Meiboom–Gill modification of the Carr–Purcell spin echo sequence.

on the relaxation. This is often not true for protons, and different relaxation rates for selective and nonselective pulses will be observed. To observe the selective relaxation in protons it is necessary to selectively excite a single proton resonance using some pulse sequence like DANTE for selective excitation (29).

2.6.1.2 Spin-Spin Relaxation The spin-spin relaxation times are sensitive to molecular motions over a wide range of frequencies and can be measured with the spin-echo pulse sequence shown in Figure 2.24a. Spin-spin relaxation measures the decay of magnetization in the xy plane, and the most basic pulse sequence for measuring the relaxation rate is the $90°–\tau–180°–\tau$ sequence. The Meiboom–Gill modification of the pulse sequence uses a loop of $[\tau–180°–\tau]_n$ where the value of τ is constant and the time variable is determined by the number of times the loop is executed. This gives rise to an exponential decay in signal intensity with a time constant of T_2. The purpose of the Meiboom–Gill modification is to remove the contribution from inhomogeneous broadening from the relaxation decay. The apparent spin-spin relaxation rate (T_2^*) has contributions both from T_2 and inhomogeneous broadening and is given by

$$\frac{1}{T_2^*} = \frac{1}{T_2} + \frac{1}{T_2^{\text{in hom o}}} \qquad (2.23)$$

where $T_2^{\text{in hom o}}$ is the inhomogeneous contribution (due to magnetic field inhomogeneities an inhomogeneous broadening) to the spin-spin relaxation, and T_2^* is the apparent relaxation, time determined from the linewidth as

$$\Delta v_{1/2} = \frac{1}{\pi T_2^*} \qquad (2.24)$$

Figure 2.25 shows a proton spin-spin relaxation rate measurement for a mobile polymer in an organic/inorganic nanocomposite.

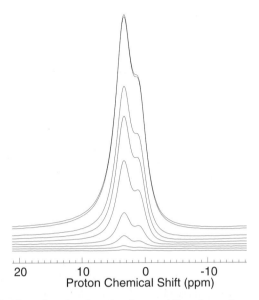

FIGURE 2.25 The spin-spin relaxation for a mobile polymer in a nanocomposite.

2.6.1.3 Nuclear Overhauser Enhancement The NOEs are very sensitive to molecular motion near the Larmour frequency and are often used in conjunction with other relaxation measurements to study the solution dynamics of polymers. Proton–proton NOEs are most commonly measured using 2D NMR and contain information about the structures of polymers, since the NOE intensities depend on the inverse sixth power of the internuclear separations.

Heteronuclear NOEs are most commonly measured by comparing the intensity for spectra gathered with proton decoupling only during acquisition and with proton decoupling during the entire experiment using the pulse sequences shown in Figure 2.5. The ratio of the observed intensities between the two experiments is the NOE. Since proton saturation requires less power than scalar decoupling, it is desirable to use lower decoupling power during the long delay time between acquisitions to reduce sample heating. As with many relaxation measurements, it is necessary to wait five times T_1 between scans for quantitative measurements.

2.6.2 NMR Relaxation in Solids

Solid-state NMR relaxation measurements are used to provide information about the chain dynamics of polymers in the solid state, and the morphology of solids over a length scale of 20–200 Å. Some of the measurements, such as T_1 and T_2, are similar to those measured in solution and provide information about the fast motions of polymers. Others, such as the rotating-frame spin-lattice relaxation times, $T_{1\rho}$, are sensitive to molecular motions in the kHz frequency regime and can in favorable cases be related to the mechanical properties (30). The solid-state proton relaxation times are dominated by proton spin diffusion and are more useful for studying the length scale of mixing in polymers rather than the molecular dynamics.

2.6 NMR RELAXATION

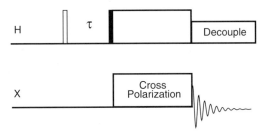

FIGURE 2.26 The pulse-sequence diagram for measuring the solid-state proton spin-lattice relaxation times.

2.6.2.1 Spin-Lattice Relaxation

The solid-state relaxation times for polymers are sensitive to the chain dynamics and are very different for polymers in the crystalline, amorphous, and rubbery phases. As with other solid-state NMR studies, the relaxation times are most commonly measured using cross polarization and magic-angle sample spinning. Separate cross polarization experiments are used to measure the spin-lattice relaxation times for the protons and the insensitive nuclei.

Figure 2.26 shows the pulse-sequence diagram for measuring the solid-state proton spin-lattice relaxation time. The pulse sequence differs from the standard cross-polarization pulse sequence in that a 180° proton pulse and a relaxation delay time τ precedes the cross-polarization and detection periods. The initial 180° pulse inverts the proton magnetization so that the cross-polarized spectrum is inverted relative to the experiment without the 180° pulse. The analysis of the relaxation times is identical to that used with the inversion recovery sequence (Equation (2.20)). The dominant relaxation mechanism for protons is magnetization exchange with nearby protons. The mathematics of the exchange follows the mathematics of diffusion and this process is known as spin diffusion (31). The rate of spin diffusion contains important information about the morphology, as discussed in Section 4.3.2.

The spin-lattice relaxation times for the insensitive nuclei can be measured using the modified cross-polarization sequence shown in Figure 2.27. Following cross

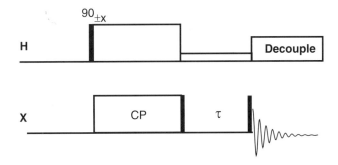

FIGURE 2.27 The pulse-sequence diagram for spin-lattice relaxation-rate measurements using cross polarization.

polarization the magnetization is rotated to the $-z$ axis with a 90° pulse to the carbons. The return to equilibrium is monitored by another 90° pulse after a variable delay time τ, and the spectrum is recorded with magic-angle spinning and proton decoupling. The implementation of this pulse sequence is slightly more complex than the solution version of the inversion-recovery experiment because the experiment begins with cross polarization, which is a nonequilibrium state. We can correct for this by inverting the phase of the proton 90° pulse in every other scan and alternately adding and subtracting the FIDs. The resulting difference spectrum is at a maximum at the shortest delay times, and the signals decay exponentially with a time constant of T_1. This pulse sequence retains all of the advantages of the cross-polarization sequence in terms of the sensitivity enhancements and the shorter waiting times between scans. This approach to measuring the spin-lattice relaxation times assumes that good signals can be observed using cross polarization. For those samples that do not cross polarize well, the relaxation can be measured using saturation recovery.

2.6.2.2 Rotating-Frame Spin-Lattice Relaxation The relaxation of protons or insensitive nuclei in a spin-locking field provides information about the dynamics of polymers in the kHz frequency regime. The rotating-frame relaxation times for insensitive nuclei can be measured using the pulse sequence shown in Figure 2.28. This sequence differs from the standard cross-polarization sequence in the delay period following cross polarization during which the proton field is turned off and the magnetization evolves under the influence of the spin-lock field of the insensitive nuclei. The maximum signal is observed for the shortest delay times and at longer delay times the signal exponentially decays with a time constant $T_{1\rho}$. In many polymers the decay is more complex than a single exponential, due both to spin dynamics and heterogeneities in many polymer systems, as will be discussed in Section 5.3.2.

The solid-state proton rotating-frame spin-lattice relaxation time $T_{1\rho}^H$ can be measured using the cross-polarization sequence shown in Figure 2.29. In this case a variable relaxation time precedes the Hartman–Hahn spin-lock during which the proton magnetization decays under the influence of the proton spin-locking field. The relaxation times measured by this procedure are dominated by proton spin diffusion and often represent an average for the entire spin system.

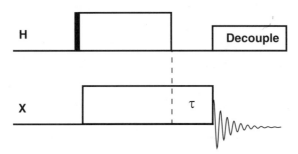

FIGURE 2.28 The pulse sequence diagram for measuring the rotating-frame relaxation times of insensitive nuclei.

2.7 MULTIDIMENSIONAL NMR

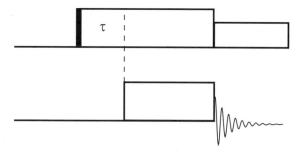

FIGURE 2.29 The pulse sequence diagram for measuring the proton rotating-frame relaxation times.

2.7 MULTIDIMENSIONAL NMR

nD NMR methods have had a revolutionary impact on the NMR of polymers because they address one of the most fundamental limitations in most NMR studies of polymers, the peak overlap that arises from the repeating-sequence nature of polymers. This overlap makes it difficult to resolve the peaks of interest that may be present at very low levels. The resolution can be greatly improved by using nD NMR experiments. These nD NMR experiments are more time-consuming and require different strategies for data acquisition and processing relative to 1D NMR experiments, but this effort is justified by the results. Among the factors that must be considered are the acquisition parameters, the time required to acquire the data, the strategies for data processing and the means to visualize extremely large data sets. The acquisition times required for nD NMR experiments would be prohibitively large if the data were to be acquired at the same digital resolution as for 1D NMR experiments. The general strategy in nD NMR is to acquire the minimum amount of data needed to visualize the interactions of interest. This often involves mathematical methods such as linear prediction (32) to extrapolate the data to longer acquisition times so that the highest resolution can be obtained in the shortest period of time. The use of pulse-field gradients for coherence selection rather than extensive phase cycling also leads to dramatic decreases in the time required for data acquisition (33). Aquiring data in the shortest amount of time requires the highest sensitivity, which can be vastly improved with proton detection in heteronuclear NMR experiments.

2.7.1 Data Acquisition

Data acquisition in nD NMR experiments requires a careful setup of experimental parameters. The greatest efficiency is achieved in a compromise between the acquisition times and the minimum digital resolution needed to identify the peaks of interest. This often involves maximizing the sensitivity by using inverse detection whenever possible, using pulse-field gradients for coherence selection, and collecting the minimum number of data points in the indirect dimensions.

TABLE 2.4 The Acquisition Times Required to Acquire nD Data Sets at Comparable Digital Resoultuions Using 1024 Data Points in Each Dimension, 16 Transients, and a 3-s Delay Time Between Scans.

nD	Time (h)	Data Size (MB)
1	0.013	0.008
2	13.65	8.3
3	13,981	8,499.2

2.7.1.1 Digital Resolution and Acquisition Times in nD NMR The digital resolution, the number of acquired data points, and the acquisition times are important parameters to consider in nD NMR experiments. The digital resolution in 1D experiments is chosen to visualize the smallest feature in the spectrum. It is generally not possible to use such a high digital resolution in the nD NMR experiments because of the time required to collect the data sets. This is illustrated in Table 2.4, which compares the time and the data sizes required to gather nD data with equal digital resolution in all dimensions. It is obvious from this table that compromises must be made in the data acquisition and digital resolution for these experiments to be acquired in a reasonable period of time.

The digital resolution required for an nD NMR experiment depends on the interactions giving rise to peaks in the nD spectrum. For simple chemical shift correlations it is common to have digital resolutions on the order of 10 to 20 Hz/point, compared with 0.25 Hz/point in 1D NMR experiments. It is often not necessary to actually collect this many data points in the indirectly detected dimension, since the digital resolution can be increased by linear prediction or zero filling. A typical heteronuclear correlation experiment may be collected with 1024 points in the directly detected dimension and 128 or 256 points in the indirectly detected dimension. Zero filling along the indirectly detected dimension would lead to a final data set size of 1024 × 1024.

An important factor to consider when setting up an nD NMR experiment is the length of the acquisition times. A simple product-operator analysis of a COSY-type experiment shows that the cross peaks do not exist at the beginning of the t_1 and t_2 time periods, but evolve into observable magnetization during these periods. The efficiency of generating cross peaks is approximately given by

$$M \propto \sin(\pi J t_1) \sin(\pi J t_1) \qquad (2.25)$$

To observe significant intensity in the 2D spectra it is important to have the acquisition times in the t_1 and t_2 dimensions be on the order of $1/2J$, which is about 0.05 s for proton–proton couplings and 0.003 s for carbon-proton couplings. Very long acquisition times are required to observe weak couplings. In most polymers, spin-spin relaxation during the delay times makes it impossible to see COSY cross peaks for coupling constants less than 1 Hz. It is for these reason that the maximum values for the t_1 and t_2 delay times (t_1^{max} and t_2^{max}) are often listed in the experimental section.

2.7.1.2 Inverse Detection The sensitivity is low in NMR compared to other spectroscopies because the separation between the energy levels is very small, resulting in an extremely small ($1/10^5$) difference in populations between the upper and lower levels. The signal-to-noise ratio for a heteronuclear 1D NMR experiment depends on the nuclei involved and the concentrations. The SNR is approximately given by

$$\frac{S}{N} \sim N \gamma_{exc} \gamma_{det} B_0^{3/2} \sqrt{NS} \frac{T_2}{T} \qquad (2.26)$$

where N is the number of nuclei in the NMR coil, γ_{exc} and γ_{det} are the magnetogyric ratios of the excited and detected nuclei, B_0 is the magnetic field strength, NS is the number of scans, and T is the temperature. Although many of the factors affecting the SNR depend on the hardware (NMR tube size, magnetic field, etc.), others can be influenced by the experiment design and the pulse sequence. The largest gain in sensitivity is observed by exciting and detecting the high-sensitivity nuclei (protons or fluorines), rather than a nucleus with lower sensitivity (carbon, nitrogen, etc.). If we compare the time required to acquire a signal with the same SNR for carbon detection and excitation compared with proton excitation and detection, the time for the carbon spectrum is 1024 times longer than for the proton spectra. These gains in the SNR are only possible if we have some means (scalar or dipolar couplings) to efficiently transfer magnetization between the high- and low-sensitivity nuclei.

Inverse detection is frequently used in nD NMR to improve the sensitivity, since the sensitivity gains in nD NMR are even larger than for 1D experiments. These experiments typically start with magnetization transfer from the protons to the less sensitive nuclei and end with transfer back to the protons for detection. The HMQC experiment begins with a proton magnetization that is transferred to the carbons through the carbon–proton one-bond J coupling. The maximum efficiency is obtained when the delay time between the proton and carbon pulses is $1/2J$, which is on the order of 3 ms. This is a very efficient transfer because $1/2J$ is usually much less than the spin-spin relaxation time. The experiment ends with a reversed sequence that transfers carbon magnetization back to the protons for detection.

The HMQC experiment detects only those protons bonded to ^{13}C carbon atoms. To specifically detect these protons it is necessary to suppress all of the signals from protons bound to ^{12}C nuclei, which make up 98.9% of the sample. This can be accomplished through phase cycling, or more efficiently using pulse-field gradients (33).

2.7.1.3 Phase Cycling Phase cycling is used in nD NMR to remove artifacts and select coherence pathways so that the peaks of interest can be visualized. In some cases the phase cycle is simple and requires only two steps, while more complex phase cycling is required to remove other types of artifacts. In many cases this can be more efficiently accomplished with pulse-field gradients (33).

Phase cycling can be used to remove a number of different types of artifacts from the nD spectrum. This can be illustrated by considering the effect of T_1 relaxation during the $90°-t_1-90°$ COSY pulse sequence. Any magnetization that relaxes during the t_1 period will be tipped into the xy plane by the second pulse. This gives rise to magnetization that is not modulated during the t_1 period, the so-called axial peaks that appear at zero frequency in the t_1 dimension after Fourier transformation. The axial peaks can be removed by inverting the phase of the second pulse and coadding the resulting FIDs.

Phase cycling is also important to select coherence transfer pathways and for quadrature detection (Section 2.7.1.4). This is illustrated by considering the effect of the pulse phases on the double-quantum coherences generated by the $90°-t_1-90°-90°$ double-quantum filtered COSY spectrum (34). If the phases of the first two pulses are shifted by $90°$ in subsequent scans, the phase of the double-quantum (but not the single quantum) signals will be inverted. The FIDs can then be subtracted to select for only the double-quantum signals. As a general rule the phase cycling for selection of p-quantum coherences are 0, 180 $/p$, 2 × 180 $/p$, ...$(2p-1)$× 180 $/p$, where p is the coherence order (double-quantum, triple-quantum, etc.). For coherence orders greater than two, phase shifts of less than $90°$ are required. Small-angle phase shifting is commonly available on modern NMR spectrometers.

2.7.1.4 Quadrature Detection

As with 1D NMR, there is a tremendous advantage to performing nD NMR experiments with quadrature detection in the indirect dimension. Quadrature detection makes it possible to place the rf offset in the center of the spectrum, rather than at one end of the spectrum, so the data can be acquired with narrower sweep widths, and higher digital resolution. In addition, quadrature detection often makes it possible to acquire the data in a pure absorption mode so that the spectra can be phased in all dimensions.

In the earliest 2D NMR experiments such as COSY, quadrature detection in the t_1 dimension was achieved with a single data set through phase cycling. Although this approach gives quadrature detection in the t_1 dimension, the peak phases were a mixture of absorption and dispersion phase. This means that it was not possible to phase the peaks, and the spectra had to be viewed in magnitude mode where the resolution is lower. Much higher resolution is possible if the spectra are acquired in absorption mode.

One strategy for acquiring absorption-mode nD NMR spectra with quadrature detection is to acquire two spectra for each t_1 point in which one or more of the pulses is phase shifted by $90°$ using the hypercomplex method (35). The spectra are stored separately and after Fourier transformation of the t_2 dimension, the parts of the data sets are combined before Fourier transformation in the t_1 dimension. Although the details of data processing depend on the pulse sequence, it is common to combine the real part of the FID from the first experiment with the real part of the FID from the second experiment into a new data set before Fourier transformation in the t_1 dimension. This allows the spectra to be phased in the normal manner. The disadvantage of this approach is that any axial peaks that are not suppressed by phase cycling appear at zero frequency in the t_1 dimension.

A second approach is the so-called time-proportional phase incrementation (TPPI) method where quadrature detection is achieved by incrementing the phases of one or more of the pulses along the t_1 delay time (34). In a COSY spectrum, for example, the phase of the first pulse is incremented by $90°$ with each t_1 point. The spectrum in the t_1 dimension is obtained by a real FT, which is sometimes more difficult to phase than the complex spectrum. It is necessary to double the sweep width in the t_1 dimension

2.7 MULTIDIMENSIONAL NMR

using TPPI, and half of the data are discarded after the Fourier transformation in the indirectly detected dimension. TPPI has the advantage that the axial peaks appear at the edge of the spectra and are not likely to overlap with the peaks of interest.

More recently, the combination of these two methods (TPPI-Hypercomplex) has been used to acquire nD NMR spectra (36). This approach combines the best features of the two approaches, a complex FT in the t_1 dimension, and the axial peaks appear at the edge rather than the center of the spectrum. As with the hypercomplex method, two separate data sets are acquired that differ in the phases of one or more of the pulses by 90°. In addition, the phases of pulse and the receiver are advanced by 180° for every t_1 point. The processing of the data is identical to the hypercomplex method.

2.7.1.5 Pulse-Field Gradients

Pulse-field gradients are extensively used in nD NMR for coherence selection and to eliminate artifacts. The gradients can be used in place of phase cycling to dramatically shorten the number of scans required for each transient in the indirectly detected dimension. This sometimes leads to a decrease in the sensitivity, but it also leads to a decrease in the noise. In inverse-detected nD NMR experiments, for example it is necessary to detect only the signals bonded to ^{13}C atoms. This can be achieved by using phase cycling, but the resulting spectrum is obtained by taking the difference between two scans and subtracting away the signals of protons bound to ^{12}C atoms. Since the ^{12}C atoms account for 98.9% of the signal, the ^{13}C spectrum results from the subtraction of two very large signals. Using pulse-field gradients, it is possible to directly detect only those protons bonded to ^{13}C atoms. A much higher receiver gain can be used in these experiments to increase the SNR.

For most NMR experiments the magnetic field is adjusted to be as homogeneous as possible to obtain the highest resolution. In pulsed-field gradient experiments magnetic field gradient is purposely introduced along one or more axes. This is shown in Figure 2.30 for a z-axis gradient. Under the gradient the signals at the center of the gradient are unaffected, while those with a positive z-axis displacement experience a higher magnetic field and those with negative z-axis displacement experience a lower magnetic field. The net effect is that during the pulse-field gradient the signals precess with different frequencies, depending on their sample position, because they are experiencing different magnetic field strengths, and the frequency as a function of position is given by

$$w(z) = \gamma B_0 + \gamma z G_z \qquad (2.27)$$

where z is the displacement along the z axis and G_z is the strength of the gradient along the z axis. In a typical experiment the difference in frequency between spins in the highest and lowest magnetic fields will be on the order of 8 kHz.

To understand why gradients are so useful in nD NMR, it is instructive to consider the recalled gradient echo pulse sequence shown in Figure 2.31a. After the 90° pulse, the spins precess under the influence of their chemical shifts. The spins then rapidly dephase under the influence of the gradient pulse (since they have different frequencies) and, even though the magnetization is in a coherent state, signals can no longer be detected because they are dephased in the xy plane. This dephasing can be

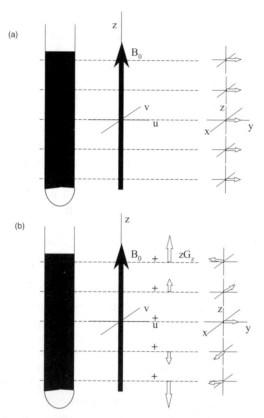

FIGURE 2.30 A diagram showing the effect of a pulse-field gradient.

reversed with the application of a second gradient pulse of the same duration but with opposite phase and signals can once again be detected. A similar result is observed for the gradient spin echo shown in Figure 2.31b. Again the magnetization is dephased by the first gradient. The direction of precession is reversed by the application of the 180° pulse, and the magnetization can be refocused with a gradient pulse of the same phase.

The utility of gradients becomes obvious when we combine these simple echo sequences with more complex pulse sequences. Consider a sequence that generates double-quantum coherences. One property of multiple-quantum coherences is that the signals precess at the sum of the frequencies of the two spins. Thus, a double-quantum proton coherence would precess at a frequency of $\omega_A + \omega_B$ and would dephase twice as fast as a single quantum coherence under the influence of a gradient. If we apply additional pulses to convert these signals to single-quantum coherences, they would precess at a frequency of ω and require a gradient that was twice as long (or strong) as the first gradient to refocus. Gradients can be applied such that only magnetization that had a frequency of 2ω during the first gradient and ω during the second gradient would be refocused. Any spins that precess with a frequency of ω or 2ω during both gradients would not be observed.

2.7 MULTIDIMENSIONAL NMR

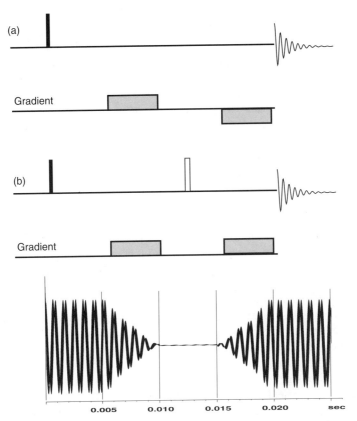

FIGURE 2.31 The pulse sequence diagram for (a) the recalled gradient echo and (b) the gradient spin-echo.

Pulse-field gradients are often used for inverse-detected NMR experiment, as illustrated in the pulse sequence for the gradient-enhanced HMQC (ge-HMQC) sequence shown in Figure 2.32. The first proton pulse, the $1/2J$ delay, and the 90° carbon pulse generates both double- and zero-quantum coherences for protons directly bonded to ^{13}C atoms. For the double-quantum coherences the first gradient pulse generates a

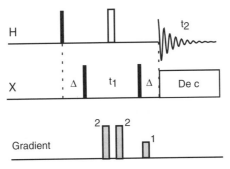

FIGURE 2.32 The pulse-sequence diagram for gradient-enhanced HMQC.

phase proportional to $G_z(\gamma_H + \gamma_C)$, the area of the gradient times the sum of the magnetogyric ratios. The proton 180° pulse converts the double-quantum coherence to zero-quantum coherence and the coherence is partially rephased by the second gradient with a phase proportional to $G_z(-\gamma_H + \gamma_C)$. This leaves a net phase of $G_z(2\gamma_C)$. Since the signal is detected as proton single-quantum coherence and γ_C equals $\gamma_H/4$, the proton signal can be refocused with a gradient pulse with half of the strength of the first two gradient pulses. The symmetric gradient pulses about the proton 180° pulse refocus any proton magnetization not bound to ^{13}C atoms (gradient spin echo) and leaves them with a net phase of zero. Therefore, they are dephased by the final gradient pulse and are not observed. The most important advantage of pulse-field gradients in nD NMR experiments is that the vast majority of the (unwanted) signal is dephased and only the signals of interest are observed. Also the noise in the nD NMR experiments is greatly reduced relative to the experiments that use phase cycling since the spectra does not result from the difference between two large signals.

The other important advantage of gradients is that they can be used in place of phase cycling to remove artifacts. In multiple-quantum experiments, for example, the phase cycling is used to select the coherence order by acquiring signals with different phases and adding or subtracting the spectra. The coherence order can be selected by adjusting the magnitude of the gradients, so multiple scans are not required. In many experiments this makes it possible to acquire a single scan for each point in the indirectly detected dimension, greatly reducing the acquisition time.

Pulsed-field gradient experiments have become very common for solution NMR studies of polymers. The gradients can be applied as square pulses or as half-sine pulse shapes. The advantage of the half-sine shape is that it minimizes the eddy currents that occur when the field gradients are rapidly changed. The ratio of the intensities of the gradient pulses are important for refocusing the coherence, and this can be achieved by either controlling the strength of the gradient pulse or by controlling the period of time that the gradient is on. Thus, the first two gradient pulses in Figure 2.31 could be twice as long or twice as strong as the final pulse. The additional equipment required for pulse-field gradient NMR studies includes a pulse-field gradient probe, a gradient amplifier, and a pulse programmer to control the strength and shape of the gradient pulses.

2.7.1.6 Decoupling
The decoupling sequences used in nD NMR are similar to those introduced in Section 2.4.3. The two main uses for composite pulse decoupling are for magnetization exchange in TOCSY-type pulse sequences and decoupling during acquisition. The TOCSY-type pulse sequences typically use the MLEV-17 (6) or DIPSI (7) decoupling for magnetization exchange.

The inverse detection experiments are particularly challenging in terms of decoupling because of the wide range of chemical shifts to be decoupled. In a proton-detected carbon experiment, the range of carbon chemical shifts with directly bonded protons is 160 ppm (20 kHz in an 11.7-T field). This can be accomplished with WALTZ decoupling, but it requires high pulse powers. The bandwidth ($\Delta f/\gamma B_2$, where Δf is the decoupled frequency range and B_2 is the decoupler field strength) covered by WALTZ decoupling is about two, so a 10-kHz proton field strength would be required. Such

2.7 MULTIDIMENSIONAL NMR

high powers lead to sample heating and noise in the nD spectra. A larger bandwidth can be decoupled using lower power with the computer-designed GARP decoupling pulse sequence that has a bandwidth of five (5).

2.7.2 Data Processing

The acquisition and processing of nD NMR data requires considerable time, energy, and expertise, so it is advisable to carefully consider experimental considerations at the start of an experiment. This includes not only the factors discussed in Section 2.4.2, but also how the acquisition parameters affect the data processing. Among the factors to consider are how the exponential multiplication or other window function can affect the data, phasing in the indirectly detected dimension, and how the experimental setup can affect the baselines and noise in the nD spectra. In certain favorable cases it is possible to design experiments to shorten the acquisition time and fill in the missing data using linear prediction methods.

2.7.2.1 Apodization The effect of window functions on the data prior to Fourier transformation was discussed in Section 2.4.4.3, where it was noted that window functions are used to increase the SNR at the expense of the resolution, or to increase the resolution at the expense of the SNR. Both of these approaches are used in nD NMR, as well as others that deal with the short acquisition times in the indirectly detected dimensions that can cause artifacts in the spectra.

The highest resolution in nD NMR is obtained for pure absorption-phase spectra that can be phased in all dimensions just like 1D NMR data. However, in some cases pure absorption-phase spectra cannot be acquired, and the lineshapes are a mixture of absorption and dispersion lineshapes. The dispersive component gives rise to broad peak bases that make it difficult observe high-resolution spectra. These bases can be removed using resolution-enhancing window functions as long as the SNR is sufficiently high.

The highest SNR is obtained when the window function matches the natural shape of the FID. In most experiments these are exponential signals, and exponential multiplication is the most commonly used window function (Equation (2.4)). However, in many nD NMR experiments, such as COSY, the signal intensity has a functional dependence on $\sin(\pi \omega t)$. Thus, there is no intensity at the start of the FID, and the FID reaches a maximum at some later time before it decays from T_2 relaxation. A window function often used for processing such data is the sine–bell function given by

$$w(t) = \sin\left(\frac{\pi t}{t_{\text{aq}}}\right) \tag{2.28}$$

This function has no adjustable parameters and leads to very strong resolution enhancement (with negative lobes on all the peaks) when applied to exponentially decaying signals. The peaks appear much sharper in the nD spectra at the cost of the

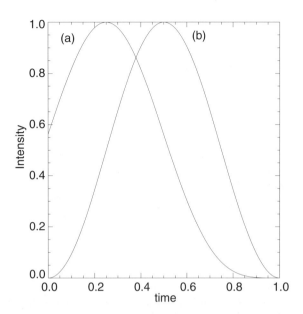

FIGURE 2.33 A plot of the (a) phase-shifted sine–bell and (b) sine-bell window functions used in nD NMR.

SNR. A slightly modified function with adjustable parameters is the sine-squared bell function that is given by

$$w(t) = \sin^2\left(\frac{\pi t}{t_{aq}} + p\right) \tag{2.29}$$

where p is the phase shift. This function is extensively used in nD Nmr. Figure 2.33 shows the window functions for the sine–bell and the phase-shifted sine-squared–bell window functions.

Window functions are also used to remove artifacts that result from the Fourier transformation of FIDs that have not decayed to zero at the end of the acquisition time. This is often observed in nD NMR spectra where the acquisition time is limited and it is not possible to gather enough spectra in the indirectly detected dimension to allow the FIDs to decay to zero. These cause distortions in the baselines near the peaks. One way to force the FIDs to decay to zero at the end of the acquisition time is to use the Hanning or Hamming windows that are given by

$$w(t) = 0.5 + 0.5\cos\left(\frac{\pi t}{t_{aq}}\right) \tag{2.30}$$

and

$$w(t) = 0.54 + 0.46\cos\left(\frac{\pi t}{t_{aq}}\right) \tag{2.31}$$

2.7 MULTIDIMENSIONAL NMR

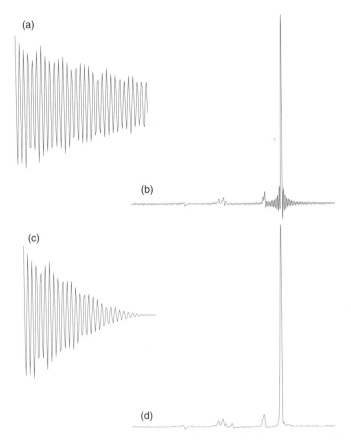

FIGURE 2.34 A plot of a FID (a) and spectrum (b) acquired with a short acquisition and the same FID (c) and spectrum (d) after application of the Hanning window function.

where t_{aq} is the acquisition time. Figure 2.34 shows a partially decayed FID before and after the application of a Hanning apodization function along with the Fourier transformed spectra. The nD NMR spectra acquired with short acquisition times in the indirectly detected dimensions can be greatly improved with the combination of the Hanning window and zero filling.

2.7.2.2 Phasing The phasing in pure absorption-phase nD NMR experiments is similar to phasing in 1D NMR experiments. The data sets contain the real and imaginary parts that differ in phase by 90°. Phasing is the process by which the real and imaginary parts are mixed such that one is purely absorptive (the real) and one is purely dispersive (the imaginary). In nD NMR the data are Fourier transformed and phased in the directly detected dimension before Fourier transformation and phasing in the indirectly detected dimension.

Phasing is not possible in some experiments where the lineshapes are a mixture of absorptive and dispersive signals, as in COSY without TPPI or State-type quadrature

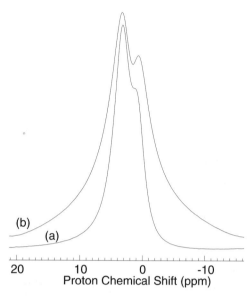

FIGURE 2.35 Comparison of the (a) absorptive and (b) magnitude lineshapes.

detection in the t_1 dimension. In such cases the magnitude spectra are plotted. The magnitude spectra $S(\omega)$ is given by

$$S(\omega) = \sqrt{S_r^2 + S_i^2} \qquad (2.32)$$

where S_r and S_i are the real and imaginary parts of the data. The problem with magnitude (or absolute value) spectra is that the lines are much broader than for the absorptive lines. This is illustrated in Figure 2.35. Note that the base of the peak is much greater in the magnitude spectrum.

In some cases it is possible to choose the parameters in nD NMR experiments such that phasing is not required in the indirectly detected dimensions, or that the values for the phase correction are known. The value for the frequency-independent (P_0) and frequency-dependent (P_1) phase corrections in the indirectly detected dimensions for an nD NMR spectrum are given by

$$P_1 = -2P_0 = \frac{x\left[\sum \delta + \sum \tau_{180} + t_1(0)\right]}{\Delta t} \qquad (2.33)$$

where $\delta = 2\tau_{90}/\pi$, τ_{90} is the 90° pulse length, τ_{180} is the 180° pulse length, $t_1(0)$ is the first delay time in the t_1 dimension, Δt is the dwell time in the t_1 dimension, and x is 360° for State-type quadrature detection or 180° for TPPI quadrature detection. The terms $\sum \delta$ and $\sum \tau_{180}$ are included to account for the time delays associated with

2.7 MULTIDIMENSIONAL NMR

90° and 180° pulses during the t_1 period. For a NOESY pulse sequence the phase correction is equal to $-x[4\tau_{90}/\pi + t_1(0)]/\Delta t$.

2.7.2.3 Baselines and t_1 Noise The processing and analysis of nD NMR spectra can be greatly simplified if the spectra are collected in a way that minimizes artifacts, such as noise in the indirectly detected dimension and baseline offsets. These artifacts can be suppressed with a careful tune-up of the NMR spectrometer, a corrected FT, and the proper choice of delay times in the nD NMR experiments.

There are a number of experimental factors that contribute to noise or ridges in the indirectly detected dimension in nD NMR experiments. Since the spectra are recorded over a long period of time (hours to days), spectrometer stability is an important factor. Variations in the rf power and phase of the pulses over the course of an experiment can lead to such noise, as can variations in the lock frequency, the temperature, or the spinning frequency (37). The noise introduced from variations in the rf pulses is often minimized on modern NMR spectrometers, but the spinning and temperature stability must be carefully monitored.

Ridges in the indirectly detected dimensions can arise from improper implementation of the FT (38). These ridges can be large in NOESY spectra where the peaks of interest are small cross peaks that must be detected in the presence of very large diagonal peaks. Baseline offsets are introduced into the Fourier-transformed data because the data acquisition generally does not begin at $t = 0$ because of the dead times after the application of a pulse. This can be corrected by scaling the first data point by 0.5 prior to the FT. Alternatively, the data can be sampled at $0.5 \Delta t$, where Δt is the dwell time in the indirectly detected dimension (39). Inserting this short delay time leads to flat baselines without correcting the first point. The initial delay time for a 2D NOESY $90°-t_1-90°-\tau_m-90°-t_2$ pulse sequence is

$$\tau = \frac{4\tau_{90}}{\pi} + t_1(0) \tag{2.34}$$

since there are two 90° pulses surrounding the t_1 period. The value of $t_1(0)$ is chosen such that $\tau = \Delta t$. For HMQC the first delay time is

$$\tau = \frac{4\tau_{90}}{\pi} + \tau_{180} + t_1(0) \tag{2.35}$$

Insertion of a short delay also affects the way that peaks are folded or aliased in the indirectly detected dimensions. Folding or aliasing occurs when the frequency of a peak is outside the sweep width (3). The frequency of the aliased peaks depends on how far the peaks are from the edge of the spectrum. It is occasionally useful to use folding in nD NMR spectra to increase the digital resolution in the indirectly detected dimensions. If the initial delay time in the indirectly detected dimension is chosen as $0.5 \Delta t$, then the aliased peaks have the opposite phase and can be easily identified (39).

2.7.2.4 Linear Prediction and Zero-Filling The digital resolution in the indirectly detected dimensions of nD NMR experiments is often limited by the data acquisition times. The spectra can be improved by artificially increasing the digital resolution in the indirectly detected dimension through the combination of linear prediction and zero-filling. Linear prediction is used to predict the future points in a FID based on the previous points (32). The future points are predicted based on a time series of the previous points as

$$X_n = C_1 X_{n-1} + C_2 X_{n-2} + C_3 X_{n-3} + \cdots + C_1 X_{n-k} \qquad (2.36)$$

where X_n are the FID points and C_n are the linear prediction coefficients to be determined. Using this approach it is often possible to extend the time series in the indirectly detected domain by 50%. Additional digital resolution can be gained by zero-filling. Linear prediction can also be used to correct the initial points in the FID that may be distorted (40). Distortions in the initial points can lead to baseline offsets in the nD NMR spectra. Linear prediction is commonly used in nD NMR and has become a standard feature on modern NMR spectrometers.

REFERENCES

1. Bax, A. *J. Mag. Res.*, 1983, *52*, 76.
2. Heatley, F. *Prog. NMR Spectroscopy*, 1979, *13*, 47.
3. Croasmun, W. R.; Carlson, M. K., eds. *Two-Dimensional NMR. Applications for Chemists and Biochemists*, VCH Publishers, New York, 1994.
4. Shaka, A.; Keeler, J.; Freeman, R. *J. Magn. Reson.*, 1983, *53*, 313.
5. Shaka, A. J.; Barker, P. B.; Freeman, R. *J. Magn. Reson.*, 1983, *64*, 547.
6. Bax, A.; Davis, D. *J. Magn. Reson.*, 1985, *65*, 355.
7. Shaka, A.; Lee, C.; Pines, A. *J. Magn. Reson.*, 1988, *77*, 274.
8. Cavanagh, J.; Rance, M. *J. Magn. Reson.*, 1992, *96*, 670.
9. Bendall, M. R.; Doddrell, D. M.; Pegg, D. T. *J. Am. Chem. Soc.*, 1981, *103*, 4603.
10. Komoroski, R. A. *High Resolution NMR Spectroscopy of Synthetic Polymers in Bulk*, Vol. 7, VCH Publishers, Dearfield Beach, 1986.
11. Brown, S. P.; Schnell, I.; Spiess, H. W. *J. Am. Chem. Soc.*, 1999, *121*, 6712.
12. Peerson, O. B.; Wu, X.; Kustanovich, I.; Smith, S. O. *J. Magn. Reson.*, 1993, *104*, 334.
13. Hediger, S.; Meier, B.; Kurur, N. D.; Bodenhausen, G.; Ernst, R. R. *Chem. Phys. Lett.*, 1994, *223*, 283.
14. Metz, G.; Wu, X.; Smith, S. O. *J. Magn. Reson. Ser. A*, 1994, *110*, 219.
15. Bennett, A. E.; Rienstra, C. M.; Auger, M.; Lakshimi, K. V.; Griffin, R. G. *J. Chem. Phys.*, 1995, *103*, 6951.
16. Spiess, H. *Coll. Polym. Sci.*, 1983, *261*, 193.
17. Pake, G. E. *J. Chem. Phys.*, 1948, *16*, 327.
18. Taylor, R.; Pembleton, R.; Ryan, L.; Gerstein, B. *J. Chem. Phys.*, 1979, *71*, 4541.

REFERENCES

19. Waugh, J. S.; Huber, L. M.; Haeberlen, V. *Phys. Rev. Lett.,* 1968, *20*, 180.
20. Mansfield, P. *Phys. Lett.,* 1970, *32A*, 485.
21. Mansfield, P. *J. Phys. C: Solid State Phys.,* 1971, *4*, 1444.
22. Mansfield, P.; Orchard, M. J.; Stalker, D. C.; Richards, K. H. B. *Phys Rev.,* 1973, *B7*, 90.
23. Rhim, W. K.; Elleman, D. D.; Schreiver, L. B.; Vaughn, R. W. *J. Chem. Phys.,* 1973, *60*, 4595.
24. Burum, D. P.; Linder, M.; Ernst, R. R. *J. Magn. Reson.,* 1981, *44*, 173.
25. Lee, M.; Goldberg, W. I. *Phys Rev. A,* 1965, *140*, 1261.
26. Bielecki, A.; Kolbert, A. C.; De Groot, H. J. M.; Griffin, R. G.; Levitt, M. H. *Adv. Magn. Reson.,* 1990, *14*, 111.
27. Bielecki, A.; Burum, D.; Rice, D.; Karasz, F. *Macromolecules,* 1991, *24*, 4820.
28. Mirau, P. A.; Heffner, S. A. *Macromolecules,* 1999, *32*, 4912.
29. Morris, G.; Freeman, R. *J. Magn. Reson.,* 1978, *29*, 433.
30. Schaefer, J.; Stejskal, E. O.; Buchdahl, R. *Macromolecules,* 1977, *10*, 384.
31. VanderHart, D. L.; McFadden, G. B. *Solid State NMR* 1996, *7*, 45.
32. Olejniczak, E. T.; Eaton, H. L. *J. Magn. Reson.,* 1990, *87*, 628.
33. Tolman, J. R.; Prestegard, J. H. *Concepts Magn. Reson.,* 1994, *7*, 247.
34. Marion, D.; Wuthrich, K. *Biochem. Biophys. Res. Commun.,* 1983, *113*, 967.
35. States, D.; Haberkorn, R.; Ruben, D. *J. Magn. Reson.,* 1982, *48*, 286.
36. Marion, D.; Bax, A. *J. Magn. Reson.,* 1989, *85*, 393.
37. Mehlkopf, A. F.; Korbee, D.; Tiggelman, T. A. *J. Magn. Reson.,* 1984, *58*, 315.
38. Otting, G.; Widmer, H.; Wagner, G.; Wuthrich, K. *J. Magn. Reson.,* 1986, *66*, 187.
39. Bax, A.; Ikura, M.; Zhu, G. *J. Magn. Reson.,* 1991, *91*, 174.
40. Marion, D.; Bax, A. *J. Magn. Reson.,* 1989, *83*, 205.

3

THE SOLUTION CHARACTERIZATION OF POLYMERS

3.1 INTRODUCTION

Solution NMR is an important method for polymer characterization. Even in the earliest studies using low magnetic field strengths it was observed that high-resolution spectra could be acquired for many polymers in solution. This observation was not expected since it was well known that broad lines are expected for high molecular weight materials. Subsequent studies have shown that atomic fluctuations on the length scale of a monomer or smaller almost completely average the dipolar interactions that lead to line broadening. This high resolution makes it possible to characterize the polymer microstructure, including the stereochemistry, regioisomerism, geometric isomerism, endgroups, branches, and other defects. It is also possible to use NMR to measure the chain conformation and the local structure using the through-bond coupling constants and the through-space dipolar interactions.

The earliest studies used proton NMR for the solution characterization because of its high sensitivity. As spectrometers improved and higher field magnets were introduced, the sensitivity improved to such a degree that NMR could be used for polymer characterization with less sensitive nuclei. Carbon is a common element in most polymers, and the carbon chemical shifts are extremely sensitive to small changes in the chemical structure and microstructure. In addition, carbon NMR can be used to characterize carbonyl groups and quaternary carbons, which could not be directly observed in the proton spectrum. More recently, silicon, phosphorus, and fluorine NMR have been used to characterize polymers containing these elements.

The progress in using NMR as a tool for microstructural characterization is due in large part to advances in NMR technology and in the methodology used to assign the peaks in the high-resolution spectra. The introduction of higher field magnets makes

---

A Practical Guide to Understanding the NMR of Polymers, by Peter A. Mirau
ISBN 0-471-37123-8   Copyright © 2005 John Wiley & Sons, Inc.

it possible to resolve increasingly finer detail in the NMR spectrum. This has led to the development of new methods for spectral assignments. Assignments were made in the earliest studies by comparing the spectra with those from model compounds. This was an effective, but very labor-intensive approach. These assignments were further aided by the development of chemical shift calculations. After the NMR spectra had been reported for a large number of compounds, it was possible to develop empirical relationships between the chemical structure and the chemical shifts that could be used to establish the assignments in polymers. In addition, it was observed that the chemical shifts depend on the conformation through the γ-*gauche* effect, and the assignments could be established in conjunction with chain conformation (1,2). This development was very important for establishing the stereosequence assignments in polymers. Additional information about the assignments was also obtained using spectral editing methods, such as DEPT, to assign the carbon types (methyl, methylene, methine, quaternary) in the carbon spectra. Finally, the assignments have been aided by the development of nD NMR, which makes it possible to correlate long sequences with their neighbors.

3.1.1 Polymer Microstructure

The microstructure of polymers has important consequences for structure–property relationships. Depending on the synthetic approach, polymer chains can range from very homogeneous to very heterogeneous. The NMR spectra of polymers are very sensitive to polymer microstructure because the chemical shifts and coupling constants can be very sensitive to structural or stereochemical changes several bonds removed. At the grossest level, changes in polymer microstructure, such as the introduction of a branch, can lead to chemical shift changes of more than 10 ppm that are readily apparent in the spectra. The introduction of some microstructures do not lead to large changes in chemical shifts, but rather to changes in distances or coupling constants that can be identified with nD NMR experiments. This section contains a brief introduction to polymer microstructure, and interested readers are referred to other introductory texts for a more complete description (3).

3.1.1.1 Regioisomerism   Regioisomerism arises from the synthesis of asymmetric monomers that may be introduced into the growing polymer chain in either a head-to-head (H–H) or a head-to-tail (H–T) fashion, as illustrated in Figure 3.1. Note that in the H–T polymers each substituted carbon has two methylene carbon neighbors, while in the polymer with regiodefects, several different types of environments are observed. At the H–H junction there are two neighboring substituted carbons that are bordered by methylene carbons, and for the tail-to-tail (T–T) junctions there are neighboring methylene carbons surrounded by substituted carbons. As noted in Section 1.3.3, the so-called β-effect resulting from changing the substitution at the nearest-neighbor carbon can give rise to induced chemical shifts in the carbon spectra as large as 10 ppm. Thus, the chemical shifts from the inverted monomer would be very different from the chemical shifts in the H–T parts of the chain. Furthermore, the chemical shifts of the H–T monomers adjacent to the inverted monomer

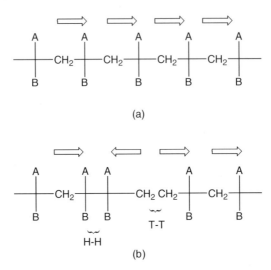

FIGURE 3.1 A diagram showing polymer regioisomerism in polymers.

would experience smaller chemical shift changes (1–5 ppm) from the so-called γ- and δ-effects. The induced chemical shifts are smaller in the proton spectrum, but very large in the fluorine spectrum for fluorine-containing polymers.

The other consequence of inserting inverted units into the chain is that the couplings between neighboring groups can be changed. Consider, for example, a polymer in which either the A or B substituent is a proton. In the H–T polymer we would see only three-bond coupling between methine and methylene protons. The H–H and T–T monomers would have a definitive signature in the COSY spectrum, as only inverted units would show methine–methine and methylene–methylene couplings.

Regioisomers are most commonly introduced into polymer chains at a low level, so the peaks from the inverted monomers are small compared to the H–T resonances. Even though they have a low intensity, they can often be studied using 1D NMR methods. If the levels are 1% or higher, they can be efficiently studied using nD NMR methods.

3.1.1.2 Stereochemical Isomerism The stereochemistry of polymers refers not to the absolute configuration, but the relative handedness of neighboring monomer units. In vinyl polymers the substituted carbon is termed the α-carbon and is said to be *pseudoasymmetric*, since if the chain ends are disregarded these carbons do not have the required four unique substituents to be classified as an asymmetric center. This relative handedness influences both the polymer properties and the NMR spectrum. The simplest is the so-called *isotactic* sequence shown in Figure 3.2a. in which, when viewed in the all-*trans* conformation, all of the substituents lie on one side of the chain. It is often more convenient to draw the stereochemical configuration using the stick-type drawings also shown in Figure 3.2. The terms *meso* (or *m*) or *racemic* (or *r*) are used to denote the relative orientation of pseudoasymmetric centers along the polymer chain. In the *m* configuration, both substituents are along the same

3.1 INTRODUCTION

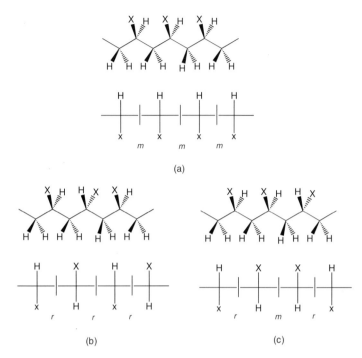

FIGURE 3.2 A diagram showing (a) isotactic, (b) syndiotactic and (c) atactic stereochemistries in vinyl polymers.

side of the chain, while the *r* configuration denotes substituents on opposite sides of the chain. The purely isotactic chain contains only a *meso* configuration of the pseudoasymmetric centers. The polymer chains in which the configuration of the pseudoasymmetric center alternates are known as a *syndiotactic* polymers and contain only racemic units. Polymer chains containing a random mixture of *meso* and *racemic* pseudoasymmetric centers are termed *atactic*.

There is a nomenclature associated with the specification of polymer stereochemistry that is based on the configuration of neighboring pseudoasymmetric centers. A *diad* (such as *m* or *r*) refers to the orientation of neighboring asymmetric centers, while a *triad* (such as *mr*, *rr*, or *mm*) refers to the relative orientation of three pseudoasymmetric centers. The NMR spectrum of vinyl polymers is often sensitive to very high-order stereochemistry, and it is not uncommon to observe hexad-level splittings.

The situation is slightly more complex for polymers of 1,2-disubstituted monomers, since the polymer chain has two interleaved systems of nonidentical pseudoasymmetric centers. The structures for *disyndiotactic, erythrodiisotactic*, and *threodiisotactic* polymers are shown in Figure 3.3.

3.1.1.3 Geometric Isomerism The NMR spectrum of diene polymers is very sensitive to geometric isomerism. The two most basic forms of isomerism, *cis* and *trans*, are shown in Figure 3.4. These designations refer to the relative orientation of the

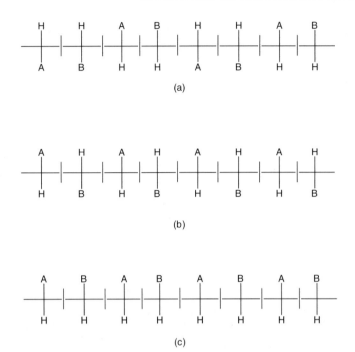

FIGURE 3.3 A diagram showing (a) disyndiotactic, (b) erythrodiisotactic and (c) threodiisotactic stereochemistry in 1,2-disubstituted polymers.

polymer main chain. The isomerism in diene chains can be very complex because of the possibilities of regio- and stereochemical isomerism in addition to geometric isomerism.

3.1.1.4 Branching and Endgroups Branching and endgroups introduce large changes into the NMR spectra of polymers because these structural features change the local magnetic environment in polymer chains, and the peaks from branches and endgroups are often well resolved from the main-chain signals. The nomenclature for branches and endgroups is shown in Figure 3.5. This nomenclature describes branches and chain ends based on their distance from the site. The carbons are labeled with Greek letters (α, β, γ, and δ) along the main chain that denote the distance from the branch site. The chemical shift perturbations from introducing branches do not usually extend for more than four or five carbons atoms. For chains with a high degree of branching it is possible for a carbon atom to be near two branches. The primed

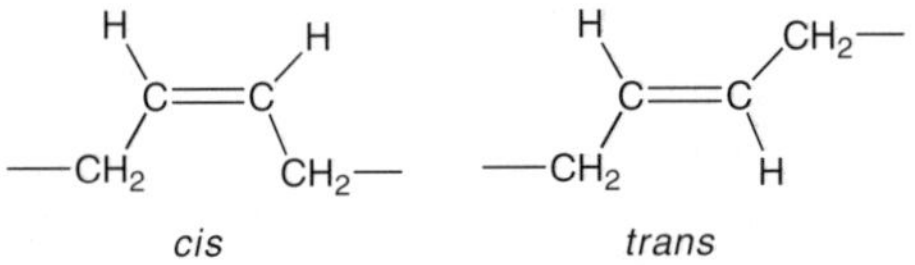

FIGURE 3.4 A diagram showing geometric isomerism in polymers.

3.1 INTRODUCTION

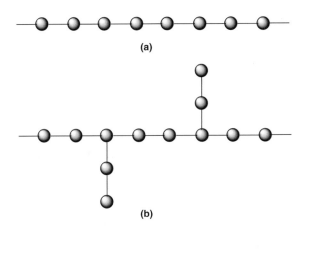

FIGURE 3.5 A diagram illustrating the nomenclature for branches and defects in polymers.

double Greek letters and *Bn* nomenclature is used to describe the branch carbons, while the *s* nomenclature is used to describe the chain ends.

3.1.1.5 Chain Architecture The NMR spectra of polymers are sensitive to chain architecture. Figure 3.6 shows some of the commonly observed chain architectures

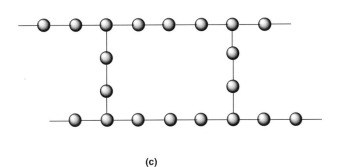

FIGURE 3.6 A diagram showing (a) linear, (b) grafted (or branched) and (c) crosslinked chain architectures in polymers.

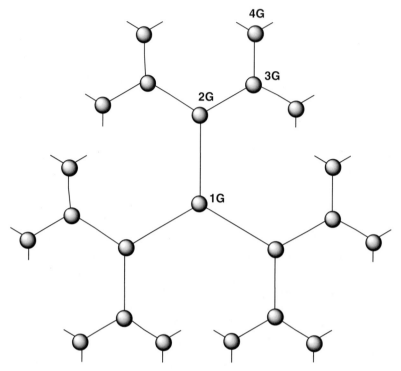

FIGURE 3.7 A diagram showing the nomenclature for dendritic polymer chains.

in polymers, including linear, branched, and cross-linked chains. Short branches can be introduced into polymers through alternate polymerization pathways, and longer branches through graft polymerization. At low densities the cross links may anchor two chains together, but at higher cross-link densities networks involving many chains can be formed. As might be expected, there are many similarities in the NMR spectra of grafted and cross-linked chains and the endgroups and branches discussed in the preceding section.

More recently dendritic polymers have been the focus of much synthetic attention. Dendritic polymers are hyperbranced polymers in which branching arms are connected to a central core, as shown in Figure 3.7. The central core (1G, the first generation) connects to three monomers (2G, the second generation). Each of these 2G monomers connects to two more monomers to give a third generation, and so on. The limitation in the number of possible generations is usually determined by monomer crowding at the exterior of the dendrimers where the density becomes very high. It can be seen that the number of monomers in each generation doubles with each generation after the second. Because of the large number of similar monomers, dendrimers remain very challenging to study by NMR.

3.1.1.6 Copolymers Copolymerization is another form of polymer microstructure that is amenable to analysis by NMR. Copolymers are available in a variety of

3.1 INTRODUCTION

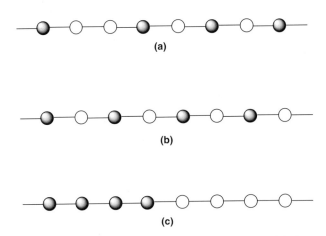

FIGURE 3.8 A diagram showing (a) random, (b) alternating and (c) block copolymer architectures.

architectures, including random, alternating, and block copolymers, as shown by Figure 3.8. NMR is very sensitive to the main-chain structure of polymers, and is thus very sensitive to the copolymer sequence. The chemical shifts can be sensitive to the substituents four and five bonds removed from the carbon of interest. The analysis of block copolymers by NMR can be very challenging because the junctions between blocks only occur one or two times per chain, depending on the block copolymer architecture (diblock, triblock, etc.). In addition to these clearly defined architectures, intermediate types of architectures, such as mixtures of random and alternating microstructures or "blocky" microstructures, can also be observed and characterized.

3.1.2 Spectral Assignments in Polymers

Establishing the chemical shift assignments for polymers remains one of the fundamental challenges of solution NMR in polymer science. With modern high-field spectrometers, very low levels of microstructural features can often be resolved from the main polymer peaks, and some means must be used to assign these peaks. This is a particularly difficult problem in polymers because of the repeating-sequence nature of polymer chains. This leads to a situation where the minor peaks are in close proximity to the much larger main-chain signals.

Over the years, a number of methods have evolved to assign polymer NMR spectra. Most methods make use of both the proton and carbon spectra, as well as any other NMR-active nuclei in the chain. The approach used to establish the assignments depends on the polymer of interest, its NMR properties, and the availability of model compounds and polymers. Model compounds and polymers have historically been used to establish the stereochemical assignments in polymers (4). The assignments of stereochemical sequences have also relied heavily on the chain statistics and chemical shift calculations (2). The assignments established with the model compounds and chemical shift calculations are empirical and cannot provide rigorous proof for the assignments. More recently nD NMR has emerged as a method able to offer definitive proof for the assignments of polymers in solution.

3.1.2.1 Model Compounds and Polymers One of the earliest methods used to establish the peak assignments in polymers was to compare the solution NMR spectra of a polymer with model compounds and other polymers. This is a very labor-intensive process since compounds must be synthesized, but it can be very effective. Model compounds such as 2,4,6-trimethyl heptane have been used, for example, to study stereochemical isomerism in polypropylene (5). The diastereomers of 2,4,6-trimethyl heptane can be chromatographically separated and the stereochemical assignments in polypropylene can be made by comparison to the model compounds. In a similar way, compounds such as 2,3,6-trimethyl hepatane could be used as model compounds for the H–H and T–T defects in polypropylene.

Model polymers are also used to make assignments in polymers. In studies of polymer stereochemistry, for example, the resonance assignments can be established by comparing the spectra to those from the isotactic or syndiotactic polymers if they are available. This is illustrated in Figure 3.9, which compares the spectra of atactic polypropylene with the isotactic and syndiotactic polymer. The isotactic (*mmmm*) and syndiotactic (*rrrr*) peaks can be directly assigned by this comparison.

The comparison with the isotactic and syndiotactic polymers leads directly to the assignments of the *mmmm* and *rrrr* stereosequences, but not many of the others that appear in Figure 3.9b. Additional assignments can be made from the epimerization of stereochemically "pure" materials (6,7). Epimerization is the inversion of the stereochemistry at methine carbons that has been observed in the presence of certain catalysts. Epimerization appears to be a random process, so the appearance of additional

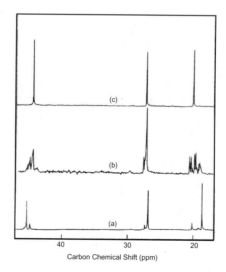

FIGURE 3.9 The carbon NMR spectra of (a) isotactic, (b) atactic and syndiotactic polypropylene.

3.1 INTRODUCTION

stereosequences from the isotactic or syndiotactic polymers can be predicted from basic statistics.

This method has been used with great success with polypropylene, where epimerization of the isotactic and syndiotactic materials give rise to new resonances in the methyl region (6). Syndiotactic polypropylene gives rise to three resonances with an intensity ratio of 2:2:1 that can be assigned to *rrmm*, *rrrm*, and *rmmr*, while in isotactic polypropylene these resonances are assigned to *mmmr*, *mmrr*, and *mrrm*. By consideration of the necessary pentad–pentad relationships

$$2rmmr + rrrm = mmrm + mmrr \tag{3.1}$$

and

$$2mrrm + mrrr = mmrr + rmrr \tag{3.2}$$

the three remaining stereosequences *mmrm*, *rmrr*, and *rmrr* can be assigned.

Model polymers can also be used to characterize other types of polymeric microstructures. This approach has also been used for poly(vinylidine fluoride), where the resonances from the H–T and T–T defects can be easily identified by comparison with the isoregic material (8).

3.1.2.2 Polymer Chain Statistics In favorable cases, the polymer chain statistics can be used to establish the chemical shift assignments for polymers. If the statistics of chain propagation are known or can be determined, then it is possible to make assignments based on the relative intensities of peaks. This is illustrated in Figure 3.10, which shows triad probabilities for polymer chain growth by Bernoullian statistics as a function of the probability of adding a *meso* configuration to the end of the growing chain. In Bernoullian trial propagation the chain end is represented as not having any stereochemical influence over the addition of the next monomer. When a monomer is added to the growing chain the relative fraction of *m* and *r* diads is determined by the addition probabilities P_m and P_r. The relative intensities of the *mm*, *mr*, and *rr* triads can be calculated from simple statistics and are given by

$$(m) = P_m \tag{3.3}$$

$$(r) = (1 - P_m) \tag{3.4}$$

$$(mm) = P_m^2 \tag{3.5}$$

$$(rm) = 2P_m(1 - P_m) \tag{3.6}$$

and

$$(rr) = (1 - P_m)^2 \tag{3.7}$$

From these equations and Figure 3.10 we can see that if the probability of adding a monomer in the *meso* configuration is 0.25 ($P_m = 0.25$), then the relative intensities of the *m* and *r* diads will be .25 and .75, and the probabilities for the *mm*, *mr/rm*,

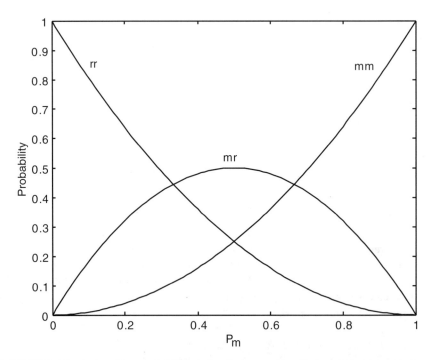

FIGURE 3.10 A plot of the probabilities of various stereochemical configurations as a function of the probability of adding a *meso* diad to the end of the growing chain, assuming Bernoullian statistics.

and *rr* triads are .0625, .375, and .5625. Therefore, if P_m is known and Bernoullian statistics are followed, the triad peaks could be assigned directly from the relative intensities.

In many cases, higher order *n*-ads can be resolved in the NMR spectrum, and these peaks can also be assigned from the chain statistics. For Bernoullian chain propagation the probabilities for the tetrad sequences are given by

$$(mmm) = P_m^3 \tag{3.8}$$

$$(mmr/rmm) = 2P_m^2(1 - P_m) \tag{3.9}$$

$$(rmr) = P_m(1 - P_m)^2 \tag{3.10}$$

$$(mrm) = P_m^2(1 - P_m) \tag{3.11}$$

$$(rrm/mrr) = 2P_m(1 - P_m)^2 \tag{3.12}$$

and

$$(rrr) = (1 - P_m)^3 \tag{3.13}$$

The chain statistics for higher order *n*-ads are easily calculated for the Bernoullian model.

3.1 INTRODUCTION

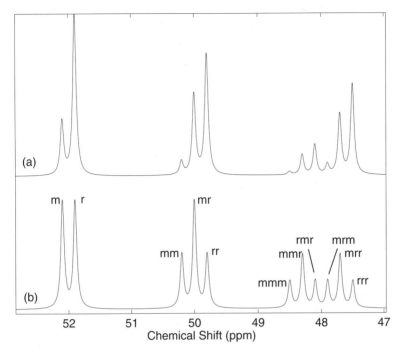

FIGURE 3.11 A simulated spectra showing the peak intensities expected for Bernoullian statistics calculated for (a) $P_m = 0.25$ and (b) $P_m = 0.5$.

To understand how to apply these statistics to polymers, consider the simulated spectra shown in Figure 3.11. This simulated spectrum shows peaks at 52, 50, and 48 ppm that are sensitive to polymer stereochemistry at the diad, triad, and tetrad levels. These two spectra are calculated by assuming Bernoullian chain propagation statistics with P_m equal to either to 0.5 or 0.25. For the spectrum with $P_m = 0.25$, many of the peaks can be assigned directly from the intensities and the chain statistics. However, for the spectrum with $P_m = 0.5$, many of the peaks have the same intensity (such as *mm* and *rr*) and other information is required to establish the peak assignments. The chain statistics are often used in combination with nD NMR to establish the stereochemical assignments for polymers.

In many cases, peak overlap prohibits the accurate measurement of all of the peak intensities, particularly for the higher-order stereosequences. Under such circumstances those peaks that are not resolved are excluded from the analysis. Another common occurrence is that combinations of peaks, rather than resolved peaks, can be measured. In some cases this can provide information about the assignments, making use of the stereosequence relationships listed in Table 3.1. For example, consider the case when the *mmm* and *mmr* tetrads are resolved, but the *mm* triad is not. The *mm* configurational sequence probability can be calculated from the triad–tetrad relationship shown in Table 3.1.

Of course, not all polymerizations follow Bernoullian statistics, so other models must be considered. If, for example, the polymerization follows first-order Markov

TABLE 3.1 The Stereochemical Relationships for Polymers.

Diad–Diad	$m + r = 1$
Diad–Triad	$m = mm + 1/2\, mr$
	$r = rr + 1/2\, mr$
Diad–Tetrad	$m = mmm + mrm + 1/2\, mmr + 1/2\, mrr$
	$r = rrr + rmr + 1/2\, mmr + mrr$
Triad–Triad	$mm + mr + rr = 1$
Triad–Tetrad	$mm = mmm + 1/2\, mmr$
	$mr = mmr + 2\, rmr = mrr + 2\, mrm$
	$rr = rrr + 1/2\, mrr$
Tetrad–Tetrad	$\sum = 1$
	$mmr + 2\, rmr = 2\, mrm + mrr$

statistics, then monomer addition depends on the stereochemistry of the chain end and the incoming monomer. This gives rise to four probabilities for the addition of new monomers, $P_{m/m}$, $P_{m/r}$, $P_{r/m}$, and $P_{r/r}$, where $P_{m/m}$ refers to the addition of an m centered monomer to a chain with an m chain end. In this case, the diad and triad probabilities are given by

$$(m) = P_{m/m} + P_{r/m} \qquad (3.14)$$

$$(r) = P_{m/r} + P_{r/r} \qquad (3.15)$$

$$(mm) = \frac{P_{m/m} P_{r/m}}{P_{m/r} + P_{r/m}} \qquad (3.16)$$

$$(mr) = \frac{2 P_{m/r} P_{r/m}}{P_{m/r} + P_{r/m}} \qquad (3.17)$$

and

$$(rr) = \frac{P_{r/r} P_{m/r}}{P_{m/r} + P_{r/m}} \qquad (3.18)$$

The use of these configurational statistics is more complex, since there are now four parameters and the values cannot be simply determined by measuring the diad peak intensities, as is common for Bernoullian statistics. If these addition probabilities are known, then the relative intensities of the triads can be easily calculated. The probabilities for larger sequences have been compiled elsewhere, along with the probabilities for other propagation statistics, including the second-order Markov and Coleman–Fox models (9,10).

It is important to note that polymer chain statistics can provide important information about the assignments, but they cannot be considered as proof of an assignment. The best that can be said is that the assignments are consistent with a particular type of chain statistics. In some cases, this is the best that can be done.

3.1 INTRODUCTION

TABLE 3.2 The Parameters for the Empirical Calculation of Carbon Chemical Shifts.

Carbon Position	A_l (ppm)
α	9.1 ± 0.10
β	9.4 ± 0.10
γ	-2.5 ± 0.10
δ	0.3 ± 0.10
ε	0.1 ± 0.10

3.1.2.3 Chemical Shift Calculations Over the past several decades the chemical shifts for a wide variety of compounds have been reported, and empirical rules have been developed to relate the chemical structure to the chemical shifts, particularly for carbon NMR (11–15). In Section 1.3.3 we discussed some of these relationships, including the α, β, and γ effects on the chemical shifts of hydrocarbons. The empirical relationship between the structure and the carbon chemical shifts for hydrocarbons are given by (15)

$$\delta_C = B + \Sigma A_l n_l + \Sigma S_l \tag{3.19}$$

where δ_C is the calculated chemical shift, B is a constant given by the chemical shift of methane (-2.3 ppm), n_l is the number of carbons at position l away from the carbon of interest, A_l is the additive shift due to carbon l, and S_l is a term included to account for branching. The shift parameters A_l are given in Table 3.2 for the α to ε carbons, and Table 3.3 contains the correction terms for branching. The correction terms depend on the degree of substitution of a given carbon and its neighbor. The nomenclature is such that the correction term 3°(2°) would be used for a tertiary carbon next to a secondary carbon.

To appreciate the value of this empirical relationship it is instructive to consider the chemical shift calculation for one of the methylene carbons (marked with *) in the polypropylene structure shown in Figure 3.12a. This methylene has two α, four

TABLE 3.3 The Parameters for Calculating the Effect of Branching on the Carbon Chemical Shifts.

	S_{lk} (ppm)
1°(3°)	-1.10 ± 0.20
1°(4°)	-3.35 ± 0.35
2°(3°)	-2.50 ± 0.25
2°(4°)	-7.5
3°(2°)	-3.65 ± 0.15
3°(3°)	-9.45
4°(1°)	-1.50 ± 0.10
4°(2°)	-8.35

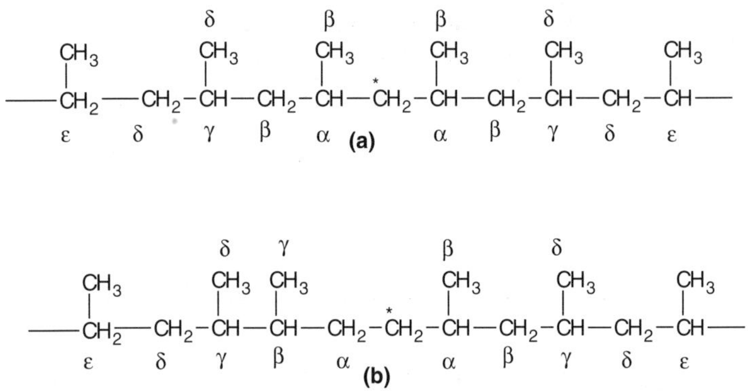

174 THE SOLUTION CHARACTERIZATION OF POLYMERS

$$\begin{array}{cccccccccc}
& & \delta & \beta & \beta & \delta & & \\
CH_3 & & CH_3 & CH_3 & CH_3 & CH_3 & & CH_3 \\
| & & | & | & | & | & & | \\
—CH_2—CH_2—CH—CH_2—CH—CH_2—CH—CH_2—CH—CH_2—CH— \\
\varepsilon & \delta & \gamma & \beta & \alpha & \alpha & \beta & \gamma & \delta & \varepsilon
\end{array}$$

(a)

$$\begin{array}{cccccccccc}
& & \delta & \gamma & & \beta & \delta & & \\
CH_3 & & CH_3 & CH_3 & & CH_3 & CH_3 & & CH_3 \\
| & & | & | & & | & | & & | \\
—CH_2—CH_2—CH—CH—CH_2—CH_2—CH—CH_2—CH—CH_2—CH— \\
\varepsilon & \delta & \gamma & \beta & \alpha & \alpha & \beta & \gamma & \delta & \varepsilon
\end{array}$$

(b)

FIGURE 3.12 The structure of polypropylene used for chemical shift calculations.

β, two γ, four δ, and two ε neighbors, and the methylene is a secondary carbon with two tertiary neighbors, so the chemical shift is given by

$$\delta_C = B + 2A_\alpha + 4A_\beta + 2A_\gamma + 4A_\delta + 2A_\varepsilon + 2(2°(3°))$$
$$= 44.9 \text{ ppm} \tag{3.20}$$

This chemical shift can be compared with the methylene carbon in a defect structure containing a T–T and a H–H structure like that shown in Figure 3.12b. The methylene carbon in the T–T structure has two α, three β, three γ, four δ, and two ε neighbors, and the methylene is a secondary carbon with a tertiary neighbor, so the chemical shift is given by

$$\delta_C = B + 2A_\alpha + 3A_\beta + 3A_\gamma + 4A_\delta + 2A_\varepsilon + (2°(3°))$$
$$= 35.3 \text{ ppm} \tag{3.21}$$

This comparison illustrates several important points. The coefficients for A_α and A_β are quite large, so changes in the α, β, or γ neighbors gives rise to large chemical shift changes. The changes due to differences in the δ or ε neighbors are more subtle, but the chemical shift changes are still large enough that they can be easily detected in the carbon spectrum.

It should also be noted that the uncertainties in the coefficients in Tables 3.2 and 3.3 are large enough that it is not possible to quantitatively predict the chemical shifts using these simple rules. Furthermore, other factors such as the chain conformation (through the γ-*gauche* effect) or the solvent can also affect the chemical shift. These empirical relationships can be used as a guide in establishing the chemical shift assignments, and they are particularly useful for understanding the differences in chemical shift as the structure is changed. The predicted difference in chemical shift for the methylene carbons in Figures 3.12a and 3.12b is given by

$$\Delta\delta_C = A_\beta - A_\gamma + (2°(3°))$$
$$= 9.4 \text{ ppm} \tag{3.22}$$

3.1 INTRODUCTION

TABLE 3.4 The Chemical Shift Parameters for Halogens in Hydrocarbons.

Carbon Position	Cl (ppm)	Br (ppm)	I (ppm)
α	31.2	20.0	−6.0
β	10.5	10.6	11.3
γ	−4.6	−301	−1.0
δ	0.1	0.1	0.2
ε	0.5	0.5	1.0

The relationship between the chemical structure and the carbon chemical shift has been studied for many types of molecules, including those with halogens, double-bonded carbons, cyclic structures, and aromatic groups. Table 3.4 lists the effect of halogen atoms on the carbon chemical shifts. Bromine and chlorine give rise to very large chemical shifts for the substituents at the α and β positions. Iodine has a smaller (and negative) α effect, but a large β effect. The reader is referred to other texts for a more complete listing of the factors affecting the carbon chemical shifts (15). It should also be noted that there are commercial programs available to calculate the carbon and proton chemical shifts from the structure.

3.1.2.4 The γ-Gauche Effect

Conformational effects on the chemical shifts of polymers are important for understanding polymer structure–property relationships and for establishing the chemical shift assignments in polymers. The most useful of the conformational effects is the so-called γ-gauche effect. It has been observed that the carbon chemical shifts in polymers and other molecules are sensitive to the substituent of its γ neighbor. The γ-gauche effect is thought to arise from a through-space interaction between a carbon and its γ neighbor along the chain. Depending on the conformation about the intervening bonds, the γ neighbor can be in either a *gauche* or *trans* conformation relative to the carbon of interest as shown in Figure 3.13 for butane and 2-methyl butane and 2,2-dimethyl butane. The distance between the carbon and its γ neighbor is shortest in the *gauche* conformation and this leads to an induced shift in the carbon spectrum. This effect depends on the substituent at the γ substituent, and the full γ-gauche effect can be as large as 5 ppm. Since *gauche–trans* isomerization in most polymers is fast on the time scale of the difference in chemical shift, the magnitude of the γ-gauche effect depends on the time average of the fraction of bonds in the *gauche* conformation. To use the γ-gauche effect to make polymer resonance assignments it is necessary to calculate the average conformation of the polymer using a rotational isomeric state model. All *gauche* conformations in Figure 3.13 contribute to the induced chemical shift, so the induced shift from each γ-gauche interaction is given by

$$\Delta\delta_C = \chi_g \cdot \Delta\delta_C^{gg} \quad (3.23)$$

where $\Delta\delta_C^{gg}$ is the shift induced by the γ-gauche effect, and χ_g is the fractional population of the *gauche* conformers.

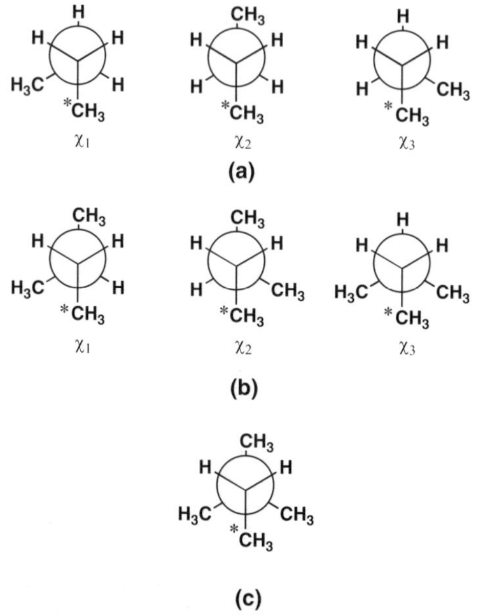

FIGURE 3.13 A drawing of the *gauche* and *trans* conformations of (a) butane, (b) 2-methyl butane and (c) 2,2-dimethyl butane.

The chemical shift induced by the γ-*gauche* effect results from the sum of all γ-*gauche* interactions. Two of the conformations for butane (χ_1 and χ_3) each have one *gauche* interaction, so the induced shift is given by $\Delta\delta_C = \chi_1 \cdot \Delta\delta_C^{gg} + \chi_3 \cdot \Delta\delta_C^{gg}$. For 2-methyl butane two conformations have one *gauche* interaction while the third has two, so the induce shift is given by

$$\Delta\delta_C = \chi_1 \cdot \Delta\delta_C^{gg} + \chi_2 \cdot \Delta\delta_C^{gg} + 2\chi_3 \cdot \Delta\delta_C^{gg} \qquad (3.24)$$

The three conformations of 2,2-dimethyl butane are equivalent and each has two γ-*gauche* interactions, so the induced chemical shift is given by

$$\Delta\delta_C = 2\Delta\delta_C^{gg} \qquad (3.25)$$

The magnitude of the induced shift also depends on the substituent in the γ position, and the relative shifts for methyl, hydroxyl, and chlorine are listed in Table 3.5. Each

TABLE 3.5 The Magnitude of the γ-*gauche* Effect for Several Substituents.

Group	$\Delta\delta_C$ (ppm)
$-CH_3$	-5.2
$-OH$	-7.2
$-Cl$	-6.8

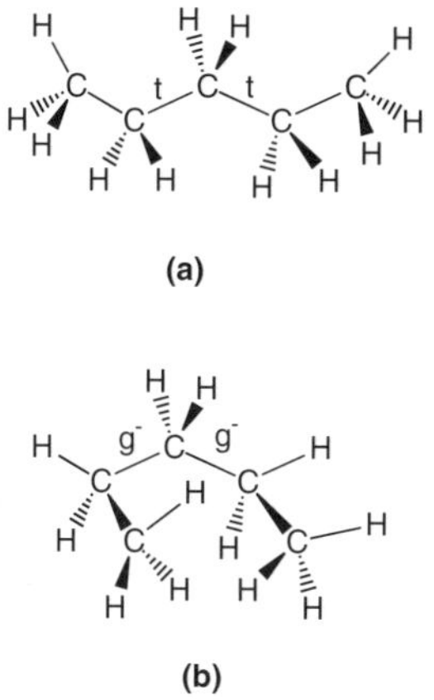

3.1 INTRODUCTION

FIGURE 3.14 Drawings of n-pentane in the (a) *trans-trans* and (b) *gauche'-gauche'* conformations.

γ-*gauche* interaction with a methyl group leads to an induced chemical shift of -5.3 ppm, while hydroxyl, and chlorine atoms lead to induced chemical shifts of -7.2 and -6.8 ppm in the carbon spectrum.

The γ-*gauche* effect is useful for making chemical shift assignments in polymers because the conformational probabilities depend on the local structure, including the stereochemistry. One consequence of the chain connectivity is that not all conformations are equally probable. This is illustrated in Figure 3.14, which shows some of the conformations for *n*-pentane. *n*-Pentane has four rotatable bonds that can affect the energy. The lowest energy conformation is the all-*trans* state. The all-*trans* conformation is much more probable than the g^+g^- (or g^-g^+) conformations that have unfavorable steric interactions between the methyl groups. The methyl groups are on opposite sides of the chain in the g^+g^+ conformation and strong repulsion is not a problem. It is clear from this illustration that the energy (and conformational probability) depends on the conformation of the nearest-neighbor bond along the chain. Longer range interactions can also affect the probabilities.

The chemical shifts assignments can often be established in polymers by combining the γ-*gauche* effect with some model to calculate the conformational probabilities. One very effective model is the so-called rotational isomeric state (RIS) model (2,16). The RIS model is based on the assumption that the average chain conformation for a polymer can be represented by a small number of conformations about each bond. The rotational isomeric-state models for polymers are constructed by summing up all of the pairwise interactions for the nearest-neighbor and next-nearest-neighbor interactions for all of the stereosequences. A critical part of this process is the parameterization of the interaction energies for the conformations. The starting point is typically an energy calculation for a small molecule to measure the relative stability of the various conformers. This model is refined by comparing the calculated results from the rotational isomeric states model with some experimental data, such as the average end-to-end distance or the dipole moment.

TABLE 3.6 The Calculated Bond Probabilities for Pentad Stereosequences in Polypropylene.

Pentad Stereosequence	P_{trans}
mrmr	0.44
rrmr	0.47
mmmm	0.52
rmmr	0.54
rmmm	0.58
rrrr	0.63
mrrm	0.68
rrrm	0.71
mmrr	0.74
rmrm	0.76
mmrm	0.79

To assign the NMR spectrum for a vinyl polymer, for example, the RIS model is first used to calculate the relative conformational probabilities for all of the stereosequences. Table 3.6 shows an example of the calculated fraction of bonds in the *trans* conformation for the pentad stereosequences in polypropylene (2). Note that the average *trans* conformation varies between 0.44 and 0.79 as a function of the pentad sequence. From these bond probabilities we can make predictions about the relative chemical shifts of the peaks from the carbon spectrum based on the number of γ-*gauche* interactions. The chemical shifts calculated from the γ-*gauche* effect are not very precise, but they are extremely useful when combined with other assignment methodologies.

3.1.2.5 Spectral Editing Spectral editing has a long and successful history in polymer microstructure determination. Often the first step in successful resonance assignment strategies is to identify the atom type by the number of directly bonded protons. This can be accomplished using the DEPT pulse sequence show in Figure 2.6 (17). The DEPT method makes it possible to generate subspectra based on the number of attached protons. In the carbon spectrum this makes it possible to extract subspectra containing only the quaternary, methine, methylene, and methyl carbons.

The DEPT pulse sequence is in some ways similar to the INEPT pulse sequence introduced in Section 1.7.1.1, in that it is based on polarization transfer using heteronuclear J couplings. The two differences between DEPT and INEPT are that a variable tip-angle pulse has replaced on the proton $90°$ pulses and an extra delay time is added after the pulses to allow the multiplets to rephase so that the decoupler can be turned on during acquisition. This pulse sequence is able to distinguish between carbons with differing numbers of attached protons, because the efficiency of polarization transfer is very sensitive to the tip angle θ. The polarization transfer efficiency as a function of θ was shown in Figure 2.7 for the DEPT pulse sequence. All peaks are observed for $\theta = 45°$, but the intensity of the methylene and methyl

3.1 INTRODUCTION

peaks would be greater than for the methine signals. With $\theta = 90°$, only the methine peaks are observed, as the transfer efficiency for the methylene and methyl signals is zero. When $\theta = 135°$, polarization transfer will be very efficient for the methylene carbons, but they will be opposite in phase from the methyl and methine signals.

The DEPT pulse sequence can be used to generate subspectra containing only the methine, methyl, and methylene carbons by acquiring three spectra for θ values of 45°, 90°, and 135°. Because the intensities and phase of the peaks depend on the multiplicity, it is possible to take linear combinations of the raw data to obtain subspectra for the carbon types. For example, if we add the spectrum for $\theta = 45°$ and $\theta = 135°$, the peaks from the methylene carbons would disappear. If we take the sum of these two spectra and subtract some of the intensity from the spectrum with $\theta = 90°$, which contains only signals from carbons with a single attached proton, we would obtain a spectrum containing only the methyl signals. In practice the polarization efficiency may differ from carbon to carbon due to variations in the coupling constants, so weighting of the spectra before the subtraction must be done interactively.

Figure 3.15 shows the carbon spectra for poly(propylene glycol) acquired using the DEPT pulse sequence with values for θ of 45°, 90°, and 135°, along with the spectrum from a one-pulse experiment. Figure 3.16 shows the results of linear combinations of the spectra showing the methyl, methylene, and methine signals. DEPT is a relatively high sensitivity experiment and can be used to identify signals as small as 1% of the main chain signals.

3.1.2.6 Multidimensional NMR The most recent, but perhaps the most powerful method for assigning resonances in polymers, is nD NMR. While there are a number of advantages to nD NMR in polymer science, perhaps the most important is that it is possible to use nD NMR to prove the resonance assignments, rather than to show that they are consistent with the known data. Furthermore, the synthesis and purification of model compounds are not required. Detailed assignments are possible because the spectra are spread out into two or more frequency dimensions, so a much higher resolution is obtainable, making it possible to assign the resonances even in highly overlapped spectra. Recent advances in NMR methods and equipment has made it possible to use nD NMR to identify low levels of defects in polymers (18).

Multidimensional NMR methods use through-space dipolar interactions or through-bond scalar couplings to establish the resonance assignments. The through-space interactions (NOE) depend on the inverse sixth power of the internuclear distances and can be used to identify pairs of protons that are in close proximity (2–5 Å). The NOEs can also be used to study inter- and intermolecular interactions in polymers, including hydrogen bonding and van der Walls interactions (19,20). Heteronuclear NOEs can also be used for resonances assignments, such as those between carbonyl groups and other nearby protons.

Through-bond scalar couplings can provide important assignment information that can be easily observed in nD NMR experiments. Information about the type of carbon or proton and the number of coupled protons can be obtained through *J*-resolved experiments (21). It is possible to use 2D *J*-resolved NMR much like

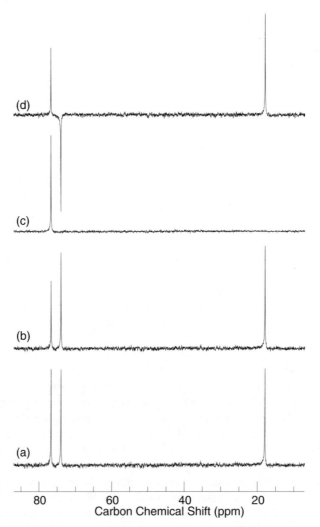

FIGURE 3.15 The DEPT spectra for poly(propylene glycol). The data show (a) the one-pulse spectrum and the spectra with (b) $\theta = 45°$, (c) $\theta = 90°$, and (d) $\theta = 135°$.

DEPT to identify the methyl, methylene, methine, and quaternary carbons and protons in polymers. These methods can also be used to make assignments in the nitrogen spectra, but this typically requires isotopically labeled compounds. nD NMR methods can also be applied to silicon-, phosphorus-, and fluorine-containing polymers.

Correlations between nearby atoms via the scalar couplings can be observed in nD NMR experiments to obtain information about the assignments using any of a number of COSY-type hetero- and homonuclear correlation experiments. The most efficient experiments are those utilizing large coupling constants such as the one-bond carbon-proton couplings. As noted in Section 1.7.1.1, magnetization is transferred through the

3.1 INTRODUCTION

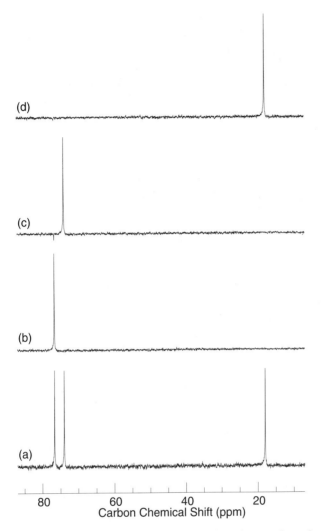

FIGURE 3.16 The (a) 1-pulse and edited DEPT spectra for poly(propylene glycol) showing the (b) methine, (c) methylene and (d) methyl carbon signals.

evolution of a coupled spin system for a time on the order of $1/2J$. During this period the signals can decay from spin-spin relaxation, so magnetization is most efficiently transferred when $1/2J$ is short compared to T_2. This makes heteronuclear correlations in polymers very efficient. The one-bond carbon–proton coupling constants are in the range of 120–160 Hz, so the $1/2J$ delay times are on the order of a few milliseconds. The same is true for the one-bond carbon–fluorine and nitrogen–proton couplings.

Multiple-bond heteronuclear correlations are also important for polymer resonance assignments. The sensitivity is lower because the coupling constants are attenuated by the intervening bonds. The two-bond carbon–proton couplings are on the order of 5–10 Hz, but the sensitivity of nD NMR is such that such correlations can be routinely observed using the or multiple-bond multiple-quantum correlation (MHBC) correlation experiments (22).

The correlations between nearby protons via the homonuclear couplings also provide important assignment information. The two-bond proton–proton couplings (as in olefins) are in the range of 12–15 Hz and are easily observed. The three-bond coupling constants are smaller (2–8 Hz), but the sensitivity of protons is very high, so these correlations are also observed routinely. The four-bond coupling constants are generally too small ($^4J_{HH} = 0$–1 Hz) to be directly observed in polymers, but longer-range correlations can be observed using relayed correlation experiments such as TOCSY. By using sequential correlations between nearby protons, it is possible to observe correlations between pairs of protons separated by four or more intervening bonds.

One of the advantages of nD NMR experiments is that the various correlated and resolved experiments can be combined to give additional information. Experiments such as HMQC–COSY or HMQC–TOCSY can combine the carbon–proton and proton–proton correlation experiments. The resolution in the proton spectrum is typically lower than in the carbon spectrum, so the proton–proton correlations can be more easily observed if they are resolved via the carbon chemical shifts in the HMQC part of the experiment before they are correlated via the proton–proton scalar couplings using COSY or TOCSY. The through-bond and through-space correlations can be combined in a similar way with HMQC–NOESY correlation experiments.

3.2 STEREOCHEMICAL CHARACTERIZATION OF POLYMERS

The NMR spectrum is very sensitive to stereochemical isomerism in a variety of vinyl and other types of polymers. The degree to which the signals from the stereosequences are resolved depends on a number of factors, including the nuclei under observation and the polymer dynamics. We noted earlier (Section 1.3) that some nuclei, notably carbon, fluorine, silicon, and phosphorus, have very wide of chemical shift ranges. As a general rule, those nuclei with the largest range in chemical shift tend to be most sensitive to stereochemical isomerism, and the stereochemical characterization of polymers is most easily accomplished using these nuclei. Furthermore, some of the polymer signals are more sensitive to stereochemical isomerism than are others. It is often observed, for example, that the methylene carbons show better resolution than the methine carbons in the same polymer. This arises mainly from differences in the γ-gauche effects. Those atoms furthest from the pseudoasymmetric center experience smaller induced shifts than the closer nuclei.

The resolution of stereoisomers depends both on the degree to which the signals are shifted away from other stereosequences and on the linewidth. The linewidth is inversely proportional to the spin-spin relaxation times (Section 2.6.1.2) and depends on the segmental dynamics through the correlation time. It is well established that polymer relaxation is usually dominated by segmental motion rather by reorientation of the entire chain, so sharp lines are often observed, even for high molecular-weight polymers (23). It is sometimes possible to obtain spectra with smaller linewidths by acquiring the spectra at higher temperature or in solvents with a lower viscosity.

3.2 STEREOCHEMICAL CHARACTERIZATION OF POLYMERS

The proton NMR spectra of polymers are frequently the easiest to measure, but they are often not as useful for stereochemical characterization as the carbon spectra. This is because protons have a small chemical shift range, and the differences in chemical shifts between the stereosequences are often as large as the linewidths. In many cases the stereochemical isomerism leads to inhomogeneous broadening in the proton spectrum, and broad unresolved lines are observed. While this makes it difficult to use proton NMR for stereochemical characterization, the proton chemical shifts can be observed indirectly using heteronuclear nD NMR.

3.2.1 The Observation of Stereochemical Isomerism

The first step in the NMR analysis of polymer stereochemistry is to acquire a 1D spectrum to observe the presence of stereochemical isomerism. The proton spectrum is the easiest to acquire, but it may not be as informative as the carbon spectra. Once the splitting from stereochemical isomerism has been observed, the peaks can be assigned by comparison of the spectra to those from model compounds and polymers, chemical shift, and conformational calculations, or by nD NMR.

The relative sensitivity of the proton and carbon NMR spectrum to stereochemical isomerism is illustrated in Figure 3.17, which compares the 500-MHz proton and 125-MHz carbon spectrum of poly(vinyl acetate). There are three groups of signals in the proton spectrum that are assigned to the methine, methylene, and methyl protons. The stereochemistry of poly(vinyl acetate) causes the methine protons in m- and r-centered sequences to be slightly shifted from each other, but the shifts are not large enough such that the peaks are resolved. Some fine structure is observed in the proton spectra for the methylene signals, but again the stereosequences are not well resolved. By comparison, the carbon spectrum has much better resolution. It can be seen in Figure 3.17b that the methine peak is split into five signals. There is also splittings from stereochemical isomerism for the methylene peak. The carbonyl peak is broadened from unresolved splittings, and the methyl signal appears to be insensitive to stereochemical isomerism.

$$\begin{array}{c} +CH_2-CH+ \\ | \\ O=C \\ \quad \backslash CH_3 \end{array}$$

Poly(vinyl acetate)

While the resolution is often better in the carbon spectrum relative to the proton spectrum of polymers, the proton spectrum offers important clues about the assignments that cannot be directly deduced from the carbon spectrum. One consequence of having m-centered stereosequences in vinyl polymers is that the environments for the methylene protons are not equivalent to each other. This is illustrated in Figure 3.18, which shows a hypothetical vinyl polymer in the m and r configurations. In the m configuration the magnetic environments for the geminal methylene protons are different from each other, as one of the methylene protons is directly adjacent to the methine substituent on both sides, while the other methylene proton has protons for neighbors.

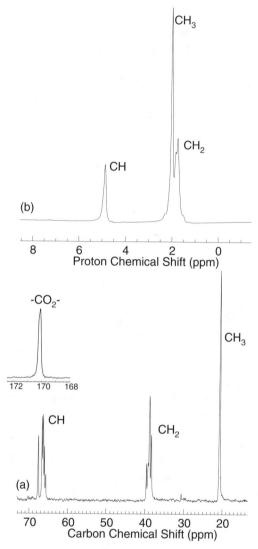

FIGURE 3.17 The (a) 500-MHz proton and (b) 125-MHz carbon spectra of poly(vinyl acetate).

In some polymers (such as poly(methyl methacrylate)) the chemical shift difference between the methylene protons can be as large as 1.5 to 2 ppm in the m-centered stereosequences. In the r-centered stereosequences both methylene protons are bordered on one side by the methine substituent and on the other side by a proton. Thus both methylene protons have similar magnetic environments and similar chemical shifts.

The difference in the magnetic environment between the m- and r-centered stereosequences is clearly observable in the high-resolution spectrum of predominantly isotactic (m) and syndiotactic (r) poly(methyl methacrylate) shown in Figure 3.19.

3.2 STEREOCHEMICAL CHARACTERIZATION OF POLYMERS

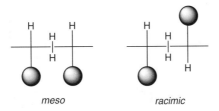

FIGURE 3.18 A schematic diagram of the effect of stereochemistry on the magnetic environment for the methylene protons.

The largest differences between the spectra are the signals for the methylene protons that appear between 1.25 and 2.0 ppm. In the syndiotactic polymer, a single group of peaks is observed for the methylene protons near 1.5 ppm, while the methylene protons is split into two peaks at 1.25 and 2.0 ppm in the isotactic polymer. Note also that the chemical shift for the main chain methyl signal is also very sensitive to the stereochemistry.

Poly(methyl methacrylate)

The stereosequence distributions can be measured from the relative intensities of the peaks if they are resolved and have been assigned. We calculate from triad

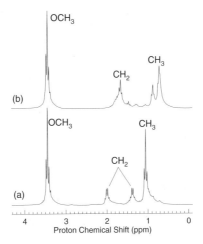

FIGURE 3.19 The 500-MHz proton spectrum for (a) isotactic, (b) syndiotactic, and (c) atactic poly(methyl methacrylate).

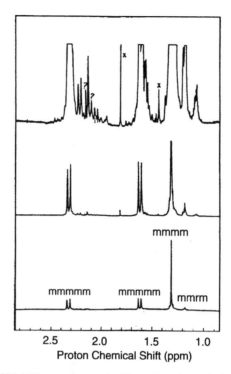

FIGURE 3.20 The 500-MHz proton spectra of isotactic poly(methyl methacrylate) at different gain settings.

distribution from the main-chain methyl group intensities between 0.5 and 1.0 ppm. This analysis shows that the predominantly syndiotactic poly(methyl methacrylate) has 5% *mm*, 30% *mr*, and 65% *rr* stereosequences. If the peaks are more highly resolved, as for the carbon spectrum, then higher-order stereosequence can be measured. If enough stereosequences can be quantified, then it is possible to study the polymerization mechanism. In this way it is often possible to distinguish between a polymerization following Bernoullian statistics from other statistical models.

Figure 3.20 shows an expanded plot of the proton spectra of predominantly isotactic poly(methyl methacrylate) and illustrates the level to which proton NMR can provide information about the polymer mircrostructure. At the lowest gain settings the polymer appears to be nearly purely isotactic poly(methyl methacrylate), but expansion of the gain shows many stereosequences are present at the 1% level. The assignments for the peaks shown in Figure 3.20 cannot be directly established from the 1D spectra, but are made on the basis of chain statistics, conformational and chemical shift calculations, and nD NMR studies.

Figure 3.21 shows the carbon spectrum of a predominantly syndiotactic poly(methyl methacrylate) obtained at 100°C in 1,2,4-trichlorobenzene. As with the poly(vinyl acetate), the chemical shifts of some resonances are much more sensitive to polymer microstructure than are others. The main-chain methyl signals (18 to 22 ppm)

3.2 STEREOCHEMICAL CHARACTERIZATION OF POLYMERS

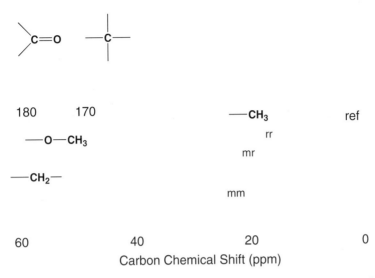

FIGURE 3.21 The 125-MHz carbon NMR spectrum of predominantly syndiotactic poly(methyl methacrylate).

are very sensitive to stereochemistry and the peaks from the *rr*, *mr*, and *mm* sterosequences are well resolved. In contrast to the poly(vinyl acetate), the carbonyl signals are extremely sensitive to polymer microstructure. The methoxyl carbon signals are the least sensitive to stereochemistry and the stereochemical splittings for the methylene carbons lead to inhomogeneous broadening of these signals rather than well-resolved peaks. The quaternary carbon signals are sensitive to stereochemistry to the triad level.

Figure 3.22 shows the carbon and proton spectrum of polystyrene. It might have been intuitively expected that the proton spectrum of polystyrene would be more highly resolved than other vinyl polymers because of ring current shifts from the aromatic ring. This is not observed, but rather the spectrum is broadened from inhomogeneous broadening from microstructures that do not give rise to well-resolved peaks. Two peaks are observed in the aromatic region of the proton spectrum that are assigned to the *ortho* and *meta/para* aromatic protons. The carbon spectrum is also not well resolved and only broad peaks are observed for the methine and methylene main carbons. Only the C1 carbon of the aromatic ring (146 ppm) is resolved from the other aromatic signals (125 ppm), and none of the aromatic carbon signals are particularly sensitive to polymer microstructure.

Figure 3.23 shows the proton and carbon solution spectra for poly(vinyl chloride). The chlorine atom is strongly electron withdrawing and the methine and methylene signals are well resolved from each other in both the carbon and proton spectrum. Triad-level splittings are resolved for the methine protons, while well-resolved peaks for the methylene protons are not observed. The carbon spectrum again shows increased resolution relative to the proton spectrum and tetrad or higher sequences are resolved for the methylene carbons.

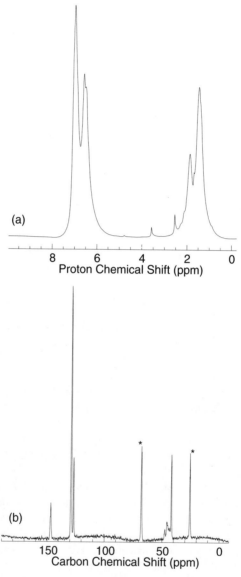

FIGURE 3.22 Comparison of the (a) 500-MHz proton and (b) 125-MHz carbon spectra of polystyrene.

Nuclei other than carbons and protons are also useful for the characterization of polymer microstructure. This is illustrated in Figure 3.24, which shows the 188-MHz ^{19}F spectrum of poly(vinyl fluoride) prepared by two synthetic routes (8). Three main peaks are resolved between 179 and 183 ppm from the *mm*-, *mr*-, and *rr*-centered triads, and each of these peaks is further split into tetrad peaks. The large chemical

3.2 STEREOCHEMICAL CHARACTERIZATION OF POLYMERS

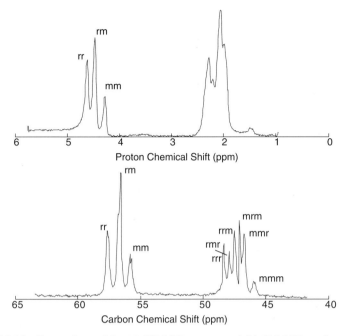

FIGURE 3.23 Comparison of the (a) 500-MHz proton and (b) 125-MHz carbon spectra of poly(vinyl chloride).

shift range of fluorine makes this an excellent tool to characterize the stereochemistry in fluorine-containing polymers.

Silicon NMR is also a useful probe of polymer microstructure, but relatively few polymers have silicon near a pseudoasymmetric center. One group of polymers that has been studied by solution NMR is the asymmetrically substituted polysilanes, such as poly(methyl-n-propyl silane) (24). The NMR results show that the silicon signal is sensitive to stereochemistry, and at least six signals are resolved. Much sharper

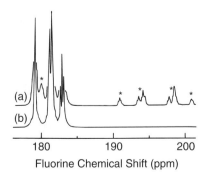

FIGURE 3.24 The 188-MHz ^{19}F NMR spectra of (a) commercial and (b) isoregic poly(vinyl fluoride).

signals are observed for the symmetric poly(di-*n*-hexyl silane), demonstrating that the splitting in the asymmetric polysilanes must be attributed to tacticity.

3.2.2 Resonance Assignments for Stereosequences

Multinuclear NMR is a very effective tool for the observation of polymer stereochemistry. However, for a complete interpretation of the NMR spectrum it is necessary to assign all of the resolved peaks. A number of methods have been used to establish these assignments, including comparisons of the spectra with model compounds and polymers, intensity measurements for chains with known chain statistics, conformational and chemical shift calculations, and nD NMR. These methods have evolved historically following the first studies using model compounds. With the exception of some nD NMR measurements, these methods provide evidence for, but not proof of, the stereochemical assignments.

3.2.2.1 Assignments of Stereosequences Using Model Compounds In the earliest days of polymer NMR, the assignments were established by comparing the spectra to those from model compounds and polymers. This involves the synthesis and separation of diastereomers for small molecules, and the preparation of stereochemically pure polymers. While the synthesis of model compounds can often be used to establish the assignments, it must be kept in mind that the chemical shifts are also sensitive to temperature and solvent, so comparisons must be made with care.

2,4,6-Trichlorheptane has been used as a model compound to establish the chemical shift assignments for stereochemical sequences in poly(vinyl chloride). 2,4,6-Trichlorheptane has three asymmetric centers and serves as a model for the *mm* (isotactic), *mr* (heterotactic or atactic), and *rr* (syndiotactic) sequences. Two sets of signals are observed in the carbon spectra for the model compounds that are assigned to the methine carbons (internal and external), while a single set of signals is observed for the methylene and methyl carbons. In addition, to the *mm* and *rr* signals, the two heterotactic sequences (*mr* and *rm*) also gave resolved signals.

$$H_3C-CH-CH_2-CH-CH_2-CH-CH_3$$
$$|||$$
$$ClClCl$$

2,4,6-Trichloroheptane

The results from the model compound studies showed that the chemical shifts for 2,4,6-trichlorheptane are very sensitive to stereosequence, but also to solvent and temperature. Figure 3.25 shows a simulated spectrum of the reported chemical shifts for 2,4,6-trichlorheptane (1). Several peaks are observed in the carbon spectra in the methine (54 to 60 ppm), the methylene region (47 to 49 ppm), and the methyl region (not shown). Seven peaks are observed in the methine region for neat 2,4,6-trichlorheptane. The three peaks of most interest are the central methine signals labeled S_c, H_c, and I_c to denote the assignments for the syndiotactic, heterotactic, and isotactic sequences, and the four methylene signals. It is expected that the central

3.2 STEREOCHEMICAL CHARACTERIZATION OF POLYMERS 191

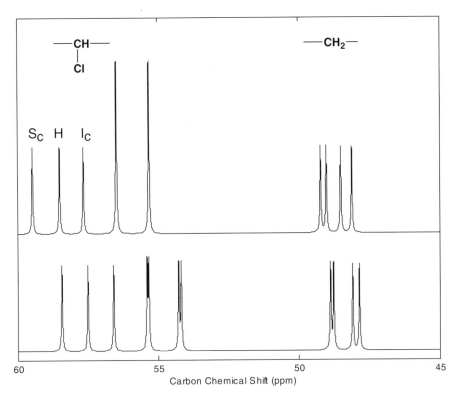

FIGURE 3.25 The simulated carbon spectra from the reported chemical shifts for 2,4,6-trichlorheptane (a) neat and (b) in acetone.

part of 2,4,6-trichlorheptane is a better model for poly(vinyl chloride), since the other signals are perturbed by end effects. The results show that the shifts from stereochemical isomerism are large and much of the chemical shift difference can be attributed to conformational effect through the γ-gauche effect (1).

Caution must be used when using model compounds to make assignments in polymers because the chemical shifts can be very sensitive to the solvent. This is illustrated by comparing the spectrum of neat 2,4,6-trichlorheptane with the spectrum acquired in acetone, which is also shown in Figure 3.25b. The conformational properties of 2,4,6-trichlorheptane are different in the neat solution and in acetone, resulting in a change in both the absolute values of the chemical shifts, and, for the methylene signals, the relative ordering of the chemical shifts. Note that using 2,4,6-trichlorheptane in acetone as a model compound would lead to incorrect assignments for poly(vinyl chloride). Nonetheless, model compounds have historically been important for establishing the stereosequence assignments in polymers.

Model polymers, if they are available, are an excellent source of chemical shift assignments for polymer microstructure. This is illustrated in Figure 3.26, which compares the carbon NMR spectra of isotactic, atactic, and syndiotactic polypropylene. The spectrum for the isotactic polymer shows that the isotactic material is extremely

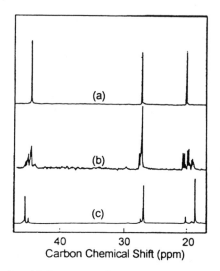

FIGURE 3.26 The carbon NMR spectra of (a) isotactic, (b) atactic, and (c) syndiotactic polypropylene.

pure. More peaks are observed in the spectrum of the syndiotactic material, but the largest peaks can be used to assign the syndiotactic resonances. The spectrum for the atactic material is much more complex, particularly for the methyl region, but the peaks from the isotactic and syndiotactic sequences can clearly be assigned from the comparison with the stereochemically pure materials.

Additional information is required for the complete assignments of the polymer spectrum. One approach to making these assignments is to use the epimerization of stereoregular polymers. It has been shown that certain catalysts cause some vinyl polymers to undergo configurational inversion at their pseudoasymmetric centers (6,7). In stereoregular polypropylene these catalysts lead to random inversions of monomer configuration at low conversion ratios. If we start from the *mmmm* pentad configuration, a single epimerization gives rise to *mmmr*, *mmrr*, and *mrrm* stereosequences in a 2:2:1 ratio. If we start with the syndiotactic polymer, the *rrrr* pentad gives rise to *rrmm*, *rrrm*, and *rmmr* configurational sequences in a 2:2:1 ratio. Thus, the stereochemically pure polymers and the epimerization results give the assignments for seven of the ten pentad sequences. The remaining sequences *mmrm*, *rmrr*, and *rmrm* can be assigned from known pentad–pentad relationships that are given by

$$\Sigma \text{pentads} = 1 \qquad (3.26)$$

$$2rmmr + mmmr = mmrm + mmrr \qquad (3.27)$$

and

$$2mrrm + mrrr = mmrr + rmrr \qquad (3.28)$$

The assignments can be made by measuring the intensities for resolved peaks and for groups of peaks that are not resolved (2).

3.2 STEREOCHEMICAL CHARACTERIZATION OF POLYMERS

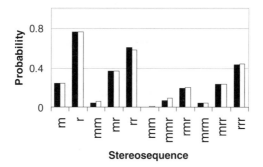

FIGURE 3.27 The peak intensities for poly(methyl methacrylate) comparing the experimental intensities (filled bars) with the intensities calculated assuming Bernoullian trail statistics (unfilled bars).

3.2.2.2 Assignments of Stereosequences Using Polymerization Statistics

The stereosequence peak assignments can sometimes be established by comparing the peak intensities with those calculated based on a particular type of chain propagation statistics, as discussed in Section 3.1.2.2. This approach is illustrated in Figure 3.27, which shows a comparison of the experimental data for poly(methyl methacrylate) prepared using a free-radical polymerization and the calculated peak intensities using Bernoullian trail statistics calculated with $P_m = .24$ (25). As expected with a predominantly syndiotactic polymer, the *racemic* signals are most intense, and the peaks from some of the *meso* tetrads are close to zero. The excellent agreement between the experimental and calculated peak intensities provides evidence both for the assignments and the polymerization statistics.

For those polymers that cannot be assumed to proceed by Bernoullian statistics, the situation is more complex, and it is difficult to make assignments based on the intensities alone. This is illustrated in Figure 3.28, which compares the experimental data (filled bars) for poly(methyl methacrylate) prepared by anionic polymerization with the intensities calculated from Bernoullian, first-order Markov, second-order Markov, and the Coleman–Fox models for chain propagation (25). This figure shows that the Bernoullian statistics do not accurately predict the peak intensities for any

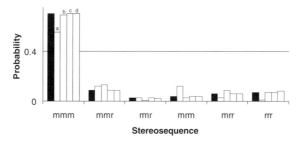

FIGURE 3.28 Comparison of the poly(methyl methacrylate) peak intensities (filled bars) with the values calculated assuming (a) Bernoullian, (b) first-order Markov, (c) second-order Markov, and (d) Fox–Coleman chain statistics.

of the stereosequences, except *rmr*, using a $P_m = .82$. The first-order Markov model accurately predicts some of the intensities, but not those for the *mmr*, *mrm*, and *mrr*. The best fit is for the second-order Markov and Coleman–Fox models.

As noted previously, caution must be used when the assignments are made using the polymerization statistics. The fit of the experimental data to a particular polymerization model can be considered as evidence, but not proof, for the stereochemical assignments.

3.2.2.3 Assignments of Stereosequences Using Chemical Shift and Conformational Calculations
Chemical shift calculations using the γ-*gauche* effect are a powerful tool for making stereochemical assignments in vinyl and other polymers. As noted in Section 3.1.2.4, the through-space interactions between substituents separated by three bonds can lead to large changes in the carbon chemical shifts. The magnitude of the shift depends on the atom at the γ position and the fraction of bonds with a *gauch* relationship to the carbon of interest. Each γ-*gauche* interaction between a carbon and its methyl neighbor, for example, leads to a 5.3-ppm shift in the carbon spectrum. The induced chemical shift from the γ-*gauche* effect therefore depends on the chain conformation, which can be calculated through the RIS model (2).

To understand the relationship between chain conformation and the carbon chemical shift we can consider the case of polypropylene. An RIS model for polypropylene was developed to predict the fractional probabilities of the bond conformations (26). The results from these calculations showed that the bond probabilities were extremely sensitive to the polymer stereosequence. This was shown in Table 3.6, which lists the fraction of bonds at the center of pentad stereosequences that are in the *trans* conformation. Note that the differences in conformation for the pentad stereosequences are relatively large and may give rise to resolved peaks for many of the stereo sequences. In addition, it is important to note that the chemical shifts for the stereosequences are not simply ordered relative to the bond probabilities. The γ-*gauche* calculations predict that the *mrmr* and *mmrm* stereosequences would be the most separated from each other, rather than having the *mmmm* and *rrrr* stereosequences at opposite ends of the spectrum.

As shown in Figure 3.26, the carbon spectrum of polypropylene is extremely sensitive to polymer stereochemistry. The spectra for the isotactic and syndiotactic polymers are very simple for the methyl, methylene, and methine signals. In the atactic material the methyl signals are most sensitive to polymer stereochemistry and many peaks are resolved.

Figure 3.29 shows the 90-MHz carbon spectrum of polypropylene along with the chemical shift for the 36 heptad stereosequences predicted from the RIS modeling (27). Since there is some uncertainty both in the magnitude of the γ-*gauche* effect and the bond probabilities, the assignments cannot be unequivocally established using this approach. To get a better idea of the quality of the RIS calculations, the chemical shifts can be combined with the peak intensities to simulate the spectrum. This is shown in Figure 3.30, which shows the spectrum obtained from combining the chemical shifts from the RIS calculation and the peak intensities from the experimental spectrum. In the case of atactic polypropylene the intensities could not be described by simple polymerization models such as the Bernoullian or the first-order Markov models.

3.2 STEREOCHEMICAL CHARACTERIZATION OF POLYMERS

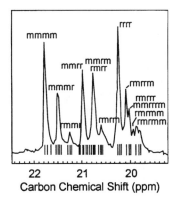

FIGURE 3.29 The 90-MHz carbon spectrum of polypropylene along with the chemical shifts calculated from the γ-*gauche* effect.

Atoms in addition to carbon also give rise to large γ-*gauche* effects, making these calculations valuable for establishing the chemical shift assignments in a wide variety of vinyl polymers. Poly(vinyl chloride) has been extensively studied, and the chemical shifts for the methine and methylene carbons have been assigned by a number of methods. Since the assignments have been established by other means, poly(vinyl chloride) is an excellent polymer to test the accuracy of RIS methods and the γ-*gauche* effect for making assignments in polymers.

Given the structure of poly(vinyl chloride), several types of γ-*gauche* interactions are possible, and each of these may affect the chemical shift differently. The chemical

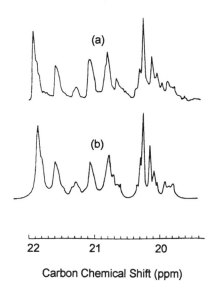

FIGURE 3.30 Comparison of the (a) experimental 90-MHz carbon spectrum of polypropylene with (b) the spectrum calculated from the chain statistics (intensities) and the γ-*gauche* effect (chemical shifts).

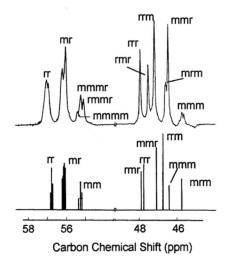

FIGURE 3.31 Comparison of the (a) experimental and (b) calculated spectra for poly(vinyl chloride).

shifts are calculated from the bond probabilities in the RIS model (28) using values of −2.5 ppm for CH_2—CH, −5 ppm for CH—CH_2, and −3 ppm for CH—Cl γ-*gauche* interactions. Figure 3.31 compares the chemical shifts calculated from the γ-*gauche* model with the experimental spectrum. Good agreement is observed for the methine part of the spectrum, with the proper ordering of the *rr*, *mr*, and *mm* groups of peaks. The agreement is not as good for the methylene carbons, where even the ordering of the peaks does not match the experimental spectrum. Such behavior might have been expected based on the observation that the methylene carbon spectrum is solvent dependent (29).

In summary, the γ-*gauche* effect can provide important information about the stereochemical assignments in polymers. However, the assignments from the γ-*gauche* effect must be combined with other information to establish the assignments, because of uncertainties in the RIS models, the measured intensities, the magnitudes of the γ-*gauche* effects, and solvent effects on the spectrum.

3.2.2.4 Assignments of Stereosequences Using nD NMR nD NMR has emerged as an important method for establishing the chemical shift assignments for stereochemical and other polymer microstructures. This approach differs from other methods in that it is often possible to prove the assignments using these techniques, rather than to demonstrate that the data are consistent with the assignments. One of the most obvious advantages of nD NMR is that the spectrum is spread out into two or more frequency dimensions, where the resolution is greatly improved.

The first step in the establishing the stereosequence assignments in polymers is to identify *meso* and *racemic* configurations. NMR is very valuable for this task because, as noted in Section 3.1.1.2, the magnetic environments for geminal pairs of protons are very sensitive to the configuration. In *racemic* sequences in vinyl polymers, each

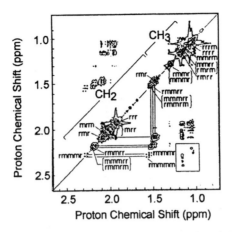

FIGURE 3.32 The 500-MHz proton COSY spectrum for atactic poly(methyl methacrylate).

geminal methylene proton is on the same side of the chain as one of the neighboring vinyl substituents, and therefore has a similar environment (Figure 3.18). In a *meso* configuration one of the geminal protons is on the same side of the chain as two of the vinyl substituents, while the other proton has two methine proton neighbors on the same side of the chain. This leads to a chemical shift difference between the methylene protons in *meso* diads. Thus, the NMR spectral signature for *meso* configurations is a pair of nonequivalent methylene protons. These protons can be identified in homonuclear nD NMR experiments because they are coupled with a much larger two-bond coupling constant ($^2J_{HH} = 12 - 15$ Hz) than most of the other three-bond couplings ($^3J_{HH} = 2 - 8$ Hz) and they are separated by only 1.77Å. *Meso* configuration stereosequences can also be identified in heteronuclear nD experiments, since a single carbon may give rise to cross peaks to two proton chemical shifts.

The differences in the chemical shifts for the methylene protons in *meso* and *racemic* sequences is particularly obvious in stereochemically pure polymers. This was illustrated in Figure 3.19, where the proton NMR spectra for isotactic was compared with syndiotactic poly(methyl methacrylate). The isotactic polymer can be easily identified by the two peaks that are observed for the methylene protons. This task is considerable more complicated in the atactic polymer, as the *meso* and *racemic* peaks overlap and the couplings nonequivalence of the geminal methylene protons cannot be observed by the visual inspection of the one-dimensional spectra. These couplings can be identified from the nD spectra.

Figure 3.32 shows the 500-MHz 2D COSY spectrums for atactic poly(methyl methacrylate) as an example of a polymer with a complex proton spectrum. Although the methoxyl signal at 3.4 ppm appears to be insensitive to polymer stereochemistry (not shown), both the methyl and methylene protons show a number of resolved peaks. The nonequivalent methylene protons in *meso* configurations will be coupled by a large two-bond coupling and can be easily identified as the only off-diagonal peaks in the methylene region of the COSY spectrum of poly(methyl methacrylate). The assignments shown in Figure 3.32 are the result of a large number of studies, and it should be noted that although the *meso*-centered configurations can be identified from the COSY spectrum, they cannot be directly assigned from this information alone.

As with other polymers, the carbon spectrum for poly(methyl methacrylate) is more highly resolved than the proton spectrum and can be used to establish the stereosequence assignments. This was illustrated in Figure 3.21, which showed well-resolved lines for the methyl and carbonyl stereosequence peaks, and poor resolution for the quaternary and methylene carbons. The higher resolution in the carbon spectrum can be used as a starting point for establishing the stereosequence assignments using nD heteronuclear NMR.

Figure 3.33 shows the HMQC spectrum for atactic poly(methyl methacrylate). This spectrum alone does not lead to an assignment of the stereosequence peaks, but does provide valuable information about the assignments. If some of the assignments are known from other studies (such as the studies of isotactic and syndiotactic poly(methyl methacrylate)), then the assignments in the proton spectrum can be directly inferred from the heteronuclear correlations. The *meso*-configuration methylene carbons can also be identified from the correlation of the carbon chemical shifts with two peaks in the proton spectrum.

In order to establish the stereosequence assignments using nD NMR, it is necessary to have longer range correlations along the polymer main chain. In vinyl polymers it is possible to use the combination of proton–proton couplings and heteronuclear correlations, because the proton–proton couplings can correlate several pseudoasymmetric centers together. This is not possible in poly(methyl methacrylate), because the

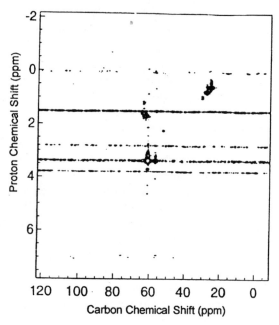

FIGURE 3.33 The HMQC spectrum of poly(methyl methacrylate).

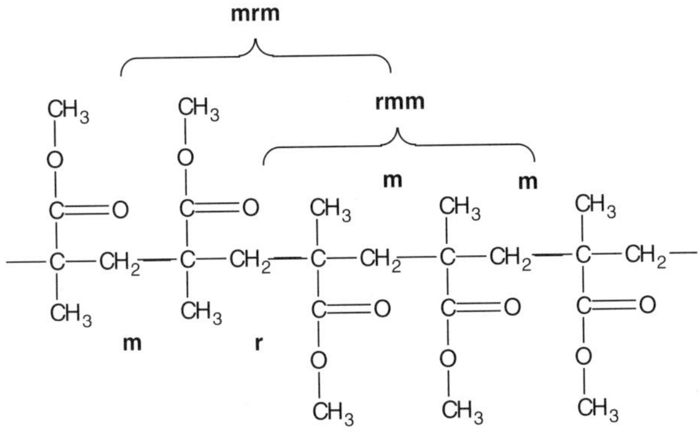

3.2 STEREOCHEMICAL CHARACTERIZATION OF POLYMERS

FIGURE 3.34 A schematic drawing of the correlation of stereochemical sequences in poly(methyl methacrylate).

coupling does not propagate between pseudoasymmetric centers due to the quaternary main-chain carbon, so another approach is required.

The assignments for poly(methyl methacrylate) can be established using longer range correlations, such as the two-bond correlation between the carbonyl and the methylene carbon signals. This is illustrated in Figure 3.34, which shows a drawing of the *mrmm* pentad, which contains the *mrm* and *rmm* tetrads. The carbonyl chemical shift in poly(methyl methacrylate) is extremely sensitive to stereosequence and each carbonyl in the pentad will be correlated with two tetrad methylene peaks. As expected, the *m*-centered sequences will show a splitting in the proton dimension because of the nonequivalent methylene protons. These assignments have been published using ^{13}C labeled poly(methyl methacrylate) and heteronuclear correlation (30). On a more modern spectrometer, these chemical shift assignments could be more efficiently measured using HMBC (22).

Poly(vinyl chloride) is a commercially important polymer that has been extensively studied by NMR, and Figure 3.23 showed the carbon and proton spectrum for poly(vinyl chloride). While the carbon spectrum is resolved to the tetrad level, little resolution is observed in the proton spectrum, particularly for the methylene protons. This low resolution makes it difficult to identify the *meso*-centered methylene signals from the 1D proton spectrum. The *m*-centered stereosequences can be identified as the only signals showing off-diagonal peaks in the methylene region in the proton COSY spectrum, as illustrated in Figure 3.35. Correlations can be observed between the methylene and methine protons, but it is difficult to identify all of the *m*-centered sequences because of the poor chemical shift resolution in the proton spectrum for the methylenes. The cross peaks between nonequivalent methylene protons are sometimes difficult to identify because there is little chemical shift dispersion for the methylene protons and the cross peaks may overlap with the intense diagonal peaks. The methine protons are resolved to the diad level, and several triad–tetrad (methine–methylene) correlations can be observed in homonuclear experiments. These include two correlations for the *mm* triads (*mm-mmm* and *mm-mmr*) and the two for the *rr* triads (*rrr* and *rrm*). The remaining signals will be correlated with the *mr* triad peak.

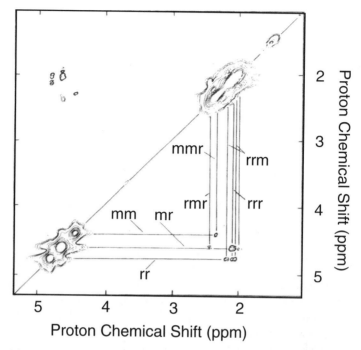

FIGURE 3.35 The 500-MHz double-quantum filtered COSY spectrum of poly(vinyl chloride).

A 2D homonuclear single-quantum–double-quantum correlation experiment (known as INADEQUATE for incredible natural abundance double-quantum transfer experiment (31)) can be used to identify the nonequivalent protons, because the INADEQUATE spectrum does not have the intense diagonal peaks that obscure the COSY correlations. The original INADEQUATE experiment was used to correlate neighboring ^{13}C atoms using the one-bond carbon–carbon coupling constants. This is a potentially powerful tool to establish resonance assignments in polymers because nearest-neighbor carbon atoms can be identified, but it is not extensively used because the sensitivity is low, since the probability of having two ^{13}C atom neighbors in the polymer chain is very low, ($1/10^4$). This approach has been used at natural abundance to study regioisomerism in polymers (32,33).

Figure 3.36 shows the pulse-sequence for INADEQUATE and a schematic diagram of the correlation of single- and double-quantum coherence. The $90°-\tau-180°-\tau$ portion of the pulse sequence creates double-quantum coherences that evolve during the t_1 period. The creation of double-quantum coherences is most efficient for $\tau = 1/2J$, although a shorter delay time may be chosen if signal loss from spin-spin relaxation is a problem. If the primary purpose is to identify nonequivalent methylene protons, a short value of τ ($1/2J \sim 0.033$ s) can be used to maximize the signals resulting from the geminal couplings.

Since the data for the INADEQUATE experiment is the correlation of single- and double-quantum coherences, no diagonal signals are observed. This is the most

3.2 STEREOCHEMICAL CHARACTERIZATION OF POLYMERS

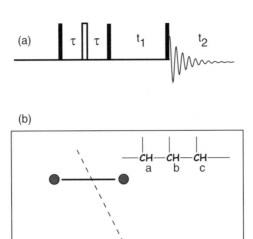

FIGURE 3.36 The (a) pulse-sequence diagram for the homonuclear INADEQUATE experiment, and (b) a schematic drawing of the single-double-quantum correlation.

important advantage of INADEQUATE, since the diagonal signals in the COSY experiment often obscure correlations between nonequivalent methylene protons that have only a small difference in chemical shift. The cross peaks appear at their normal single-quantum frequency (ω) in the t_2 dimension and at the double-quantum frequency (2ω) in the t_1 dimension. The pseudodiagonal appears at the center of the spectrum with a slope of two. Since the double-quantum frequencies are twice as large as their single-quantum counterparts, the sweep width in the t_1 dimension should be twice as large as in the t_2 dimension.

Figure 3.37 shows the 2D proton INADEQUATE spectrum for poly(vinyl chloride) obtained with τ values of 8 and 18 ms at 65°C. The τ value was chosen to be shorter than $1/2J$ (33 ms) because of signal losses due to spin-spin relaxation during the double-quantum creation period. The correlations in the methylene region can only arise from *m*-centered sequences. As with the COSY spectrum, the INADEQUATE spectrum also shows the connectivity between the methine and methylene protons in poly(vinyl chloride). The intensity of the methine–methylene peaks relative to the methylene–methylene cross peaks increases as τ is increased. In principle the correlations between nonequivalent methylene protons can also be observed in NOESY experiments from the very strong dipolar couplings between them. However, like COSY, the NOESY cross peaks are often obscured by the intense diagonal peaks.

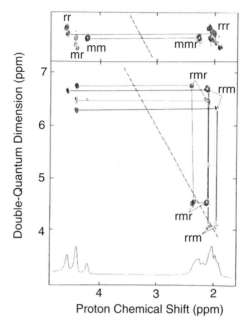

FIGURE 3.37 The 2D proton INADEQUATE spectrum for poly(vinyl chloride).

Longer range correlations can be observed in TOCSY experiments where multiple-bond correlations can be observed. This is illustrated in Figure 3.38, which shows the TOCSY spectrum of poly(vinyl chloride). Note that correlations are now observed in the methine region between the *mm* and *mr* peaks and the *rr* and *mr* peaks. Such cross peaks establish unambiguously the assignment for the *mr* peak.

The combinations of the INADEQUATE, COSY, and TOCSY experiments lead to a partial assignment of the poly(vinyl chloride) proton spectrum. The triad methine peaks can be directly assigned by combining the information from the three experiments. Since the *m*-centered methylene protons identified in the INADEQUATE spectrum are correlated with the two higher-field methine peaks, those must be the *mm* and *mr* peaks. The long-range TOCSY correlations show that the central methine peak has correlations to the other methine signals and must be assigned to the *mr* triad. The highest-field peak must therefore be assigned to the *mm* triad sequence, and the lowest-field methine peak must be assigned to the *rr* triad sequence.

For a complete assignment of the carbon and proton signals, it is often necessary to combine both homonuclear and heteronuclear NMR experiments. The correlation of the proton and carbon spectrum is most efficiently observed using the HMQC experiment, and Figure 3.39 shows the spectrum for poly(vinyl chloride) observed at 65°C. Since the methine and methylene protons and carbons are well resolved from each other, the correlations are well resolved in the 2D spectrum. The relative chemical shifts are the same in the carbon and proton spectrum in the methine region, and the peak order in both spectra are *rr*, *mr*, and *mm*. The methylene correlations are more complex, with some of the carbons showing two correlations. These are

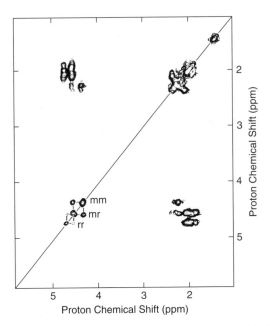

FIGURE 3.38 The TOCSY spectrum for poly(vinyl chloride).

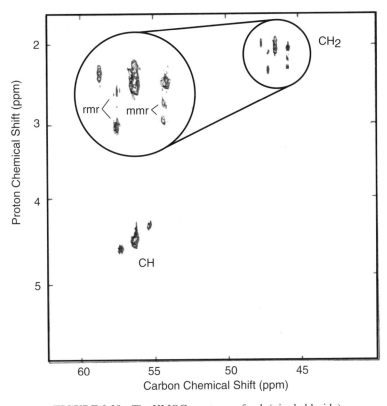

FIGURE 3.39 The HMQC spectrum of poly(vinyl chloride).

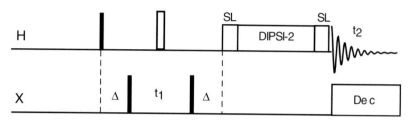

FIGURE 3.40 The pulse-sequence diagram for the HMQC–TOCSY experiment.

immediately identified as *m*-centered sequences. This sample of poly(vinyl chloride) is predominantly syndiotactic, and the correlations to the *mmm* tetrads are too low in intensity to be directly observed.

The stereosequence assignments can be directly assigned using the combination of homo- and heteronuclear correlation using the HMQC–TOCSY experiment shown in Figure 3.40. The HMQC–TOCSY experiment is simply a combination of the two 2D experiments into another 2D experiment. The HMQC is performed first, so the carbon chemical shift evolves during the t_1 period. After magnetization is transferred back to the proton spectrum, a TOCSY-type mixing sequence is applied to the protons to promote magnetization exchange over several bonds via the scalar coupling during the spin-lock period. The final spectrum shows the correlation between a carbon and its directly bonded protons as well as any protons that are scalar coupled to it.

The HMQC–TOCSY is an extremely useful experiment and it has the potential to establish all of the triad and tetrad assignments from the relayed methine–methylene correlations. The strategy for the assignments is illustrated by the data in Table 3.7, which shows the triad–tetrad correlations expected from the HMQC–TOCSY experiment. The *mmm* and *rrr* tetrads show only one correlation each, to the *mm* and *rr* triads, which can be distinguished from each other by the nonequivalent methylene protons in the *m*-centered sequences. The *mmr* and *rrm* sequences show correlations to the *mr* triad and either the *mm* or *rr* triads. The remaining sequences, *rmr* and *mrm*, show only correlations to the *mr* triad, and the *rmr* tetrad can be distinguished by nonequivalent methylene protons. Thus, for those polymers where there is good spectral resolution, the spectrum can be definitively assigned using HMQC–TOCSY experiments.

TABLE 3.7 The Diad–Triad Correlations That Can Be Observed from the HMQC–TOCSY Experiment.

Triad – Diad	*mm*	*mr*	*rr*
mmm	●		
mmr	●	●	
mrm		●	
rmr		●	
rrm		●	●
rrr			●

3.2 STEREOCHEMICAL CHARACTERIZATION OF POLYMERS

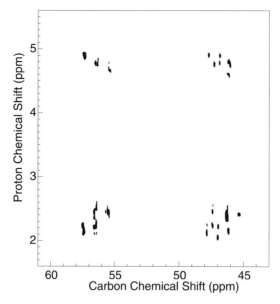

FIGURE 3.41 The HMQC–TOCSY spectrum for poly(vinyl chloride) with a mixing time of 35 ms.

Figure 3.41 shows the HMQC–TOCSY spectrum for poly(vinyl chloride). Two groups of peaks are observed for the direct correlations between carbons and protons, and two groups of peaks are observed for relayed correlations. The *rrr* peak can be assigned as the signal with a single correlation to the *rr* diad. The *mmr* and *rrm* peaks can be identified as coupled to both the *mr* and *mm* or *rr* and *mr* triads. The *mrm* and *rmr* peaks are only correlated with the *mr* triad, and again the *m*-centered sequence can be identified by the correlation to nonequivalent methylene protons.

In this discussion of poly(vinyl chloride), we noted that the HMQC–TOCSY was an excellent experiment for establishing the stereosequence resonance assignments. In the case of poly(vinyl chloride), the lines are relatively well resolved and a 2D experiment is sufficient to resolve all of the peaks. In other cases, a 3D TOCSY–HMQC experiment is better suited toward the assignment problem. Figure 3.42 shows the pulse-sequence diagram for the 3D HMQC–TOCSY experiment. As with most

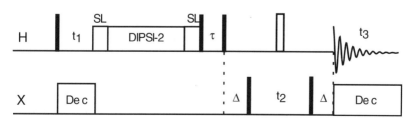

FIGURE 3.42 The pulse sequence diagram for 3D HMQC–TOCSY.

3D experiments the pulse sequence begins with proton excitation and ends with proton detection. After the initial proton 90° pulse the spin system evolves under the influence of the proton chemical shifts. A spin-locking sequence (DIPSI-2 in this case) is used to promote exchange among all of the *J*-coupled spins (34) before the magnetization is transferred by HMQC to the carbon spectrum for the second evolution time. The experiment ends with the reverse HMQC transfer back to the protons for detection under carbon decoupling. The final spectrum has two proton chemical shift dimensions and a carbon dimension. Planes through the 3D matrix at a particular carbon chemical shift give the 2D proton TOCSY spectrum for all the protons directly and indirectly coupled to the carbon of interest.

In a similar way the homo/heteronuclear correlations can be applied to polymers with more complex microstructures and polymers with other NMR-active nuclei. This was illustrated in poly(vinyl fluoride) using ^{19}F NMR (35). The fluorine spectrum for poly(vinyl fluoride) is well-resolved, as shown by Figure 3.24. The three groups of peaks observed in the isoregic material are assigned to the *mm*, *mr*, and *rr* triads, and the pentad-level fine structure is observed within each of these triads. The triad assignments were established with a combination of synthesis methods and RIS/γ-*gauche* calculations (36). Assignments of the pentads can be directly obtained by 2D COSY spectra utilizing the four-bond ^{19}F—^{19}F scalar couplings. The magnitude of the four-bond coupling is on the order of 7 Hz, but it depends on conformation, so there are situations where the expected couplings are not observed.

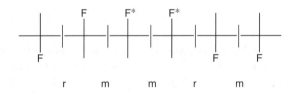

The strategy for establishing the assignments in poly(vinyl fluoride) can be understood by considering the hexad structure just shown. This hexad contains the *rmmr* and *mmrm* pentads. The central fluorines (shown by *) are separated from each other by four bonds and are scalar coupled to each other, so cross peaks are expected in the COSY spectrum. The pentads can be assigned by a process of elimination from the fluorine COSY spectrum. This approach is illustrated in Table 3.8, which shows the cross peaks expected between the pentads in a typical polymer. Some of the cross-peak patterns are unique and can be used as a starting point for the assignments. The *rrrr* peak, for example, is only coupled to the *rrrm* peak, and the *rrrm* and *mrrm* peaks can be identified by cross peaks to the *mr* region. By default, the *mrrm* peak can be identified as the one not coupled to the *rrrr* peak.

Figure 3.43 shows the fluorine COSY spectrum for poly(vinyl fluoride) obtained at 188 MHz (36). The strategy outlined previously is used to identify the *rrrr* peak as the lowest field peak in the *rr* region, since it is only coupled to one other peak in the *rr* triad region. The peak that it is coupled to is assigned to the *rrrm* pentad signal, and the remaining peak in the *rr* triad region must be assigned to the *mrrm* pentad. The peaks in the *mm* region can be assigned in a similar way, although the resolution is much poorer than in the *rr* region. In fact, the three *mm* peaks cannot be easily

3.2 STEREOCHEMICAL CHARACTERIZATION OF POLYMERS

TABLE 3.8 Pentad Couplings in Poly(Vinyl Fluoride) That May Be Observed by, ^{19}F NMR.

	rrrr	rrrm	mrrm	mmrm	rmrm	mmrr	rmrr	rmmr	mmmr	mmmm
rrrr	▓	●								
rrrm	●	▓	●			●	●			
mrrm		●	▓			●	●			
mmrm				▓	●	●		●	●	
rmrm				●	▓		●	●	●	
mmrr			●	●		▓		●	●	
rmrr				●	●		▓			
rmmr				●		●		▓	●	
mmmr				●		●		●	▓	●
mmmm									●	▓

visualized from the 1D spectrum because of the severe overlap, but the three peaks can be resolved on the diagonal of the 2D spectrum.

There are four resolved lines in the *mr* region, two of which (*mmrm* and *mmrr*) are coupled to the *mm* region, and two of which (*mmrr* and *rmrr*) are coupled to the *rr* region. The *mmrr* peak can be identified immediately as the only peak coupled to both the *mm* and *rr* region. Thus, the remaining *mr* peak with coupling to the *mm* region is assigned to *mmmr* and the remaining *mr* peak coupled to the *rr* region is assigned to *rmrr*. The final peak, *mrmr*, can be identified as the one not coupled to either the *mm* or *rr* regions.

Fluorine NMR is an excellent means to probe the stereochemistry of polymers, since ^{19}F has a high sensitivity and a wide chemical shift dispersion. As with

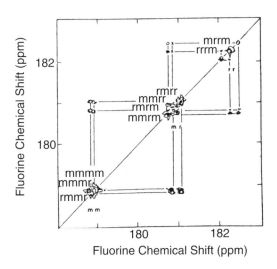

FIGURE 3.43 The 188-MHz ^{19}F COSY spectrum for poly(vinyl fluoride).

proton–proton couplings, the magnitude of the fluorine–fluorine couplings depend on the torsional angle between the fluorines via a Karplus-type relationship (Section 1.4.4). It is therefore possible that, there may be conformations with fluorine–fluorine couplings that are too small to observe in the COSY spectrum. For example, it was noted in the COSY spectrum of poly(vinyl fluoride), that the cross peaks *mmrr–mmrr* were larger than the *mmrr–mrrr* cross peaks, and this was ascribed to a conformational preference (35).

Once the fluorine spectrum is assigned for poly(vinyl fluoride), the carbon and proton spectrum can be directly assigned from the heteronuclear proton–fluorine and fluorine–carbon HMQC or HSQC spectrum. The pulse sequences are the same as for carbon–proton experiments, but the delay times are changed to account for differences in the magnitudes of the proton–fluorine and carbon–fluorine coupling constants. In the carbon–fluorine experiments the fluorine–carbon one-bond coupling constants are approximately twice as large (250–300 Hz) as the carbon–proton coupling constants, so shorter delays for the $1/2J$ can be used. This, of course, increases the SNR because there is less time for spin-spin relaxation during this delay period. In addition, the two-bond fluorine–carbon coupling constants are larger, so better sensitivity is observed in HMBC experiments. In many probes it is possible to tune the proton channel to the fluorine frequency to observe the fluorine–carbon HMQC spectrum.

The proton–fluorine correlation experiments require a more specialized probe with two high-frequency inputs. The fluorine and proton resonance frequencies are near each other (500 and 470 MHz in an 11.7 T magnet) and rf filters are required to keep the fluorine and proton channels from mixing. The three-bond proton–fluorine coupling constants are on the order of 8–15 Hz, so longer delays for $1/2J$ are required. Shorter delay times are required for two-bond correlation experiments where the proton–fluorine coupling constants are on the order of 50 Hz.

In many cases the spectral assignments can be established using 3D NMR methods when the 2D analysis is complicated by peak overlap. In polymers containing only carbons and protons as NMR-active nuclei, the experiments are typically performed such that the experiment has two proton dimensions and one carbon dimension. In cases where there are three NMR-active nuclei it is possible to establish the assignments using 3D experiments where each dimension corresponds to a different nuclei.

Poly(1-chloro-1-fluoroethylene)

3D NMR has been used to establish the stereosequence assignments in poly(1-chloro-1-fluoroethylene) using a $^{13}C/^{19}F/^{1}H$ NMR experiment (37). The poly(1-chloro-1-fluoroethylene) main chain consists of alternating disubstituted quaternary carbons and methylene carbons, and the stereochemistry is defined by the relative orientation of the chlorine and fluorine atoms. Figure 3.44 shows the 600-MHz proton, 564-MHz fluorine, and 150-MHz carbon spectrum of poly(1-chloro-1-fluoroethylene). The fluorine spectrum shows the best resolution, with three

3.2 STEREOCHEMICAL CHARACTERIZATION OF POLYMERS   209

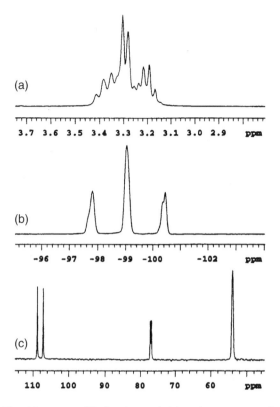

FIGURE 3.44 The (a) proton, (b) fluorine and (c) carbon spectra of poly(1-chloro-1-fluoroethylene).

well-resolved peaks with a 1:2:1 intensity ratio. No well-resolved peaks from stereosequence effects are visible from the carbon or proton spectrum. The splitting in the carbon peak at 108 ppm is from the one-bond carbon–fluorine coupling, as only proton decoupling was used during acquisition.

A modified version of an experiment used to assign the spectra of proteins (the so-called HNCA experiment (38)) is used for the $^{13}C/^{19}F/^1H$ correlation experiment, and the pulse sequence for the modified experiment is shown in Figure 3.45. The pulse sequence begins with an HSQC-type magnetization transfer from the methylene protons to the carbons. This a very efficient transfer since the one-bond J_{CH} coupling constant is on the order of 140 Hz. After the ^{13}C evolution period, magnetization is transferred from the carbons to the fluorines using an HMQC-type transfer with the delay time δ and the 90° pulses. The delay time δ is chosen as $1/2^2J_{CF}$ using a value for $^2J_{CF}$ of 40 Hz, and the magnetization evolves under the influence of the fluorine chemical shift during the t_2 period. During the remaining part of the pulse sequence the magnetization is transferred back to the carbons and finally to the protons using reverse HMQC and HSQC pulse-sequence elements. The signal is recorded during the t_3 period with GARP decoupling on both the carbon and fluorine channels.

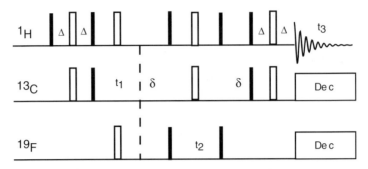

FIGURE 3.45 The pulse-sequence diagram for the 3D carbon–fluorine–proton heteronuclear correlation experiment.

The correlations are most conveniently viewed as cross sections through the 3D data matrix at the frequencies of the three resolved fluorine signals at −98.2, −99.4, and −100.7 ppm. Cross sections through the data matrix at the fluorine frequencies are planes in the carbon and proton dimensions, and show the ^{13}C–^{1}H correlations for methylene carbons near the three resolved fluorines. At least two different peaks are expected in each plane for the two methylene groups that are two bonds removed from the fluorines.

Figure 3.46 shows the planes through the 3D matrix for poly(1-chloro-1-fluoroethylene). As in other forms of heteronuclear correlation, the methylene carbons

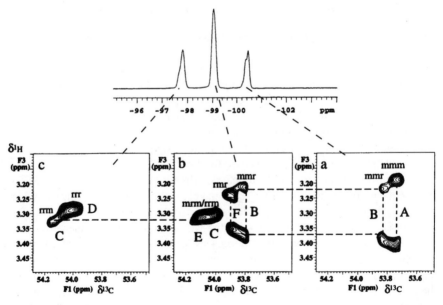

FIGURE 3.46 Cross sections through the 3D data matrix for poly(1-chloro-1-fluoroethylene) at the frequencies of the three resolved fluorine peaks at (a) −100.7 ppm, (b) −99.4 ppm and (c) −98.2 ppm.

from *m*-centered stereosequences will be correlated with two peaks in the proton dimension, since the methylene protons are nonequivalent. Two proton correlations are observed at the fluorine frequency of −98.2 ppm for each of the resolved carbon peaks, while the cross section at −99.4 ppm shows at least four correlations, two of which are correlated to doublets in the proton spectrum. The correlations at the fluorine frequency of −100.7 ppm show only singlets in the proton spectrum. From these data we can assign the fluorine peaks at −98.2, −99.4, and −100.7 ppm to the *mm*, *mr*, and *rr* triad signals. We can make further assignments from the triad–tetrad correlations, as noted above for poly(vinyl chloride). The tetrads *mmr* and *rrm* can be identified by their correlations to both the *mm* and *rr* regions. Conversely, the *mmm* and *rrr* tetrads can be identified as those peaks that do not have correlations in the *mr* region. The remaining peaks, *mrm* and *mrm*, can be identified as the remaining peaks that show one and two correlations in the proton spectrum in the cross sections of the *mr* region. Thus, using a single 3D experiment all of the signals in the proton, fluorine, and carbon spectra can all be assigned.

3.3 REGIOISOMERISM IN POLYMERS

Regioisomerism in polymers has been extensively studied, because the bulk properties of polymers are often related to the level of regio defects. NMR is a powerful tool to study regioisomerism, because these defects give rise to large chemical shift changes, and because the defects sometimes have a unique signature in the 2D NMR spectrum.

Polypropylene is an example of a commercially important polymer in which the regiodefects have been extensively studied by NMR (32). Figure 3.47 shows a schematic drawing of H–T, H–H, and T–T defects in a polypropylene chain. From the previous discussion of the factors that affect the carbon chemical shifts (Section 1.3.3), large changes are expected in the chemical shifts for the inverted monomers relative to the H–T monomers. The methine carbon in a H–T structure has three α, two β, four γ, and two δ carbon neighbors, and the methine is a tertiary carbon with two secondary neighbors, so the calculated chemical shift is given by (Section 3.1.2.3)

$$\delta_C^{H-T} = B + 3\alpha + 2\beta + 4\gamma + 2\delta + 2(3°(2°)) \quad (3.29)$$
$$= 27.1 \text{ ppm}$$

The methine carbon in a H–H defect has three α, three β, two γ, and four δ carbon neighbors and is a tertiary carbon next to a tertiary and a secondary carbon, and the

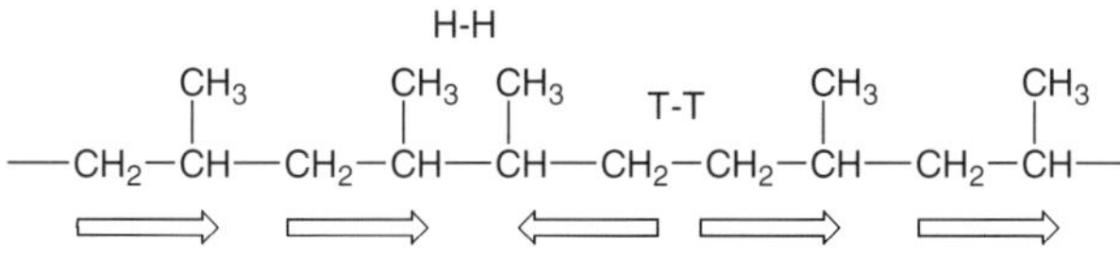

FIGURE 3.47 A schematic diagram of the structure of H–H and T–T defects in polypropylene.

TABLE 3.9 The Carbon Chemical Shifts for Polypropylene in H–H, H–T, and T–T Configurations.

	Chemical Shift (ppm)		
Carbon	H–T	H–H	T–T
CH	28.5	37.0	—
CH_2	46.0	—	13.3
CH_3	20.5	15.0	—

chemical shift is given by

$$\delta_C^{H-H} = B + 3\alpha + 3\beta + 2\gamma + 4\delta + (3°(2°)) + (3°(3°)) \quad (3.30)$$
$$= 36.3 \text{ ppm}$$

Since the resolution in the carbon spectrum is often on the order of 0.05 ppm or better, such large chemical shift differences are clearly resolved. The calculated chemical shifts are in excellent agreement with reported values for the methine chemical shifts, which are listed in Table 3.9, (28.5 and 37.0 ppm) (2). In addition, the methylene chemical and methyl chemical shift difference between the H–T and the inverted units are 14.7 and 5.5 ppm.

Large chemical shift changes are often observed in regiodefects, but the spectra, especially for saturated hydrocarbons, are often extremely complex. In vinyl polymers like poly(vinyl chloride) the methine and methylene signals are well resolved. The methine, methylene, and methyl carbons are well resolved in H–T polypropylene, but the induced chemical shifts from the regiodefects can shift the resonances such that they now overlap with the more numerous H–T structures. These spectra are often already complex from stereochemical isomerism, and the regioisomerism adds another level of complexity. In cases where the regioisomers are present at low concentrations, they can be difficult to identify and assign.

In cases where complex spectra are observed for polymers with regioisomerism, the first step in the assignment procedure is to identify the carbon types using DEPT or INEPT spectra (Section 2.4.7). This strategy has been used to assign the regiodefects in polypropylene (32). After the carbon types were assigned, the same authors used 2D carbon INADEQUATE spectra to establish many of the resonance assignments. Although the SNR is low in these experiments (because of the 1% natural abundance of carbon), many of the inverted units could be identified.

Poly(vinyl fluoride) is another commercially important polymer in which the properties are very dependent on the fraction of regiodefects. Fluorine NMR has been used

3.3 REGIOISOMERISM IN POLYMERS

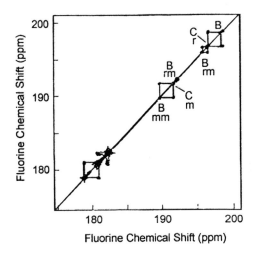

FIGURE 3.48 The fluorine COSY spectrum of poly(vinyl fluoride) containing regio defects.

to identify and assign the peaks from the defect resonances, since it has a high sensitivity and a good chemical shift dispersion (35). There are four types of peaks to be expected in the fluorine spectra, as shown by the preceding digram. Peak A and D correspond to an isolated H–T monomer and H–T monomers next to a T–T defect. Peaks B and C correspond to fluorines in H–H monomers next to H–H and T–T monomers. The peaks from the defects can be directly observed by comparing the spectra for the isoregic and regiodefective materials. The peaks from B and C appear at a higher field and show splittings from stereosequence effects. As might be expected, the H–T peak D falls near the normal H–T peak from peak A (35).

Some of the regiodefect peaks can be directly assigned from the fluorine COSY spectrum using the three- and four-bond fluorine–fluorine couplings. Figure 3.48 shows the fluorine COSY spectrum for the commercial poly(vinyl fluoride), which contains a significant level of defects (35). Six peaks are observed in the higher-field portion of the spectrum that can be assigned to the H–H sequences B and C. The large number of peaks is due to the fact that the defect peaks also show splittings from stereochemistry. The cross peaks between −190 and −193 ppm show the correlations between the C diads in the *m* configurations and the *rm* and *mm* triads, while the correlations between −195 and −200 ppm show the correlations between the C diads in the *r* configuration and the *rr* and *rm* B triads. Correlations between A and D might be expected on the basis of the stereochemical correlations observed in isoregic materials, but these were not observed. Apparently the chain conformation is such that these coupling constants are too small to be observed in COSY.

The proton COSY and NOESY spectra are also sensitive to the presence of regiodefects, since three-bond proton–proton couplings and short internuclear distances are expected between methine peaks in H–H defects and between methylene peaks in T–T defects. This is illustrated in the 2D NOESY spectrum for poly(styrene-*co*-2-phenyl-1,1-dicyanoethene) shown in Figure 3.49 (39). Note that in addition to the

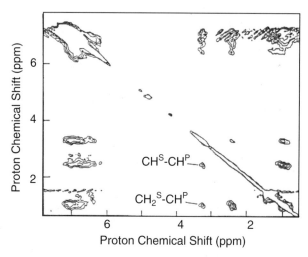

FIGURE 3.49 The 2D NOESY spectrum of poly(styrene-*co*-2-phenyl-1,1-dicyanoethene).

normal methine–methylene peaks, well-resolved methine–methine cross peaks are observed.

In summary, regioisomerism results in large changes in the 1D and nD NMR spectra of polymers. Regiodefects give rise to large chemical shift changes and correlations in the nD spectra that are not observed in the isoregic material. The spectra are often complex due to the splittings from both stereochemical isomerism and regioisomerism, and a variety of means (DEPT, *J*-resolved NMR, etc.) are used to identify the carbon type before the peaks are assigned by nD NMR. In addition to the polymers discussed previously, regioisomerism has been studied in a variety of polymers, including poly(3-hexyl thiophene) (40), poly(propylene oxide) (41), and poly(vinylidine fluoride) (36,42,43).

3.4 DEFECTS IN POLYMERS

A variety of defects, including branches and endgroups in polymers are known to affect the properties of the bulk material, even at very low levels. Because of the low levels of these defects, they are often investigated with traditional 1D NMR methods. With advances in NMR methods, including high-temperature pulse-field gradient probes and new pulse sequences, it is becoming increasingly feasible to investigate the structure of defects at relatively low levels by nD NMR (44). This leads to a better understanding of both polymer structure–property relationships and polymerization mechanisms.

3.4.1 Branching

Polymer chain branching leads to changes in the primary structure and is often accompanied by large changes in chemical shift. We noted in the Section 1.3.3 that

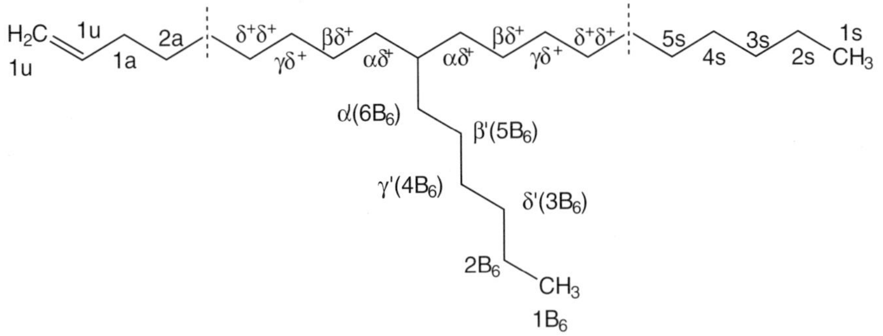

3.4 DEFECTS IN POLYMERS

FIGURE 3.50 The nomenclature for branching in polymer chains.

the carbon chemical shift is extremely sensitive to the substituents several bonds removed, and that the chemical shifts from α and β effects can be as large as 9 ppm. For simple polymers like polyethylene we expect the branched peaks to be well resolved from the main-chain signals.

Depending on the catalyst and polymerization mechanism, a number of different types of branches can be incorporated into the growing polymer chain. Many of these defects can be identified by NMR, and the nomenclature for chain branching and endgroups was introduced in Section 3.1.1.4. Figure 3.50 shows a schematic structure of a polyethylene chain with a hexyl branch and saturated and unsaturated endgroups. This drawing shows a nomenclature that is sometimes used for chains with a high density of branches. Instead of a single Greek letter to show the distance along the main chain from the nearest branch, this nomenclature shows the distance to the nearest branch in each direction. Thus, a methylene carbon between two branches would be denoted as an $\alpha\alpha$ carbon. The δ^+ symbol is used to show that a particular carbon is four or more carbons removed from a branch site.

The chemical shifts for main-chain carbons are affected both by the distance away from the branch and the branch length. This is illustrated by comparing the chemical shifts for the $\delta^+\delta^+$ (i.e., the main-chain polyethylene) with the carbons at the $\alpha\delta^+$, $\beta\delta^+$, and $\gamma\delta^+$ positions. Using the rules introduced in Section 3.1.2.3, the chemical shift for the $\delta^+\delta^+$ carbon is given by

$$\delta_C^{\delta^+\delta^+} = B + 2\alpha + 2\beta + 2\gamma + 2\delta \tag{3.31}$$
$$= 30.3 \, \text{ppm}$$

If a hexyl branch is introduced into the chain, the carbon nearest the branch experiences a large chemical shift change because the number of β, γ, and δ neighbors are changes, as well as the branching for the nearest-neighbor carbon. The chemical shift for the $\alpha\delta^+$ carbon is given by

$$\delta_C^{\alpha\delta^+} = B + 2\alpha + 3\beta + 3\gamma + 3\delta + (2°(3°)) \tag{3.32}$$
$$= 35.0 \, \text{ppm}$$

In a similar way the chemical shifts for the $\beta\gamma^+$ and $\gamma\delta^+$ carbons are given by

$$\delta_C^{\beta\delta^+} = B + 2\alpha + 2\beta + 3\gamma + 3\delta \qquad (3.33)$$
$$= 25.6\,\text{ppm}$$

and

$$\delta_C^{\gamma\delta^+} = B + 2\alpha + 2\beta + 2\gamma + 3\delta \qquad (3.34)$$
$$= 28.1\,\text{ppm}$$

Since the linewidths for polyethylene are relatively narrow ($\sim$0.05 ppm), such chemical shift changes can easily be observed.

Since the induced chemical shift depends on the number of α, β, γ, and δ neighbors, the chemical shifts will also be sensitive to the branch length. The chemical shift for a carbon α to an ethyl branch is given by

$$\delta_C^{a\delta^+} = B + 2\alpha + 3\beta + 3\gamma + 2\delta + (2°(3°)) \qquad (3.35)$$
$$= 34.7\,\text{ppm}$$

which can easily be resolved from the chemical shift position of the α carbon in the hexyl branch.

A variety of polyethylene samples have been studied by NMR, mostly in an effort to understand structure–property relationships in polyethylenes prepared with different catalysts. For the most part these studies rely on the assignments established in the earlier studies using model compounds and chemical shift calculations (45–47), and report on the number and different types of branches present in polyethylene sample. The results are often reported as the number of branches per 1000 backbone carbons, and some typical results are shown in Table 3.10 for a high-pressure polyethylene sample (45).

TABLE 3.10 The Frequency of Branch Formation in a High-Pressure Polyethylene Sample.

Branch	Branches per 1000 backbone carbons
—CH_3	0.0
—CH_2—CH_3	1.0
—CH_2—CH_2—CH_3	0.0
—CH_2—CH_2—CH_2—CH_3	9.6
—CH_2—CH_2—CH_2—CH_2—CH_3	3.6
Hexyl or longer	5.6
Total	19.8

3.4 DEFECTS IN POLYMERS

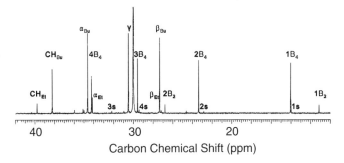

FIGURE 3.51 The 188.6-MHz carbons spectrum of poly(ethylene-*co*-hexene-*co*-butene) acquired at 120°C.

More recently, improvements in NMR spectrometers and methods have made it possible to assign branched structures using nD NMR. Figure 3.51 shows the carbon NMR spectrum of a poly(ethylene-*co*-hexene-*co*-butene) acquired at 120°C at 188.6 MHz that can be used to assign the peaks for the hexane and butene branches (44). The SNR is high enough in these spectra that HMBC studies can be performed. The HMBC pulse sequence (Figure 3.52) is a slightly modified version of the HMCQ experiment that provides the heteronuclear correlation via the 2-bond carbon–proton coupling. The main difference in the HMQC and HMBC pulse sequences is the first carbon pulse that is applied after the delay Δ_1, which is set equal to $1/2\,^1J_{CH}$, where $^1J_{CH}$ is the one-bond carbon–proton coupling constant. The effect is to remove most of the signal from the one-bond coupling so that the weaker two-bond and three-bond correlation peaks can be observed. The two-bond and three-bond couplings evolves during the Δ_2 period, which is set equal to $1/2\,^2J_{CH}$. Since the two-bond and three-bond couplings are in the range of 3–10 Hz, this delay time is quite long, and severe S/N loss can occur due to T_2 relaxation. For this reason the signal is detected directly after the t_1 period without a refocusing delay and carbon decoupling.

Figure 3.53 shows the HMBC spectrum for a polyethylene sample in which two well-resolved groups of peaks are observed (44). Few peaks are resolved at the proton chemical shift of the main-chain peaks, in part because of noise in the t_1 dimension that is associated with the extremely intense signal from the main-chain protons. Several well-resolved peaks are observed in the methyl proton region, including the methine and $2B_2$ peaks from the ethyl branches at 38.7 and 26.7 ppm. The strongest correlations to the methyl region are for the $2B_4$ and $3B_4$ peaks from the butyl branches at 23.3

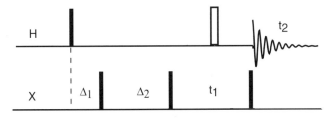

FIGURE 3.52 The pulse sequence diagram for HMBC.

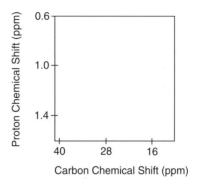

FIGURE 3.53 The HMBC spectrum of poly(ethylene-*co*-hexene-*co*-butene).

and 29.5 ppm. The authors of this study point out that suppression of the main-chain polyethylene signals is crucial for this experiment and is best accomplished using pulse-field gradients to suppress the unwanted signals.

Through the combination of model compounds, chemical shift calculations, and nD NMR, the chemical shifts for the main-chain signals and the branched carbons have been reported (48). The chemical shift assignments for the main-chain and branched carbons are listed in Table 3.11 and 3.12.

3.4.2 Endgroups

NMR can also provide information about the polymer molecular weights and the chain ends, because, like branches, the endgroups are often chemically distinct from the main-chain atoms. The endgroups are often observed as minor peaks in the spectra and assigned using model compounds, chemical shift calculations, and nD NMR. Table 3.13 shows the chemical shifts for the endgroups in polyethylene that have been assigned by using a combination of approaches (48).

The signals from endgroups are often present at low concentration in high molecular-weight polymers, but they can be easily identified if polymers with different molecular weights are available. This is illustrated in Figure 3.54, which compares the 100-MHz carbon spectra of poly(propylene glycol) with molecular weights of 400 and 4000 g/mol. If only the molecular weight in changed in this experiment, the endgroup signals can be identified as those peaks that grow in intensity as the molecular weight is decreased.

TABLE 3.11 The Chemical Shift Assignments for the Main-Chain Carbons Signals in Branched Polyethylenes.

Branch Length	C_{10}	C_8	C_6	C_4	C_2
$\alpha\delta^+$	34.653	34.646	34.639	34.626	34.145
$\beta\delta^+$	27.355	27.351	27.343	27.334	27.357
$\gamma\delta^+$	30.508	30.505	30.500	30.491	30.491
$\delta^+\delta^+$	30.00	30.00	30.00	30.00	30.00
Methine	38.293	38.286	38.283	38.243	38.790

3.4 DEFECTS IN POLYMERS

TABLE 3.12 The Chemical Shift Assignments for the Branched Carbons in Polyethylene Samples.

Branch	C_{10}	C_8	C_6	C_4	C_2
$nB_n(\alpha')$	34.612	34.615	34.623	34.226	26.797
$[n-1]B_n(\beta')$	27.326	27.325	27.286	29.573	
$[n-2]B_n(\gamma)$	30.479	30.478	30.093	23.367	
$1B_n$	14.015	14.018	14.020	14.046	11.158
$2B_n$	22.852	22.864	22.881		
$3B_n$	32.184	32.195	32.213		
$4B_n$	29.574	29.591			
$5B_n$	29.925	29.935			
$6B_n$ or $7B_n$	29.985				
$6B_n$ or $7B_n$	30.016				

In favorable cases where the molecular weight is not too high and the endgroup peaks are well resolved from the main-chain signals, the molecular weight can be calculated from the relative intensity of the main-chain signals and the endgroups. These measurements must be carefully considered, because the relaxation times for the endgroups can be very different from the main-chain signals. In poly(propylene oxide), for example, the T_1 for the endgroups were a factor of 2 longer than for the main-chain carbons (41). The value of the poly(propylene oxide) molecular weight measured by NMR in this study was 5400 g/mol, compared with the manufacturer's reported value of 4000 g/mol.

In many high molecular-weight polymers the endgroups have a very low intensity and are difficult to study with nD NMR. One strategy to assign these groups is to use the combination of model compounds and nD NMR. The idea is to synthesize low molecular-weight analogs of the endgroups in sufficient quantities that they are amenable to nD NMR. This approach has been used for the polyamide model compound just shown (49). This material is a low molecular-weight analog for both the carboxyl and amine endgroups for the polyamide, and the signals can be readily assigned from 2D NMR experiments. The chemical shifts for the endgroups in the

TABLE 3.13 The Carbon Chemical Shifts for Chain Ends in Polyethylene.

Endgroup	Chemical Shift (ppm)
$1s$	13.992
$2s$	22.846
$3s$	32.182

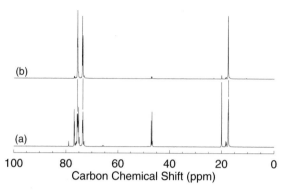

FIGURE 3.54 The 125-MHz carbon spectra of poly(propylene oxide) with molecular weights of (a) 400 g/mol and (b) 4,000 g/mol.

polymer can then be assigned by comparison to the model compound and the level of defects determined from the intensity in the 1D spectra.

In some cases the endgroups are not easily amenable to a direct NMR analysis, either because the endgroups structures do not show much chemical shift variation or because they are overlapped with the main-chain signals. In some cases it is possible to introduce NMR-active nuclei into the endgroups for the NMR analysis. This was demonstrated for the analysis of endgroups in poly(2,6-dimethyl-1, 4-phenylene oxide), where the endgroups are predominantly aromatic alcohols, which are difficult to detect as endgroup peaks separate from the main-chain signals. The endgroups can be identified from the reaction of 1,3,2-dioxaphospholanyl chloride with poly(2,6-dimethyl-1,4-phenylene oxide), as shown below (50). The phosphorus chemical shift is extremely sensitive to the structure of the endgroup and a large number of model compounds were characterized and compared with the structures in poly(2,6-dimethyl-1,4-phenylene oxide).

Tin is another NMR-active nuclei that has been incorporated into chain ends to facilitate endgroup analysis in polymers (51). Tin has a number of favorable properties, including a spin quantum number of $1/2$ and a natural abundance of 8.6%. In addition, the carbon–tin coupling constants are large over one to three bonds, so both HMQC and HMBC experiments are possible. The one-bond carbon–tin coupling constants are on the order of 250–500 Hz, while the two-bond coupling constants vary between 20 and 50 Hz. Somewhat surprisingly, the three-bond coupling constants are larger than the two-bond ones, and are in the range of 50–60 Hz.

Several endgroups structures are possible for end-labeled polybutadiene. The polymer for these studies was prepared by an anionic polymerization and reacted with tri-*n*-butyl tin chloride to give the *cis*, *trans*, and vinyl endgroups (51). The chemical

3.4 DEFECTS IN POLYMERS

shift trends are not well characterized for tin, but these structures could be assigned by correlations to the carbon and proton spectra using a 3D proton–carbon–tin NMR experiment. These structures can be distinguished by observing the proton and carbon chemical shifts from the carbons one, two, and three bonds removed from the tin. The *cis* and *trans* structures show two- and three-bond correlations only to olefinic carbons, which can be identified by their proton and carbon chemical shifts. The vinyl structure shows two- and three-bond correlations to both aliphatic and olefinic carbons.

Figure 3.55 shows the pulse-sequence diagram for the proton–carbon–tin correlation experiment (51). This is a variation of the HNCA experiment used for biomolecules, and pulse-field gradients (not shown) are used for coherence selection. The sequence begins with proton magnetization that is transferred to the carbons using an INEPT-type sequence element utilizing the one-bond carbon–proton coupling. The magnetization is then transferred from carbons to tin using another INEPT sequence element. The delay time τ is chosen to select for the one-, two-, or three-bond carbon–tin coupling. After evolution of the tin chemical shifts during the t_1 period, the magnetization is transferred to the carbons for the second evolution period and back to the protons for detection.

Four peaks are observed at -14.2, -16.3, -17, and -24 ppm in the 1D tin spectrum for end-modified polybutadiene, and the 3D data matrix is most effectively viewed as cross sections through the 3D spectrum at these four frequencies, as shown in

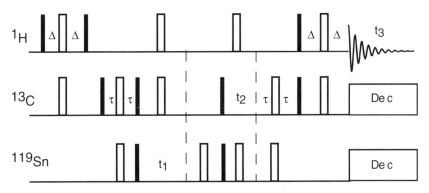

FIGURE 3.55 The pulse-sequence diagram for the carbon–proton–tin 3D heteronuclear correlation experiment.

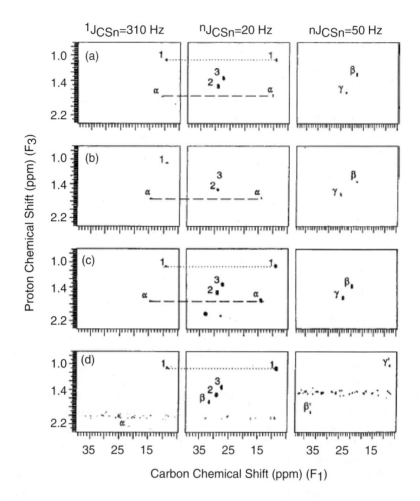

FIGURE 3.56 Cross sections through the 3D data matrix at the tin frequencies of (a) −14.2, (b) −16.3, (c) −17 and (d) −24 ppm. The cross sections are shown for three different experiments with the delay τ set for coupling constants near 310 Hz (left column), 20 Hz (middle column) and 50 Hz (right column).

Figure 3.56. At each tin frequency, the planes through the 3D data matrix show the carbon–proton correlations for those carbons coupled to the tin. The delay time τ is chosen to show the one-, two-, and three-bond correlations.

The cross sections for the one-bond coupling show peaks both for the methine carbon attachment to the polybutadiene chain and the first carbon of the *n*-propyl group. The two- and three-bond correlations show all of the peaks for the propyl groups and the α, β, and γ carbons of the endgroup structures. The *cis* and *trans* structures are distinguished by both the carbon and proton chemical shifts. The vinyl structure can be identified as the tin peak at −24 ppm, which shows two-bond correlations to both aliphatic (β) and olefinic (β′) carbons.

3.5 POLYMER CHAIN ARCHITECTURE

After the tin peaks are assigned, quantitative spectra can be acquired to measure the fraction of *cis, trans*, and vinyl groups at the chain ends. The results show that there is a higher fraction of *trans* chain ends compared to the main chain, and that the fraction of *cis* endgroups is decreased relative to the main chain, while the fraction of vinyl groups remains the same (51).

3.5 POLYMER CHAIN ARCHITECTURE

Modern synthetic methods have led to the production of polymers in a variety of chain architectures. Some of these structures include branched, grafted, crosslinked, star and hyperbranched, or dendritic polymers. Characterization of the branched, grafted, and star polymers uses the methodologies introduced to study branched defects in linear polymers and endgroups. The branch points in graft and star polymers are often chemically distinct from the main-chain polymer and show resolved peaks in the spectrum. However, as the molecular weights become large, the peaks from the branch points become small and difficult to study by nD NMR. As with branches and endgroups, the branch structures in star and graft polymers can be identified and assigned in low molecular-weight analogs.

Dendrimers are an important class of polymers that present an interesting challenge for NMR. Dendrimers can be prepared from a convergent or divergent approach, and each succeeding generation contains an increasing number of monomers (52). Each generation is unique, but many of the signals overlap, making the precise characterization of dendrimers difficult.

Figure 3.57 shows the structure for several generations of poly(propylene imine) (DAB) dendrimers. From this structure it may be expected that each succeeding generation has a different chemical environment, and hence a different chemical shift. This is observed, but the differences become difficult to observe for higher-generation dendrimers.

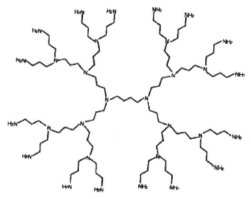

DAB-16 Dendrimer (3rd Generation)

FIGURE 3.57 The structure of the DAB-16 dendrimer.

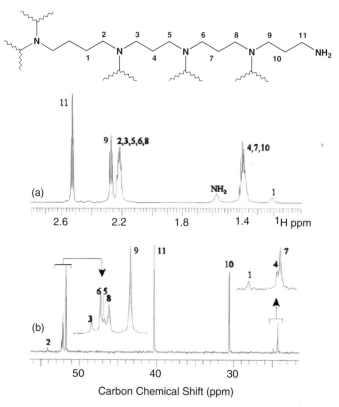

FIGURE 3.58 The (a) proton and (b) carbon NMR spectra of the DAB-16 dendrimer.

Figure 3.58 shows the carbon and proton NMR spectrum for the DAB-16 dendrimer, the third-generation dendrimer containing 16 monomer units along with the nomenclature used to describe DAB-16 (53). The assignments for the peaks were established from the expected chemical shifts and peak intensities, as well as the 3D HMQC-TOCSY spectra.

The peak intensities provide important information about the assignments in dendrimers, since the number of monomers geometrically increases with the generation number. In DAB-16, for example, the least intense peaks (1,2) are those associated with the core, while the most intense ones (9–11) are those associated with the third generation. As with most polymers the carbon spectrum is more highly resolved, and therefore more reliable for quantitative intensity measurements.

The assignments for the carbons and protons can be established using the 3D HMQC-TOCSY spectrum to correlate the carbons with both the directly bound and the nearest-neighbor protons (53). The TOCSY part of the sequence allows all of the protons in a generation to be identified, since they are scalar coupled to each other. The assignments within a generation are argued on the basis of the expected chemical shifts. The methylene next to the amine (11) in the third generation are expected at a lower chemical shift than the imine methylenes (9). The chemical shifts are less well resolved for the signals closer is to the core.

Through-space magnetization exchange (NOESY) is often valuable for establishing the assignments in polymers, but such experiments must be used with caution in dendrimers. Depending on the solvent, the arms may fold back on the core to give cross peaks between pairs of protons that are near in space, but not neighbors in the polymer chain. The NOESY data provides important information about the conformation of dendrimers.

3.6 COPOLYMER CHARACTERIZATION

NMR has long been used for copolymer characterization, because the spectra are very sensitive to copolymer sequence. In addition, it is often possible to synthesize a series of copolymers with different feed ratios of the monomers and make many of the assignments from the copolymer chain statistics. In favorable cases the NMR results provide information about any blocklike character present in the copolymer chains.

In random copolymers the fraction of monomer of each type incorporated into the polymer chain depends on the probability of adding like or unlike monomers to the growing chain. For a simple copolymer the sequence diad probabilities are given by

$$[m_1, m_1] = F_1 P_{11} \tag{3.36}$$

$$[m_1, m_2] = [m_2, m_1] = 2F_1(1 - P_{11}) \tag{3.37}$$

and

$$[m_2, m_2] = F_2 P_{22} \tag{3.38}$$

where F_1 and F_2 are overall mole fractions of monomer 1 and 2 in the chain and the P_{ij} terms correspond to the probability of adding monomer i to a chain ending with monomer j. In well-characterized systems, the reactivities are known and the mole fractions can be calculated from

$$F_1 = \frac{r_1 f_1^2 + f_1 f_2}{r_1 f_1^2 + 2 f_1 f_2 + r_2 f_2^2} \tag{3.39}$$

where the reactivity ratios r_1 and r_2 are given by

$$r_1 = \frac{k_{11}}{k_{12}} \tag{3.40}$$

and

$$r_2 = \frac{k_{22}}{k_{21}} \tag{3.41}$$

where k_{ij} is the rate constant for adding monomer i to a chain ending in monomer j.

These equations show that the statistics of copolymerization are very similar to those presented for polymer tacticity in Section 3.1.2.2. As for the tacticity, the chain statistics depend on the propagation model (Bernoullian, first-order Markov, etc.). If the polymerization statistics are known, then these equations can be used to predict the sequence distribution as a function of the monomer feed ratio. This is a useful tool for establishing the sequence assignments in copolymers.

The NMR spectra for copolymers are often complex, with many partially resolved lines. This is because in addition to the chemical shift dispersion from polymer sequence distributions, all other polymer microstructures, including tacticity and regioisomerism, are also possible. The peak intensities must be carefully evaluated because the chain statistics for copolymer sequence distribution may be different from the chain statistics giving rise to tacticity.

For the successful analysis of copolymer sequence distributions it is desirable to have at least one group of resonances from one of the monomers that is completely resolved from the other peaks in the spectrum. By integrating the resolved peak relative to the rest of the spetrum, it is possible to measure the mole fraction of each monomer incorporated into the copolymer. It is usually observed in copolymers that all of the higher-level sequence distributions are not completely resolved. In such cases, diad–triad, triad–triad, and other relationships can be used to fit the peak intensities. In an AB copolymer, for example, the triad sequence distribution probabilities can be calculated from the triad–tetrad relationships

$$P(AA) = P(BAAB) + P(BAAA + AAAB) + P(AAAA) \tag{3.42}$$

$$\begin{aligned} P(AB) = P(BBAB + BABB) + P(BBAA + AABB) \\ + P(ABAB + BABA) + P(ABAA + AABA) \end{aligned} \tag{3.43}$$

and

$$P(BB) = P(ABBA) + P(BBBA + ABBB) + P(BBBB) \tag{3.44}$$

In cases where the signals overlap, the distributions must often be calculated from groups of peaks.

The chemical shift resolution for copolymers depends on a number of factors, including the chemical shift differences between peaks in the monomers and the monomer functionalities. As with polymer tacticity, it is often observed that some signals are more sensitive to sequence distribution than are others. Fortunately, the polymer sequence distributions can be determined from one or two well-resolved peaks.

3.6 COPOLYMER CHARACTERIZATION

The resolution in the copolymer spectrum also depends on the regularity of the sequence distribution. Higher-resolution spectra are often observed for alternating and block copolymers where the magnetic environments are more uniform along the chain.

3.6.1 Random Copolymers

High-resolution solution-state NMR is a very effective means to evaluate sequence distributions in random copolymers. At the lowest level it is frequently possible to evaluate the monomer ratio from simple spectral integration even if the spectrum is heavily overlapped. In favorable cases higher-order sequence distributions and tacticity can be determined from a study of the polymer as a function of the monomer feed ratios.

Methyl Methacrylate   Methyl Acrylate

Figure 3.59 shows the carbon NMR spectra of a random copolymer prepared by free radial polymerization from a mixture of methyl methacrylate (M) and methyl acrylate (A) (54). The monomers are similar in most respects and many of the resonances overlap in the carbon spectra. The signals from the methyl methacrylate main-chain α-methyl carbons are well resolved from other peaks in the range between 16 and 22 ppm. The mole fraction of methyl methacrylate can be directly calculated by taking the integrated intensity for the α-methyl to the intensity of the carbonyl region. Alternatively, the mole fractions can be calculated by taking the ratio of the α-methyl signal intensity to the rest of the spectrum and correcting for the number of carbons in each monomer.

The α-methyl region is particularly sensitive to both sequence and stereochemistry and can be used for the analysis of these distributions. Figure 3.60 shows an expanded plot of the α-methyl region as the mole fraction of methyl methacrylate monomers is changed from 100% to 43.3%. The α-methyl region is divided into five sections (A–E). Regions A, C, and E contain the signals from the MM centered units of different stereochemistry. The assignments of these regions to the *rr*, *mr*, and *mm* peaks are known from studies of pure poly(methyl methacrylate). Regions B and D are assigned to (AMM + MMA) and AMA. These assignments are made by measuring the intensities as a function of mole fraction incorporated, and by comparison to the expected values for the reactivity ratios (2.60 and 0.27 for methyl methacrylate and methyl acrylate).

For a further analysis of the sequence distribution, it is necessary to analyze other groups of signals. The carbonyl signals are sensitive to both sequence and

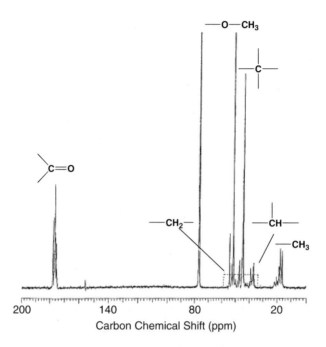

FIGURE 3.59 The carbon NMR spectrum of the methyl methacrylate/methyl acrylate copolymer containing 75 mol% methyl methacrylate.

stereochemical distributions, but they are difficult to analyze. The main-chain signals are extremely complex, but they can be analyzed using a combination of NMR methods and chemical shift calculations.

Figure 3.61 shows the DEPT spectral analysis of the main-chain spectral region for the polymer containing 43.3% methyl methacrylate monomers (54). As shown in Section 2.4.7, the DEPT method allows for the separation of subspectra for the different carbon types. Figure 3.61b shows the subspectra for the methine peaks that must arise from the methyl acrylate monomer, and Figure 3.61c shows the subspectra for the methylene signals that arise from both the methyl methacrylate and methyl acrylate monomers.

Since the methine signals from the methyl acrylate can be resolved in the DEPT spectra, they can be analyzed as a function of copolymer composition. Figure 3.62 shows the DEPT spectrum as a function of composition for the copolymer (54). Three main regions are observed that can be assigned to the AAA, AAM, and MAM copolymer sequence distribution. The AAA region is assigned based on the spectra of the homopolymer, while the assignments for the AAM and MAM signals are based on the signal intensities expected from the reactivity ratios.

The remaining main-chain signals are complex and cannot be analyzed by simple inspection. These peaks are distributed over a 20-ppm range and are sensitive to both copolymer composition and stereochemistry. The chemical shifts were assigned

3.6 COPOLYMER CHARACTERIZATION

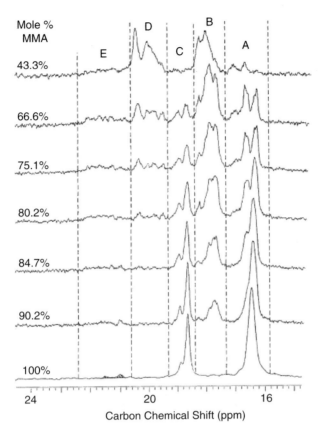

FIGURE 3.60 The carbon NMR spectra of the α-methyl region of methyl methacrylate/methyl acrylate copolymers as a function of monomer feed ratio.

in the copolymers by calculating the chemical shifts for both monomers in the ten possible tetrads (54). This gives rise to a total of 72 lines. A computer program was then used to analyze the computed chemical shifts and peak intensities as a function of composition. The copolymers with high and low acrylate content can be used to establish the assignments for some of the signals, including MMM, MMA, AAM, and AAA. The goodness of fit was evaluated by comparing eight regions of the spectra with the calculated intensities. Some of these regions are well resolved, such as the AAAA peaks from 34 ppm to 35 ppm, while others, including the region from 51.2 ppm to 55.2 ppm that contained the MMMM, AMMM, and MMMA peaks, were not. Although the assignment of peaks from chemical shift calculations and intensity variations as a function of composition are not as rigorous as those from nD NMR experiments, they are often the best that can be done on such complex materials. The analysis of many peaks ensures self-consistency and the best possible accuracy.

This example shows how we use a variety of methods for the analysis of the NMR spectra of copolymers, and the analysis is similar in many ways to the stereochemical

230   THE SOLUTION CHARACTERIZATION OF POLYMERS

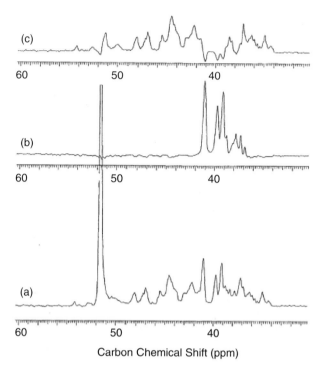

FIGURE 3.61 The DEPT spectra of the methyl methacrylate/methyl acrylate copolymer containing 43 mol% methacrylate. The spectra show (a) the methoxyl, methylene, and methyl signals, (b) the methine signals, and (c) the methylene signals.

analysis of polymers. Copolymers of ethylene and other alkanes are commercially important materials that have been extensively analyzed by NMR. In this case, the analysis is very similar to those discussed in Section 3.4.1 for branch defects in polymers. One important difference is that the comonomers in these polymers are present at much higher concentrations and many of the samples are amenable to analysis by nD NMR.

Figure 3.63 shows the carbon NMR spectrum of a 50–50 ethylene–propylene copolymer along with the nomenclature introduced in Section 3.4.1. In this case, the copolymer corresponds to an ethylene chain with a high density of methyl branches. These methyl groups are in the range of 19 ppm to 22 ppm, and the density of methyl groups is high enough that some of the minor peaks that are difficult to observe in polyethylenes with a low branching density are now easily visible. These include the methylene carbon, which is bordered on both sides by a methyl group (labeled $\alpha\alpha$) from neighboring propylene–propylene sequences. Other prominent peaks include methylene carbons that are slightly further from the methyl group such as the $\alpha\beta$ and $\beta\beta$ carbons. Also among the prominent peaks are those methylene carbons that are near only one methyl group ($\alpha\delta^+$, $\beta\delta^+$, and $\gamma\delta^+$). Since branching in ethylene carbons has been extensively studied both in polymers and model compounds, the

3.6 COPOLYMER CHARACTERIZATION

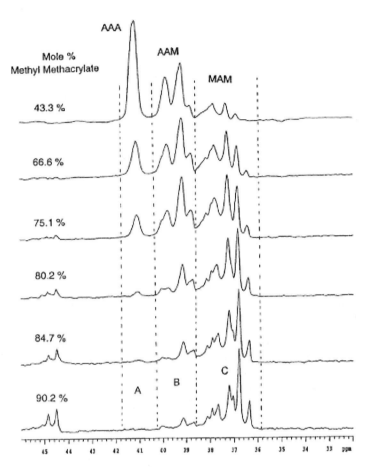

FIGURE 3.62 The DEPT methine spectra of the methyl methacrylate/methyl acrylate copolymer as a function of monomer feed ratio.

assignments for the copolymers can be made from the previously reported chemical shifts.

3.6.2 Alternating Copolymers

NMR spectroscopy can provide important information about chemical shift assignments alternating copolymers. The spectra are generally more highly resolved for alternating vs. random copolymers because the local magnetic environments are more uniform, although the spectra can still be complex due to stereochemistry or other types of isomerism.

Figure 3.64 shows the 500-MHz proton NMR spectra for random and alternating copolymers of styrene and methyl methacrylate. While it is possible to observe some features in the random copolymer, such as the aromatic signals from the polystyrene

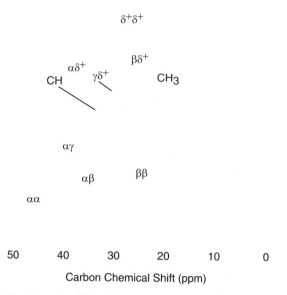

FIGURE 3.63 The carbon NMR spectrum of the ethylene–propylene copolymer.

and the methoxyl peaks from the poly(methyl methacrylate), there is extensive overlap from sequence and stereochemical isomerism. The spectrum for the alternating copolymer shows much higher resolution, and well-resolved peaks are observed for many of the signals. Almost all of the peaks in the proton spectra show multiple peaks from stereochemical isomerism.

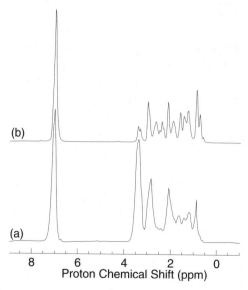

FIGURE 3.64 The 500-MHz proton NMR spectra of the alternating styrene/methyl methacrylate copolymer.

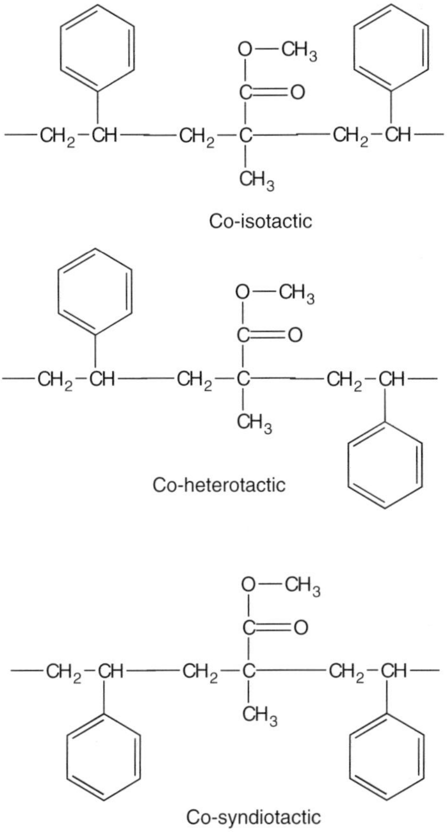

3.6 COPOLYMER CHARACTERIZATION

FIGURE 3.65 A schematic diagram showing copolymer tacticity in the styrene/methyl methacrylate copolymers.

Most of the peaks in the proton spectrum of the alternating copolymer can be assigned by nD NMR experiments and consideration of the stereochemistry between styrene and methyl methacrylate units. Figure 3.65 shows a schematic diagram of copolymer tacticity for the alternating styrene–methacrylate copolymers. Note that two of the substituents, the phenyl group of styrene and the carbonyl of methyl methacrylate, have anisotropic shieldings and are able to induce chemical shift changes for the nearby atoms. Thus it might be expected that the chemical shifts will be well resolved in the alternating copolymer. The other feature to note is that the distances between pairs of protons on the styrene and methyl methacrylate monomers depends very strongly on the tacticity. Note, for example, that the methyl methacrylate methyl protons are very close to two neighboring styrene phenyl groups in the co-syndiotactic configuration, but more distant from them in the co-isotactic configuration. In the co-heterotactic sequences the methyl group is close to one phenyl group and more distant from the other. This situation is reversed for the phenyl and methoxyl protons on the methyl methacrylate side chain. The relative distances between these and other sets of protons can be measured from NOESY experiments and provide

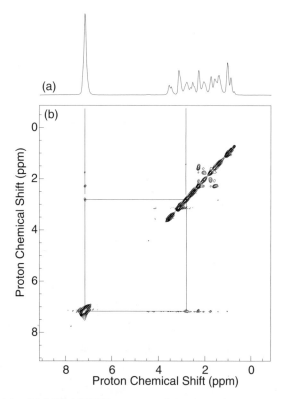

FIGURE 3.66 The 500-MHz NOESY spectrum of the alternating styrene/methyl methacrylate copolymer.

important information about the configurational assignments. The distances between the neighboring methylene protons (labeled H_e and H_t in the *meso* diads and $H_{e'}$ and $H_{t'}$ in the *racemic* ones) provides important information about the chain conformation (55).

Figure 3.66 shows the 2D NOESY spectrum for the alternating styrene–methyl methacrylate copolymer that can be used to establish the configurational assignments (56). Note the strong and medium cross peak from the phenyl protons to the methoxyl protons. The strongest peak is assigned to the co-isotactic configuration and the weaker peak to the co-hetero tactic configuration. The methoxyl peak showing no cross relaxation to the phenyl protons is assigned to the co-syndiotactic sequence. In a way the α-methyl signals can be assigned to the co-isotactic, co-heterotactic, and co-syndiotactic sequences.

3.6.3 Block Copolymers

With the exception of low molecular-weight polymers, the spectrum of block copolymers often resembles a composite spectrum of the constituent polymers. This is because the only peaks resolved from the main-chain resonances will be those within

3.7 THE SOLUTION STRUCTURE OF POLYMERS

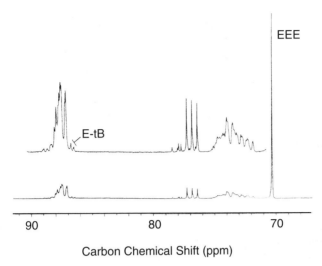

FIGURE 3.67 The carbon NMR spectrum of the poly(ethylene oxide-*b-t*-butyl ethylene oxide) block copolymer.

one or two monomer units of the junction between the blocks. Figure 3.67 shows the carbon spectrum of a of poly(ethylene oxide-*b-t*-butyl ethylene oxide) block copolymer in which some of the signals from the interface can be resolved (57). The strongest peak in the spectrum is from the ethylene oxide signals near 71 ppm. Very weak peaks from the block interface can be observed at 85 ppm.

3.7 THE SOLUTION STRUCTURE OF POLYMERS

The structure in solution has important implications for the solid-state properties of polymers. Films formed from polymer solutions will depend on whether the polymer is in a more extended or compact conformation and if there are interactions between groups on the same or different polymer chains. The hope is that controlling the structure in solution will ultimately impact on the solution properties.

The NMR chemical shifts, coupling constants, and relaxation rates depend on the local conformation of polymers, and do not in general depend on the overall chain dimensions. In some cases, however, the chain structure can be deduced from the local measures of the conformation. A polymer chain would be in an extended conformation, for example, if the measurements of local structure show that the chain is in an all-*trans* conformation.

The two measurements that are most often used to determine the local polymer chain conformation are the coupling constants between neighboring pairs of protons and the internuclear distances. As noted in Section 1.4.4, the three-bond proton–proton coupling constants depend on the torsional angle between the coupled protons. The coupling constant is at a maximum (11 Hz) for a *trans* conformation and at a minimum (2 Hz) for the *gauche* conformation. The magnitude of the coupling constants

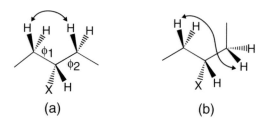

FIGURE 3.68 A schematic drawing showing the change in proton-proton distances as a function of main-chain conformation in a vinyl polymer.

can be measured from *J*-resolved nD NMR experiments and used to determine the conformation.

The distances between pairs of protons also depend on the chain conformation, as illustrated in Figure 3.68. For the main-chain protons there is a small variation of the internuclear distances in neighboring methylene protons as a function of the main-chain angles ϕ_1 and ϕ_2. These small changes in distance can be accurately measured by NOESY-type experiments because the cross-peak intensity is sensitive to the inverse sixth power of the internuclear distance. It is generally not possible to accurately determine the chain conformation from a single NOESY spectrum for a single pair of protons. However, if the interactions between several pairs of protons can be measured in NOESY experiments as a function of mixing time, then the average chain conformation can be determined.

If there are interactions between monomers that are separated in the primary sequence, distance measurements can provide important information about the solution structure. Studies of this type are of interest to determine intermolecular interactions between polymers in solution. These interactions can be identified in solution NMR experiments and can provide an understanding of why some polymer blends are miscible. The distance measurements can also be used to define the conformation of dendrimers and identify arms that are folded back on to the core.

3.7.1 Polymer Chain Conformation

The local polymer chain conformation can be measured either from the main-chain coupling constants or internuclear distances. In both cases quantitative rather than qualitative measurements are required. This requires either a measurement of the coupling constant using a *J*-resolved experiment (Section 1.7.2.5) or internuclear distances using a NOESY experiment (Section 1.7.2.4). Both of these measurements provide an average measure of the chain conformation, but the averaging is different for the *J* coupling and the distance, since they are different types of measurements.

$$—CH_2—\underset{\underset{CH_3}{|}}{CH}—O—$$

Poly(propylene oxide)

3.7 THE SOLUTION STRUCTURE OF POLYMERS

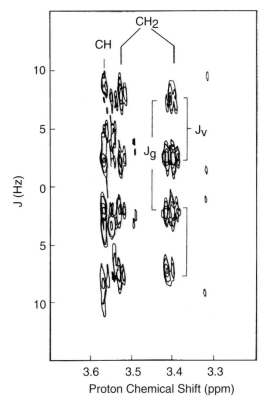

FIGURE 3.69 The 2D J-resolved NMR spectrum of poly(propylene oxide).

The coupling constants for a number of main-chain polymers have been reported. In poly(propylene oxide); for example, the coupling constants between the methine and methylene can be used as a measure of the average chain conformation. Figure 3.69 shows the 2D J-resolved spectrum for the methine and methylene protons of atactic poly(propylene oxide) (21). The analysis of the spectrum gave values for the geminal and vicinyl coupling constants of 9.9 Hz and 5.0 Hz. If we assume that the averaged vicinyl coupling is the result of averaging the *trans* (11-Hz) and *gauche* (2-Hz) coupling constants, then the measured value of 5 Hz for the methine–methylene coupling is consistent with an approximately equal weighting of the *gauche* and *trans* states. This shows that there is no large energetic difference between the *gauche* or *trans* conformations in poly(propylene oxide).

In cases where the proton spectrum is not as highly resolved as for poly(propylene oxide), higher dimensionality nD NMR experiments may be required. Figure 3.70 shows the pulse-sequence diagram for a 3D NOE–J-resolved NMR experiment for measuring the coupling constants (58). The idea behind this experiment is to spread the polymer proton magnetization out into the second dimension using the NOESY experiment before performing a J-resolved experiment. In such an experiment the cross peaks, rather than just the diagonal peaks, are modulated by the J coupling.

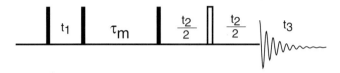

FIGURE 3.70 The pulse-sequence diagram for 3D NOE–J-resolved NMR.

This means that the coupling constant patterns can be measured from any resolved cross peak. This provides a large advantage over the 2D version of the J-resolved experiment, because there is a much greater chance that a cross peak will be resolved and the J coupling can be easily measured.

Poly(vinylidine cyanide-*alt*-vinyl acetate)

The 3D NOE–J-resolved experiment has been used to study the chain conformation in an alternating copolymer of vinylidine cyanide and vinyl acetate. The chain conformation can be measured from the methine–methylene coupling constants that are unfortunately not resolved in the 2D J-resolved spectrum. Figure 3.71 shows cross

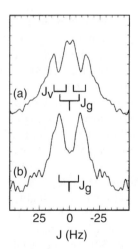

FIGURE 3.71 Cross sections through the 3D NOE–J-resolved spectrum for the alternating copolymer of vinylidine cyanide and vinyl acetate.

3.7 THE SOLUTION STRUCTURE OF POLYMERS

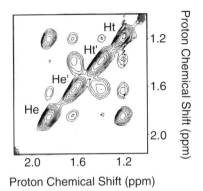

FIGURE 3.72 The NOESY spectrum of the methylene protons in the alternating styrene–methyl methacrylate copolymer.

sections through the 3D spectrum for the methylene and methine protons from which both the geminal and vicinyl coupling constants can be measured (59). The results show that the geminal coupling constant is 14 Hz and the vicinyl coupling constant is 10 Hz. This is consistent with a polymer chain conformation that is 90% *trans*.

The local polymer chain conformation can also be measured for random-coil polymers from the average distances between pairs of methylene protons along the polymer main chain. This approach has been demonstrated for the alternating copolymer of styrene and methyl methacrylate using the distances between the main-chain methylene protons (55). The methylene protons (labeled H_e and H_t in the *meso* diads and $H_{e'}$ and $H_{t'}$ in the *racemic* ones in Figure 3.67) are well resolved and the average distance between all four protons can be measured from the NOESY experiments. This is illustrated in Figure 3.72, which shows an expanded plot of the 500-MHz NOESY spectrum for the alternating copolymer with a mixing time of 0.2 s. The strongest peaks are between the geminal protons (H_e–H_t and $H_{t'}$–$H_{e'}$), which are known from the chemical structure to be separated by 1.77 Å. This is a convenient distance against which the other proton–proton distances can be measured (Equation (1.116)).

The main-chain proton–proton distances were measured from a series of NOESY experiments in which the mixing time was varied. Figure 3.73 shows the cross-peak intensities as a function of mixing time. The solid lines in Figure 3.73 were generated by measuring the peak volumes for the diagonal and cross peaks at each mixing time and solving for the relaxation rate matrix as described in Section 1.7.2.4. The conformationally sensitive relaxation rates were converted to internuclear distances using the known distance between the geminal protons, and the result was the set of distances shown in Table 3.14. The chain conformation that is most consistent with these data has a *trans* fraction of 0.53 and a 20° deviation from the normal *trans* and *gauche* states to accommodate the bulky phenyl groups (55).

In most cases the NOESY cross peaks cannot give information about the global chain conformation, because the interactions between neighboring monomers in the

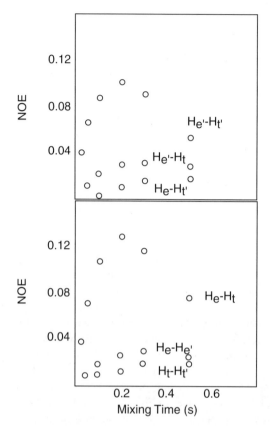

FIGURE 3.73 The NOESY cross-peak intensities as a function of mixing time for the alternating styrene–methyl methacrylate copolymer.

chain cannot be distinguished from interactions that are separated over a longer distance in the primary sequence. One important exception to this rule is for dendrimers, where the signals from higher-generation monomers can be distinguished from monomers in the first generation at the core. Figure 3.74 shows the 2D NOESY spectrum for the DAB dendrimer (see Figure 3.57 for the structure) dissolved in

TABLE 3.14 The Internuclear Distances Between the Main-Chain Methylene Protons in the Alternating Styrene–Methyl Methacrylate Copolymer.

Interaction	Distance (Å)
$H_e-H_{e'}$	2.41–2.65
$H_e-H_{t'}$	2.59–2.97
$H_{e'}-H_t$	2.32–2.40
$H_{t'}-H_t$	2.82–3.04

3.7 THE SOLUTION STRUCTURE OF POLYMERS

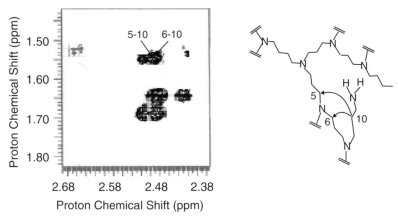

FIGURE 3.74 The 2D NOESY spectrum of the DAB dendrimer in benzene-d_6 showing interactions for a folded back conformation.

benzene-d_6 (53). Note that cross peaks are observed between the exterior protons (H_9 and H_{10}) and more interior ones (H_6 and H_5). These cross peaks can only be observed if the chain is folded back on the interior of the dendrimer. Control experiments show that the cross peaks are observed at low concentrations, demonstrating that these interactions are intra- rather than intermolecular. The folded-back conformation was not observed for the same dendrimer dissolved in chloroform.

NOESY studies are often useful for studying chain conformation in polymers with a well-defined conformation. Such well-defined conformations are often observed in biopolymers, but less frequently observed in synthetic macromolecules. NOEs can be observed in the homopolymer of amino acids and other analogs that adopt well-defined conformations. This has been quantitatively studied in poly(benzyl-*l*-glutamate), for example, where the quantitative NOESY data showed that the α-helix is preferred in 95 : 5 chloroform : trifluoroacetic acid solution (60).

In some cases, the conformation of synthetic macromolecules is determined by strong interactions between monomers along the chain. This is the case for the poly(hydroxystyrene-*r*-*N*-methyl maleimide-*r*-acetoxystryrene) copolymer. In this case, the polymer contains both hydrogen-bond donors and acceptors that form hydrogen bonds from both intra- and intermolecular interactions. Hydrogen-bond formation can be monitored because it brings the hydroxyl protons in close proximity to the methyl groups of the *N*-methyl maleimide or the acetoxy styrene. Figure 3.75 shows the 2D NOESY for a 20 wt % solution of the random terpolymer of hydroxystyrene, *N*-methyl maleimide, and acetoxy styrene (19). Strong cross peaks are observed between the stryrene hydroxy peak at 9 ppm, and the methyl group from *N*-methyl maleimide (labeled a/b) and a weaker peak is observed between the hydroxyl protons and the methyl protons from the acetoxystyrene (labeled a/c). Studies as a function of concentration showed that the hydrogen bonds to the *N*-methyl maleimide are intramolecular, while those to the acetoxystyrene are intermolecular. The strong

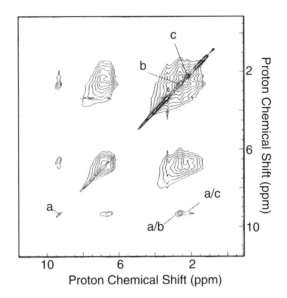

FIGURE 3.75 The 2D NOESY spectrum of the poly(hydroxystyrene-*r*-*N*-methyl maleimide-*r*-acetoxystryrene) copolymer.

intramolecular hydrogen-bonding pattern has a large effect on the solubility of the random terpolymer in base solutions.

3.7.2 Intermolecular Interactions in Polymers

The association of polymers in solution often brings pairs of protons on the different chains in close contact, and this association can be monitored by nD NMR NOESY studies. Intermolecular interactions are often expected in polymers with strongly interacting groups, such as hydrogen bonds or ionic interactions, but the interactions that promote molecular-level miscibility in other cases are not so obvious. The intensity of the intermolecular cross peaks as a function of concentration can be used as a rough measure of the strength of the intermolecular interactions.

3.7 THE SOLUTION STRUCTURE OF POLYMERS

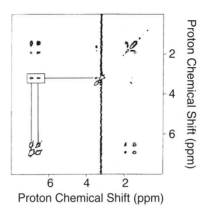

FIGURE 3.76 The 2D NOESY spectrum of the solution mixture of polystyrene and poly(vinyl methyl ether).

Intermolecular interactions in polymers were first observed for solution mixtures of poly(methyl methacrylate-co-4-vinyl pyridine) and poly(styrene-co-styrenesulfonic acid) (61), where weak cross peaks were observed between the styrene aromatic and the methoxyl peak of the methyl methacrylate. This was not observed for the polymers in the absence of strongly interacting groups, and must be a consequence of the intermolecular interactions between the 4-vinyl pyridine and the sulfonated styrene.

Since these initial observations, 2D NOESY NMR has been used to study a variety of interacting polymers. Figure 3.76 shows the 2D NOESY spectrum of a solution mixture of polystyrene and poly(vinyl methyl ether) (62). Polystyrene and poly(vinyl methyl ether) are miscible, but neither polymer contains functional groups that are expected to be strongly interacting. The results from the NOESY studies show a weak interaction between the styrene aromatic group and the methoxyl group of the poly(vinyl methyl ether). This is a very weak interaction and is only observed at very high polymer concentrations (63).

2D NOESY studies have also been used to probe the miscibility in mixtures of copolymers and silsesquioxanes that are used for low dielectric constant hybrid materials (64). In this application, the polymer phase separates from the silsesquioxanes matrix as the inorganic phase condenses, but the size of the phase-separated domains

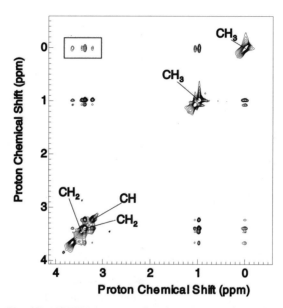

FIGURE 3.77 The 2D NOESY spectrum for the mixture of a poly(ethylene oxide-*b*-propylene oxide-*b*-ethylene oxide) triblock copolymer mixture with a methyl silsesquioxane prepolymer.

depends on the interactions between the copolymers and the methyl silsesquioxanes matrix.

Figure 3.77 shows the 2D NOESY for a poly(ethylene oxide-*b*-propylene oxide-*b*-ethylene oxide) triblock copolymer mixture with a methyl silsesquioxane prepolymer containing ~50% hydroxyl groups. Both the methyl silsesquioxane prepolymer and the triblock copolymers are low molecular-weight materials, so it is possible to obtain a solution spectrum for the neat mixture. The area enclosed in the box shows the cross peaks between the methyl protons of the silsesquioxanes at 0 ppm and the main chain methine and methylene protons of the propylene oxide and the methylene protons of the ethylene oxide. The results show that cross peaks are observed between both blocks and the methyl silsesquioxanes prepolymer. This shows that the miscibility in the polymer mixtures is a consequence of both blocks of the triblock copolymer interacting with the methyl silsesquioxanes prepolymer.

The NOESY studies can also be used to study intermolecular interactions between polymers and solvents. Cross peaks between the solvent and the polymer are usually not observed because most polymers do not strongly interact with the solvent, and most proton NMR studies use deuterated solvents. Figure 3.78 shows the 2D NOESY spectra for the DAB-16 dendrimer (see Figure 3.57 for structure) in 4:1 mixtures of either $CHCl_3/CDCl_3$ or C_6H_6/C_6D_6. For the chloroform experiment there are clear cross peaks between the solvent and several peaks (but not the core) from the dendrimer, while these peaks are missing from the experiment in benzene. A long mixing time (2 s) was used for these experiments because the solvent relaxation is

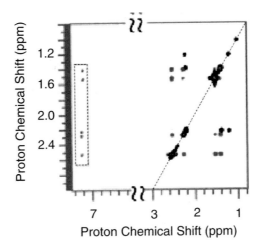

FIGURE 3.78 The 2D NOESY spectrum of the DAB-16 dendrimer in a 4:1 mixture of protonated and deuterated chloroform. The box shows the polymer-solvent cross peaks.

relatively inefficient. These results show that the chloroform penetrates the dendrimer more efficiently than the benzene and shows that the conformation is fundamentally different in the two solvents.

REFERENCES

1. Tonelli, A. E.; Schilling, F. C.; Starnes, W. H.; Shepherd, L.; Plitz, I. M. *Macromolecules*, 1979, *12*, 78.
2. Tonelli, A. *NMR Spectroscopy and Polymer Microstructure: The Conformational Connection*, VCH Publishers, New York, 1989.
3. Bovey, F. *Chain Structure and Conformation of Macromolecules*, Academic Press, New York, 1982.
4. Zambelli, A.; Locatelli, P.; Bajo, G.; Bovey, F. A. *Macromolecules*, 1975, *8*, 687.
5. Zambelli, A.; Locatelli, P.; Bajo, G. *Macromolecules*, 1979, *12*, 154.
6. Stehling, F. C.; Knox, J. R. *Macromolecules*, 1975, *8*, 595.
7. Dworak, A.; Freeman, W. J.; Harwood, H. J. *Polym. J.*, 1985, *17*, 351.
8. Cais, R. E.; Kometani, J. M. In *NMR and Macromolecules: Sequence, Dynamic and Domain Structure*, Vol. 247, Randal, J. C., ed., American Chemical Society, Washington, DC, 1983, 153 pp.
9. Bovey, F. A. *High Resolution NMR of Macromolecules*, Academic Press, New York, 1972.
10. Bovey, F. A.; Mirau, P. A. *NMR of Polymers*; Academic Press, New York, 1996.
11. Grant, D. M.; Paul, E. G. *J. Am. Chem. Soc.*, 1964, *86*, 2984.
12. Grant, D. M.; Cheney, B. V. *J. Am. Chem. Soc.*, 1967, *39*, 5315.
13. Cheng, H. N.; Bennett, M. A. *Makromol. Chem.*, 1987, *188*, 135.

14. Cheng, H. N.; Bennett, M. A. *Makromol. Chem.*, 1987, *188*, 2665.
15. Breitmaier, E.; Voelter, W. *Carbon-13 NMR Spectroscopy*, 3rd ed, VCH Publishers, Weinheim, Germany, 1987.
16. Flory, P. J. *Statistical Mechanics of Chain Molecules*, Wiley-Interscience, New York, 1969.
17. Bendall, M. R.; Doddrell, D. M.; Pegg, D. T. *J. Am. Chem. Soc.*, 1981, *103*, 4603.
18. Saito, T.; Rinaldi, P. L. *J. Magn. Reson.*, 1998, *132*, 41.
19. Heffner, S. A.; Galvin-Donoghue, M. E.; Reichmanis, E.; Gerena, L.; Mirau, P. A. In *Microelectronics Technology: Polymers for Advanced Imaging and Packaging*, Vol. 614, Reichmanis, E. R.; Ober, C. K.; MacDonald, S. A.; Iwayanagi, T.; Nishikubo, T., eds., American Chemical Society, Washington, DC, 1994, 166 pp.
20. Kogler, G.; Mirau, P. *Macromolecules*, 1992, *25*, 598.
21. Bruch, M.; Bovey, F.; Cais, R.; Noggle, J. *Macromolecules*, 1985, *18*, 1253.
22. Tokles, M.; Keifer, P. A.; Rinaldi, P. L. *Macromolecules*, 1995, *28*, 3944.
23. Heatley, F. *Prog. NMR Spectrosc.*, 1979, *13*, 47.
24. Schilling, F. C.; Bovey, F. A.; Zeigler, J. *Macromolecles*, 1986, *19*, 2309.
25. Frisch, H. L.; Mallows, C. L.; Heatley, F.; Bovey, F. A. *Macromolecules*, 1968, *1*, 533.
26. Suter, U. W.; Flory, P. J. *Macromolecules*, 1975, *8*, 765.
27. Schilling, F. C.; Tonelli, A. E. *Macromolecules*, 1980, *13*, 270.
28. Williams, A. D.; Flory, P. J. *J. Am. Chem. Soc.*, 1969, *91*, 3118.
29. Ando, I.; Kato, Y.; Nishioka, A. *Makromol. Chem.*, 1976, *177*, 2759.
30. Moad, G.; Rizzardo, E.; Solomon, D. H.; Johns, S. R.; Willing, R. I. *Macromolecules*, 1986, *19*, 2494.
31. Bax, A.; Freeman, R.; Frenkiel, T. *J. Am. Chem. Soc.*, 1981, *103*, 2102.
32. Asakura, T.; Nakayama, N.; Demura, M.; Asano, A. *Macromolecules*, 1992, 25, 4876.
33. Asakura, T.; Nakayama, N. *Polym. Commun.*, 1992, *33*, 650.
34. Cavanagh, J.; Rance, M. *J. Magn. Reson.*, 1992, *96*, 670.
35. Bruch, M.; Bovey, F.; Cais, R. *Macromolecules*, 1984, *17*, 2547.
36. Tonelli, A. E.; Schilling, F. C.; Cais, R. E. *Macromolecules*, 1982, *15*, 849.
37. Li, L.; Rinaldi, P. L. *Macromolecles*, 1996, *29*, 4808.
38. Kay, L. E.; Ikura, M.; Tschudin, R.; Bax, A. *J. Magn. Reson.*, 1990, *89*, 496.
39. Kharas, G. B.; Mirau, P. A.; Watson, K.; Harwood, H. J. *Polym. Int.*, 1992, *28*, 67.
40. Chen, T.-A.; Rieke, R. D. *J. Am. Chem. Soc.*, 1992, *114*, 10087.
41. Schilling, F. C.; Tonelli, A. E. *Macromolecules*, 1986, *19*, 1337.
42. Cais, R. E.; Kometani, J. M. *Macromolecules*, 1985, *18*, 1354.
43. Tonelli, A. E.; Schilling, F. C.; Cais, R. E., *Macromolecules*, 1981, *14*, 560.
44. Liu, W.; Ray, D. G.; Rinaldi, P. L.; Zens, T. *J. Magn. Reson.*, 1999, *140*, 482.
45. Bovey, F.; Jelinski, L. *J. Phys. Chem.*, 1985, *89*, 571.
46. De Pooter, M.; Smith, P. B.; Dohrer, K. K.; Bennett, K. F.; Meadows, D. D.; Smith, C. G.; Schouwenaars, H. P.; Geerards, R. A. *J. Appl. Polym. Sci.*, 1991, *42*, 399.
47. Randall, J. C. *Macromol. Sci., Rev. Macromol. Chem. Phys.*, 1989, *C29*, 201.
48. Liu, W.; Ray, D. G.; Rinaldi, P. L. *Macromolecules*, 1999, *32*, 3817.
49. de Vries, K.; Linssen, H.; Velden, G. V. D. *Macromolecules*, 1989, *22*, 1607.

REFERENCES

50. Chan, K. P.; Argyropoulos, D. S.; White, D. M.; Yeager, G. W.; Hay, A. S. *Macromolecules*, 1994, *27*, 6371.
51. Liu, W.; Saito, T.; Li, L.; Rinaldi, P. L.; Hirst, R.; Halasa, A. F.; Visintainer, J. *Macromolecles*, 2000, *33*, 2364.
52. Tomalia, D. A.; Naylor, A. M.; Goddard, W. A. I. *Angew. Chem., Int. Ed. Engl.*, 1990, *29*, 138.
53. Chai, M.; Niu, Y.; Youngs, W. J.; Rinaldi, P. L. *J. Am. Chem. Soc.*, 2001, *123*, 4670.
54. Kim, Y.; Harwood, H. J. *Polymer*, 2002, *43*, 3229.
55. Mirau, P.; Bovey, F.; Tonelli, A.; Heffner, S. *Macromolecules*, 1987, *20*, 1701.
56. Heffner, S.; Bovey, F.; Verge, L.; Mirau, P.; Tonneli, A. *Macromolecules*, 1986, *19*, 1628.
57. Heatley, F.; Yu, G. E.; Lawrance, J.; Booth, C. *Eur. Polym. J.*, 1994, *11*, 1249.
58. Mirau, P.; Heffner, S.; Bovey, F. *J. Magn. Reson.*, 1990, *89*, 572.
59. Mirau, P.; Heffner, S.; Bovey, F. *Macromolecules*, 1990, *23*, 4482.
60. Mirau, P.; Bovey, F. *J. Am. Chem. Soc.*, 1986, *108*, 5130.
61. Natansohn, A.; Eisenberg, A. *Macromolecules*, 1987, *20*, 323.
62. Mirau, P.; Tanaka, H.; Bovey, F. *Macromolecules*, 1988, *21*, 2929.
63. Mirau, P.; Bovey, F. *Macromolecules*, 1990, *23*.
64. Yang, S.; Mirau, P. A.; Pai, C.; Nalamasa, O.; Reichmanis, E.; Lin, E. K.; Lee, H.-J.; Gidley, D. W.; Sun, J. *Chem. Mater.* 2001, *13*, 2762.

4

THE SOLID-STATE NMR OF POLYMERS

4.1 INTRODUCTION

Solid-state NMR has emerged as an important method for polymer characterization because most polymers are used in the solid state, and these NMR methods provide the link between the chemical structure and microstructure of polymers and their mechanical, electrical, and optical properties. In addition, the chains are not mobile enough to average the local interactions, so the chemical shifts can provide important conformational information, and the proton relaxation rates can give information about the morphology. Solid-state NMR can also be used to monitor the reactivity of polymers.

Solid-state NMR developed more slowly than solution NMR, in part because the lines are broader in solids and it is more difficult to observe a high-resolution spectrum. In the carbon spectrum, for example, the lines are broadened by the combination of chemical shift anisotropy and dipolar broadening. The width of the chemical shift anisotropy depends on the carbon type, but it can be as large as several kHz. For a polymer with several carbon types, the chemical shift anisotrophy lineshapes would all overlap, resulting in a broad, featureless spectrum. The dipolar broadening can be as large as 50 kHz, which is much larger than the chemical shift range. The broad-line spectra can be used to study the chain dynamics, but they cannot provide high-resolution structural information.

In order to obtain molecular-level information about the structure of polymers in the solid state, it is necessary to average the local interactions (chemical shift anisotropy, dipolar coupling) to obtain a high-resolution spectrum. This can be accomplished with the combination of magic-angle sample spinning (Section 2.5.1) to average the chemical shift anisotropy and high-power decoupling (Section 2.5.3) to remove the

A Practical Guide to Understanding the NMR of Polymers, by Peter A. Mirau
ISBN 0-471-37123-8 Copyright © 2005 John Wiley & Sons, Inc.

4.1 INTRODUCTION

dipolar broadening. High-resolution spectra can also be observed when the spinning rate is fast compared to the breadth of the chemical shift anisotropy lineshape and the dipolar broadening.

Since the molecular dynamics of polymers are much slower in solids compared to solutions, large differences in relaxation times are often observed for solid polymers and polymers in solution. For carbon, nitrogen, and silicon atoms this frequently leads to extremely long relaxation times. This lowers the SNRs, since long waiting times are required between scans. It is often possible to overcome this limitation by using cross polarization between the protons and the nuclei of interest (Section 2.5.2), since the time between acquisitions depends on the proton relaxation times, which are typically only a few seconds. There is also a sensitivity enhancement for cross polarization, since the experiment begins with proton magnetization, which has the highest sensitivity. One of the limitations of cross polarization is that it is generally not quantitative, since the rate of cross polarization may be different for different parts of the molecule.

Another difficulty in interpreting the NMR spectra of solids is that the samples are much more likely to be heterogeneous. In semicrystalline polymers, for example, there may be crystalline, amorphous, and interfacial polymer. There are often only small differences in chemical shift for the different polymer morphologies, and some means other than the chemical shift must be used to distinguish between them. The different morphologies can often be distinguished on the basis of differences in their molecular dynamics. The rate of cross polarization, for example, depends on the strength of the dipolar coupling. If we consider a carbon with directly bonded protons, it will cross polarize much faster in the crystalline domains relative to the amorphous or rubbery domains. We can acquire a spectrum of the crystalline signals by using a short cross-polarization time. A spectrum with a longer cross-polarization time will have signals mainly from the more mobile parts of the sample. It is usually not possible to cleanly separate the signals from different morphologies simply by choosing the cross-polarization time.

The differences in relaxation times can in favorable cases be used to separate signals from different morphologies. As a general rule, the more rigid phase has a longer relaxation time than the more mobile phase. In many cases, the spin-lattice relaxation times are so long for the crystalline phase that very long delay times between scans are required to obtain a spectrum. In such cases, it is possible to obtain the spectrum of the more mobile phase by acquiring the spectrum with a short relaxation time using a normal one-pulse (or direct polarization) experiment.

The high-resolution spectrum of the crystalline fraction obtained using cross polarization and magic-angle sample spinning contains important information about the polymer conformation, because the chemical shifts are not averaged as they are in solution. We noted in Section 3.1.2.4 that the carbon chemical shifts are strongly influenced by the γ-*gauche* effect. In solution, the γ-*gauche* effect leads to induced chemical shifts that are less than the maximum because of conformational averaging. This conformational averaging is not observed in crystals, so the chemical shifts for solid-state samples can be very different from their solution counterparts. In some crystal structures, such as for syndiotactic polypropylene, the same carbon has two different magnetic environments in the crystal structure, and two separate peaks are observed in the spectrum with cross polarization with magic-angle spinning (1). The spectrum is very different in the isotactic polypropylene structure where only a single peak for the methylene carbons is observed.

The lack of chemical shift averaging and the uniform magnetic environment in the crystalline phase of polymers often lead sharp lines for crystalline polymers. The lines are typically much broader in amorphous polymers because the polymer chains can adopt a variety of conformations. Since molecular motion is slow below the glass transition temperature, there is also no averaging of the chemical shifts, and each conformation may have its own chemical shift. This leads to inhomogeneously broadenend lines from the distribution of carbon chemical shifts. The linewidths are not reduced with faster spinning, better decoupling, or even running the experiments at higher magnetic fields.

The spectral resolution for polymers above their glass transition (T_g) temperatures is often much better than below the glass transition temperatures. Above T_g there is sufficient molecular motion to average the chemical shifts and sharper lines are observed, and the spin-lattice and spin-spin relaxation times are greatly reduced. In many cases, the motion is such that the dipolar couplings are greatly reduced and it becomes feasible to observe the proton spectrum with high resolution. This is an advantage because the proton spectrum can be detected with a high sensitivity.

In addition to the chain conformation, NMR experiments can be used to probe the structure of polymers on longer length scales, mostly through the exchange of proton magnetization. The strategy behind such experiments is to apply a series of pulses to create a nonequilibrium distribution of magnetization, and to monitor the spin system as it returns to equilibrium. Differences in the chemical shifts and dynamics of complex polymer systems make it possible to selectively excite or saturate one component of a multicomponent system. The rate at which magnetization is exchanged between the components depends on the physical properties of the components and the length scale of phase separation. Such NMR experiments can be used to measure phase separation on the 1–20-nm length scale.

nD NMR is also used to study the structure, conformation, and dynamics of polymers in the solid state. The experiments are different than those for solution in that dipolar couplings and spin exchange are primarily used for magnetization transfer rather than the through-bond J couplings used in solutions. These experiments are used more frequently to measure the conformation and morphology rather than to establish the chemical shift assignments. Another general feature of the solid-state nD NMR experiments is that the resolution is much lower than for the solution spectra, particularly in the proton dimension.

4.2 CHAIN CONFORMATION IN POLYMERS

The NMR spectra of solids provide important information about the chain conformation of polymers, since the chemical shifts depend both on the structure and conformation. As for polymers in solution, the chemical shifts are determined by the local

4.2 CHAIN CONFORMATION IN POLYMERS

magnetic environment, which is due, in part, to the chemical structure. Unlike the solution spectra, the chains are restricted and the chemical shifts are not averaged. This means that the carbon chemical shifts can be used to determine the solid-state structure.

Since the chemical shift depends on the conformation, we can often use solid-state NMR to distinguish between polymers in the crystalline and amorphous morphologies. This gives us the ability to characterize complex materials, such as semicrystalline polymers that may contain crystalline, amorphous, and interfacial phases. However, the structure cannot usually be directly determined using the chemical shifts alone, since there is no direct, unambiguous relationship between the chemical shift and the conformation. In isotopically labeled polymers it is possible to determine the conformation by measuring the rate and angular dependence of magnetization exchange between isotopically labeled sites.

As with solution NMR, changes in chemical structure can often be directly observed in the NMR spectrum. This makes solid-state NMR a valuable tool to study reactivity and curing in polymers. In some cases, the final product is not soluble, and NMR is one of the few methods that can be used to follow the cure kinetics.

4.2.1 Semicrystalline Polymers

Semicrystalline polymers have been extensively studied by solid-state NMR methods, since the spectrum and relaxation times are very sensitive to the morphology. Most polymers are only semicrystalline and contain amorphous and interfacial material in addition to the crystallites. While it is not usually possible to characterize the material from a single spectrum, a variety of methods can be used to identify and quantify the various morphologies.

In order to separate the signals from the crystalline, amorphous, and interfacial phases in semicrystalline polymers, it is often necessary to take advantage of differences in the molecular dynamics and relaxation times in the phases. In carbon–proton cross-polarization experiments, for example, the signal intensity builds up as a consequence of strong carbon–proton dipolar interactions and decays away from proton $T_{1\rho}$ relaxation. This was illustrated in Section 2.5.2 where Figure 2.11 showed the simulated buildup and decay of magnetization for the (rigid) crystalline and (mobile) amorphous phases. A spectrum enriched in the crystalline phase can be recorded with a short cross-polarization contact time, while a predominantly amorphous phase spectrum can be recorded with a long contact time.

The effect of chain conformation on the solid state NMR spectrum is illustrated in Figure 4.1, which shows the carbon spectrum of polyethylene acquired in a cross polarization (CP) and single-pulse (or direct-polarization (DP)) experiments. In both cases, magic-angle sample spinning and high-powered decoupling during acquisition was used for line narrowing. The largest peak in the cross-polarized spectrum is a sharp resonance at 33.6 ppm that is assigned to the crystalline polyethylene in the all-*trans* conformation, and the broad shoulder located at 2.5 ppm to higher field is assigned to the amorphous material. Based on the cross-polarized spectrum, the sample appears to be predominantly crystalline. A very different result is observed

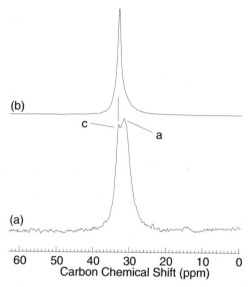

FIGURE 4.1 The solid-state carbon spectra of polyethylene acquired (a) without and (b) with cross polarization.

for the direct polarization spectrum, where the amorphous accounts for more than half of the spectral intensity. This very simple comparison of the peak intensities in solid-state spectra shows that care must be taken in the analysis of the spectra.

It is instructive to consider why the polyethylene gives such different spectra in the cross-polarization and direct-polarization experiments. The cross-polarized spectrum is not quantitative, because the signal intensities build up (from T_{CH}) and decay (from the proton $T_{1\rho}$) in the crystalline and amorphous phases at different rates. The dipolar couplings are stronger in the crystalline phase, so the intensity from the crystalline phase is greater with short cross-polarization times. The direct polarization experiment does not give an accurate measure of the crystalline and amorphous phases because the spin-lattice relaxation time for the crystalline phase of polyethylene can be as long as 5000 s, and the spectrum was acquired with only a 10-s delay time between scans.

While it usually is not feasible to accurately measure the crystallinity from a single direct polarization or cross polarization spectrum, these two methods can be combined with some other simple methods to accurately measure the crystallinity. This is illustrated in Figure 4.2, which shows several spectra acquired for an ultradrawn polyethylene sample (2). The strategy for measuring the crystallinity is to quantitatively measure the carbon spectrum of the amorphous phase from a direct polarization spectrum with a delay time that is much longer than the amorphous T_1. The amorphous phase may be present at a low concentration, and many scans are averaged. This is illustrated in Figure 4.2a and 4.2b, which compares the spectra from 512 accumulations with a delay between experiments of 20 s and the spectrum from 2048 scans with a 50-s delay time. Since the SNR depends on the square root of the number of scans (Equation 1.25), the spectra with a different number of acquisition and delay times can be scaled and compared. The intensities for the amorphous phases

4.2 CHAIN CONFORMATION IN POLYMERS

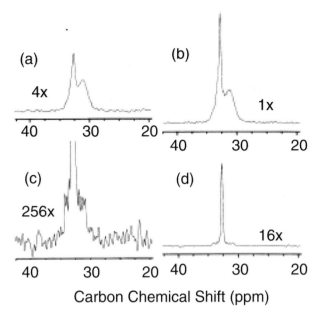

FIGURE 4.2 The direct polarization spectra of polyethylene showing how the crystallinity can be measured. The spectra were acquired using (a) a delay of 20 s and 512 scans, (b) a delay of 50 s and 2048 scans, (c) a delay of 10,000 s and 8 scans. Spectrum (d) shows the data from (c) replotted. The relative gain settings are also shown.

are similar, showing that the delay times of 20 and 50 s are sufficient for the complete spin-lattice relaxation of the amorphous phase.

The absolute intensity of the crystalline phase is measured using a small number of scans with an extremely long delay time. Figure 4.2c shows the spectrum acquired using eight scans and a delay time between scans of 10,000 s (total acquisition time: 22.2 h) using the same adjusted vertical gain as for the amorphous spectra shown earlier. The spectrum of the crystalline phase with a lower vertical gain is shown in Figure 4.2d (2).

As noted in Section 2.4.5, it is necessary to wait five times longer than the longest T_1 to have a quantitative spectrum. Although the delay times for the crystalline spectrum in Figure 4.2c are very long, they still may not fulfill this criterion. For this reason, it is necessary to measure the T_1 for the crystalline fraction and scale the spectrum accordingly.

The long T_1 for some polymers in the solid-state make it difficult to measure the relaxation times using the inversion-recovery pulse sequence introduced in Section 2.6.1.1. Progressive saturation is sometimes used to measure the polymer T_1, but this does not have the highest sensitivity. The T_1 are most efficiently measured using the cross polarization-based experiment (CPT1) shown in Figure 2.26 (3).

The CPT1 pulse sequence begins with cross polarization, and thus has the higher sensitivity associated with cross-polarization experiments. Instead of directly observing the signals after the cross-polarization period, a 90° pulse is applied on the carbon channel. If the phase of this pulse is along the x axis, then the magnetization spin-locked along the y axis is rotated to the $-z$ position, as in the inversion-recovery experiment. After the relaxation delay time τ the magnetization is sampled with another 90° pulse on the carbon channel.

One potential problem with this pulse sequence is that it begins with sensitivity-enhanced (or nonequilibrium) magnetization and relaxes toward equilibrium. To remove this potential complication, the CPT1 experiment is performed as a difference experiment. On every other scan the phase of the first 90° carbon pulse is cycled between $+x$ and $-x$, and the resulting signals are subtracted from each other. The maximum signal is observed at short τ values (no relaxation) and the difference signal decays to zero. For a single exponential relaxation process, the magnetization as a function of τ is given by

$$M(\tau) = M_0 e^{-\tau/T_1} \qquad (4.1)$$

In an attempt to determine if the crystalline ethylene signals are completely relaxed in Figure 4.2.c, the T_1 were measured using the CPT1 experiment. The results show that the delay time of 10,000 s is sufficient to relax 90% of the signal. The intensity is therefore increased by 10% to account for incomplete relaxation during the 10,000 s delay between scans. Using this correction, the data in Figure 4.2 can be combined to show that the ultradrawn polyethylene sample has a crystallinity of $87.5 \pm 1\%$ (2).

To characterize the crystalline and amorphous fractions it is often desirable to obtain the pure lineshape of the phase under study. Several methods can be used for this purpose, which generally distinguish the phases based on some differences in relaxation times. For polyethylene, for example, there is a large difference in the spin-lattice relaxation times for the amorphous and crystalline phases. In the CPT1 experiment, the spectra with short delay times have contributions from both phases. However, when the delay time is much longer than the amorphous T_1, the amorphous signals have decayed to zero and only the spectrum of the crystalline phase is observed. This experiment allows for the precise measurement of the chemical shifts and lineshapes for the crystalline phase.

A number of methods can be used to obtain the lineshape for the amorphous fraction. The spectrum obtained with a very short delay time and a large number of scans is often representative of the amorphous fraction. Another approach is to use the so-called dipolar dephasing experiment shown in Figure 4.3 (4). The key part of

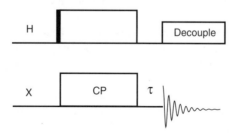

FIGURE 4.3 The pulse-sequence diagram for dipolar dephasing.

4.2 CHAIN CONFORMATION IN POLYMERS

this pulse sequence is the delay time following the cross-polarization period, during which the decoupling is turned off. The signals decay from spin-spin relaxation during this period, and the signals with the strongest dipolar couplings decay most quickly. This is a general means for editing the spectra to remove the signals from the most rigid component. The dipolar couplings are very strong for polyethylene, and a 16-μs delay is sufficient for all of the crystalline signals to decay (2). In both of these examples, it is the relative relaxation rates that are important. These experiments yield the lineshapes for the phases, but the peak intensities are not quantitative.

The chemical shifts for the crystalline and amorphous fractions of polyethylene are in general agreement with those calculated from the γ-*gauche* effect (5). In polyethylene melts and solutions, about 40% of the bonds are expected to be in the *gauche* conformation, so an upfield shift of about 5 ppm is expected for the amorphous phase. While this prediction is in the right direction, the magnitude of the shift is considerably smaller than expected. The discrepancy between the calculated and observed chemical shifts may be due to a smaller fraction of chain in the *gauche* conformation, or other factors, such as intermolecular interactions, that may also affect the chemical shifts in the solid state. It has been reported that the chemical shifts for methylene carbons in *n*-alkanes can vary as much as 1.3 ppm even though they have an all-*trans* conformation (6). In any case, while the chemical shifts can be very informative, they are not sufficient to determine the chain conformation of polymers in the solid state.

The effect of chain conformation on the solid-state NMR spectra of polymers is perhaps most dramatically shown by the comparison of isotactic and syndiotactic polypropylene shown in Figure 4.4 (1). Both polymers are crystalline, but the isotactic polymer adopts a ... *gtgtgt* ... 3_1 helical conformation, while the syndiotactic polymer

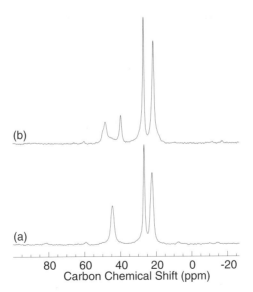

FIGURE 4.4 The carbon spectra of (a) syndiotactic and (b) isotactic polypropylene acquired with cross polarization and magic-angle sample spinning.

forms a 2_1 helix with a ... *ggttggtt* ... conformation. This difference in conformation gives rise to a large difference in the magnetic environments for the methylene carbons in the isotactic and syndiotactic structures. In syndiotactic polypropylene half of the methylene groups lie along the interior of the helix and are in a *gauche* arrangement with both of their γ neighbors, while half are on the exterior of the helix and are *trans* to their γ neighbors. In isotactic polypropylene, the methylene groups have an equivalent environment: they are *trans* to one γ neighbor and *gauche* to another. For this reason, a single resonance is observed for the methylene groups in the isotactic polymer, while two resonances separated by 8.7 ppm are observed for the syndiotactic material (5). This difference in chemical shift for the methylene carbons in the syndiotactic polymer is approximately as large as two γ-*gauche* effects. The methylene resonance for isotactic polypropylene appears midway between the two peaks in syndiotactic polypropylene, as expected for a methylene group that has one γ-*gauche* interaction.

The high resolution observed for the crystalline fraction in semicrystalline polymers makes solid-state NMR an excellent method for the evaluation of polymer conformation. This is illustrated in Figure 4.5, which shows the spectrum for poly(diethyl oxetane) crystallized at three different temperatures (7). Poly(diethyl oxetane) is known to crystallize in multiple conformations. Form I is obtained at high temperature and is associated with an all-*trans* conformation. Form II is characterized by a ... *ttggttgg* ... conformation and is observed at lower temperature. The NMR spectra show that large chemical shift differences are observed between the Form I and Form II conformations. The chemical shifts for the quaternary and methyl carbons are close to those calculated from the γ-*gauche* effect in both forms, but other carbons

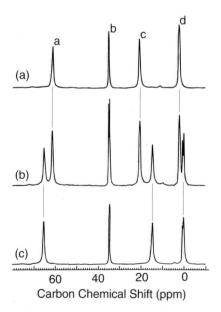

FIGURE 4.5 The cross-polarization magic-angle spinning spectra of poly(diethyl oxetane) crystallized at (a) 60°C, (b) 35°C, and (c) 0°C.

4.2 CHAIN CONFORMATION IN POLYMERS

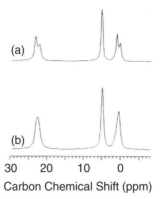

FIGURE 4.6 The cross-polarization magic-angle spinning spectra of the (a) α and (b) β forms of polypropylene.

show significant deviations. This shows that factors other than chain conformation can affect the chemical shifts.

$$\left[CH_2-\underset{\underset{CH_2-CH_3}{|}}{\overset{\overset{CH_2-CH_3}{|}}{C}}-CH_2-O \right]$$

Poly(diethyl oxetane)

Thus far we have emphasized the effect of chain conformation on the chemical shifts for the crystalline phase of polymers. While these are often the largest effects, other factors can also affect the carbon chemical shifts. This is illustrated in Figure 4.6, which compares the carbon spectra for the α and β crystalline forms of polypropylene (8). Both crystalline forms adopt a ... *tgtgtg*... conformation, so differences in the carbon spectrum must arise from chain packing effects. Helicies of different handedness pack in alternate rows in the α form, while helicies of the same handedness pack together in the β form. The 2:1 intensity ratios for the methyl and methine carbons correlate with the number of nonequivalent sites in the α-form crystal structure. The two sites have interchain packing distances of 5.28 and 6.14 Å in the α form, and 6.36 in the β form.

$$\left[\underset{R}{\overset{R}{\underset{|}{\overset{|}{Si}}}}-\underset{R}{\overset{R}{\underset{|}{\overset{|}{Si}}}} \right]$$

Poly(dialky silane)

While carbon NMR is frequently used to study semicrystalline polymers, many other nuclei may also be used. Silicon, for example, has a sensitivity similar to carbon

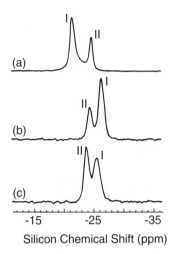

FIGURE 4.7 The solid-state silicon spectra of (a) poly(di-*n*-hexyl silane), (b) poly(di-*n*-pentyl silane) and (c)) poly(di-*n*-butyl silane).

and a wide chemical shift dispersion. Figure 4.7 shows the silicon NMR spectrum of a series of di-*n*-alkyl silanes with increasing side-chain length (9). Two crystalline forms of the silanes are often observed, a well-ordered Form I, and a higher temperature Form II with disordered sidechains. In all cases the disordered Form II signals appear near −23–24 ppm. The poly(di-*n*-hexyl silane) adopts an all-*trans* conformation for Form I and shows a peak near −22 ppm. Poly(di-*n*-pentyl silane) and poly(di-*n*-butyl silane) both adopt a 7_3 helical conformation and show signals in the silicon spectrum near −25 ppm. The silicon relaxation times in crystalline polymers can be very long, and the spectra are most efficiently acquired with cross polarization and magic-angle sample spinning.

Nylon-6

The nitrogen NMR spectrum is very sensitive to the local structure, but difficult to observe in the absence of isotopic labeling with ^{15}N. Figure 4.8 shows the cross-polarization magic-angle spinning ^{15}N NMR spectrum for nylon-6 and the nanocomposite with montmorillonite (10). Two sharp peaks are observed in the nitrogen spectra of the nylon-6 that are assigned to the α and γ crystalline forms. In addition, a broad weak signal from the amorphous regions is observed for the pure nylon-6 sample. The large chemical shift difference between the α and γ forms makes ^{15}N NMR an excellent method to measure which crystal form is favored under different crystallization conditions. The results show that the α form is favored in bulk nylon-6, and

4.2 CHAIN CONFORMATION IN POLYMERS

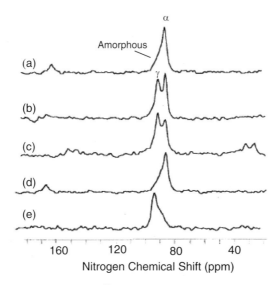

FIGURE 4.8 The 75 MHz solid-state ^{15}N spectra of nylon-6 and montmorillonite composites. The spectra are shown for (a) nylon-6, (b) the nylon-6/montmorillonite composite, (c) the reprecipitated nylon-6/montmorillonite composite, (d) annealed nylon-6 and (e) the annealed nylon-6/montmorillonite composite.

the γ form is favored for the nylon-6 synthesized in the presence of montmorillonite clay. In the bulk nylon-6, annealing increases the concentration of the α form, while the γ form is increased by annealing in the nanocomposite.

Poly[bis(4-ethylphenoxy)phosphazine]

Phosphorus is a high-sensitivity nucleus with good chemical shift dispersion that can be used to study semicrystalline polymers. Figure 4.9 shows the cross polarization magic-angle spinning ^{31}P spectrum of poly[bis(4-ethylphenoxy)phosphazine] at several temperatures (11). A complex spectrum is observed at ambient temperature with contributions from the crystalline, amorphous and interfacial phases. This polymer

260 THE SOLID-STATE NMR OF POLYMERS

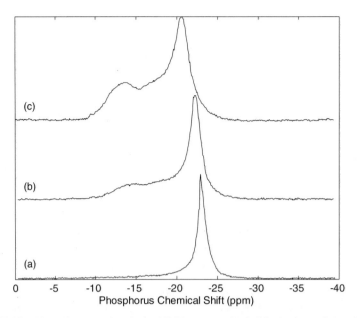

FIGURE 4.9 The solid-state phosphorus NMR spectra of poly[bis(4-ethoxyphenoxy) phosphazine] at (a) 23°C, (b) 80°C, and (c) 120°C. The spectra were acquired with cross polarization and magic-angle spinning.

undergoes a crystal-to-liquid crystal phase transition at 100°C, and a single sharp line is observed above this temperature.

$$\mathrm{+CF_2-CF_2 \mathrm{+}\mathrm{+}CF_2-CF\mathrm{+}}$$
$$\mathrm{O-CF_2-CF-O-CF_2-CF_2-SO_3^-H^+}$$
$$\mathrm{CF_3}$$
Nafion

Fluorine has the highest sensitivity of a nucleus except for protons, good chemical shift dispersion, and is an excellent probe of the crystal structure in fluorine-containing polymers. Figure 4.10 shows the solid-state fluorine spectrum of Nafion, a semicrystalline copolymer with possible applications in fuel cells (12). The spectrum was obtained without cross polarization and by using fast (35-kHz) magic-angle sample spinning for line narrowing. Fluorine has a large chemical shift anisotropy, so fast spinning is required to remove the spinning sidebands. For samples containing both fluorines and protons, the dipolar couplings must also be removed, either by proton dipolar decoupling or fast magic-angle sample spinning. The peak assignments for Nafion and other fluorinated polymers can be made by comparison to the solution spectra, model compounds, or nD NMR. The large peak at -120 ppm in the Nafion spectrum is assigned to CF_2 groups in the main chain, and the peaks at -146 and -139 ppm are assigned CF groups in the main chain and CF_2 groups near the sulfate. The peak at -115 ppm is assigned to CF_2 groups adjacent to the CF group in the

4.2 CHAIN CONFORMATION IN POLYMERS

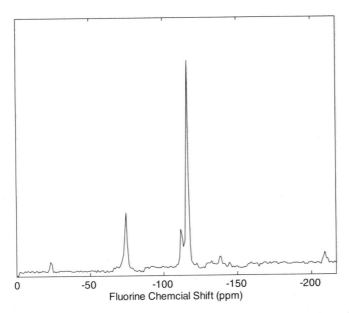

FIGURE 4.10 The solid-state fluorine spectra of Nafion acquired with magic-angle sample spinning.

polymer main chain, and the peak at –80 ppm is assigned to CF_2 and CF_3 groups in the side chain. This copolymer contains 13 mol% of the ionic monomer.

Proton NMR has the highest sensitivity, but is not often used to study crystalline polymers. A variety of methods can be used for line narrowing in protons (including decoupling and fast magic-angle sample spinning), but the resolution is typically about 1 ppm. Since most proton resonances in polymer fall between 1 and 7 ppm, many of the signals are overlapped in the proton spectrum.

4.2.1.1 Solid-Solid Phase Transitions

Section 4.2.1 shows that the chemical shifts for a variety of nuclei are sensitive to the crystal structure in polymers. This makes NMR an excellent means to study the solid-solid phase transitions in polymers. Figure 4.11 shows the carbon NMR spectrum of the low- and high-temperature form of a polyacetylene, poly(ETCD) (13). This polymer undergoes a transition from a blue phase at ambient temperature to a blue phase at 115°C. The NMR results show that the largest change in the spectrum is the 4-ppm downfield shift for the acetylic carbon and a 2-ppm downfield shift for the β and γ carbons on the side chain. These data are consistent with a transition in which the polymer undergoes rotations along the main chain away from a planar structure.

$$\begin{array}{c} CH_2-CH_2-CH_2-CH_2-CH_2-CH_3 \\ | \\ -\!\!\left[-Si-\right]\!\!- \\ | \\ CH_2-CH_2-CH_2-CH_2-CH_2-\!\!-CH_3 \end{array}$$

Poly(di-*n*-hexyl silane)

The ^{29}Si chemical shifts are extremely sensitive to chain conformation and have been used to study both silane and siloxane polymers. Two crystalline forms of polysilanes have been reported, and the chain conformation is sensitive to the side-chain groups. Figure 4.12 shows the effect of temperature on the ^{29}Si spectra for poly(di-*n*-hexyl silane), which has two forms, an ordered Form I and a disorderd Form II. The Form I for poly(di-*n*-hexyl silane) has an all-*trans* conformaiton, while the side chains are disordered in Form II. The ^{29}Si spectrum shows the transition between the two forms has an apparent midpoint near 40°C. These spectra are not quantitative because they were observed with cross polarization and magic-angle sample spinning. The relative intensities of the Form I and Form II peaks depend on the populations of the two crystalline forms, the cross polarization rates, and the proton $T_{1\rho}$ relaxation times. The dynamics of Forms I and II are very different from each other, so large differences are expected in the cross-polarization dynamics. Furthermore, the Form I to Form II conversion takes place over a very small temperature range, so the temperature must be carefully calibrated. This is always an issue, since the heating

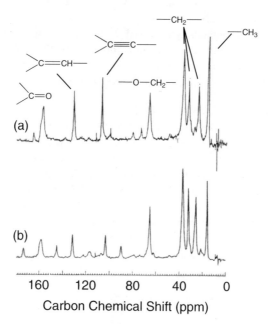

FIGURE 4.11 The solid-state carbons spectra of poly(ECTD) in the (a) blue and (b) red phase.

4.2 CHAIN CONFORMATION IN POLYMERS

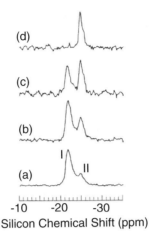

FIGURE 4.12 The solid-state silicon spectra of poly(di-n-hexylsilane) at (a) 25°C, (b) 35.9°C, (c) 41.5°C, and (d) 44.3°C.

from magic-angle sample spinning can increase the sample temperature above that measured in the nitrogen gas flowing over the sample (14).

4.2.2 Amorphous Polymers

The NMR spectra of amorphous polymers are generally not as well resolved as those from crystalline polymers, since the magnetic environments are more heterogeneous. In the amorphous phase, polymers exist in a variety of conformations that are not interconverting rapidly on the NMR timescale. This leads to a distribution in chemical shift from the γ-gauche effect (Section 3.1.2.4) and relatively broad lines in the carbon spectrum. This broadening is not greatly diminished by performing the experiments at higher magnetic fields or temperatures. The molecular dynamics are restricted in the glassy state and the signals are most effectively observed using cross polarization and magic-angle sample spinning.

Poly(n-butyl methacrylate)

Figure 4.13 shows the carbon NMR spectrum for poly(n-butyl methacrylate) and illustrates the resolution that can be observed in favorable cases. The carbonyl peak at 176 ppm is well resolved from all other main-chain and side-chain peaks. The peak from the main-chain methylene is broader than the other signals, presumably because

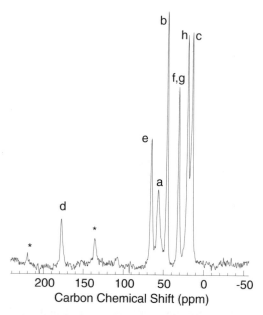

FIGURE 4.13 The cross-polarization magic-angle spinning carbon NMR spectrum of poly(n-butyl acrylate).

of unfavorable chain dynamics. If the chain dynamics are on the same time scale as the magic-angle spinning or the decoupling, the polymer dynamics can interfere with the line narrowing (Section 5.3.2). The lines from the side-chain methylene carbons are sharper than for the main chain and all of the peaks are resolved. The signals from the main-chain and side-chain methyl carbons can also be resolved.

Poly(styrene-co-methyl methacryalte)

Figure 4.14 shows the cross polarization magic-angle spinning carbon spectrum of a more poorly resolved amorphous material, the random copolymer of styrene and methyl methacrylate. The lines for both monomers are broad from the γ-gauche effect, since there is a distribution of conformation in the amorphous glassy state. Studies as a function of spinning speed are required to resolve the isotropic chemical shifts from the spinning sidebands. The peaks at 146 and 135 ppm are assigned to the nonprotonated and protonated signals from the styrene, and the overlapping methine/methylene peaks from the styrene main chain appear at 40 ppm. The carbonyl

4.2 CHAIN CONFORMATION IN POLYMERS

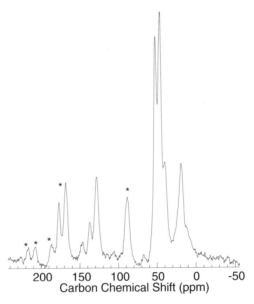

FIGURE 4.14 The 100 MHz Carbon NMR spectrum obtained with cross polarization and magic-angle sample spinning of the random copolymer of styrene and methyl methacrylate. The spinning sidebands are marked (*).

from the methyl methacrylate appears at 176 ppm and at this spinning speed (4 kHz) is resolved from the sidebands from the styrene aromatic carbons. The quaternary, methoxyl, and methyl signals from the methyl methacrylate are resolved at 53, 57, and 26 ppm.

Poly(vinyl phenol)

Proton NMR has a high sensitivity, but poor resolution for investigating the structure of amorphous polymers below their glass transition temperatures. This is illustrated in Figure 4.15, which shows the solid-state proton NMR spectrum of poly(vinyl phenol) obtained with combined multiple-pulse decoupling and magic-angle sample spinning (15). The resolution for this polymer is on the order of 2 ppm and only the broad peaks for the aromatic and main-chain protons are resolved. A study of the heteronuclear correlation spectrum for poly(vinyl phenol) in a blend showed that the hydroxyl protons appear near 5 ppm, but this peak is not resolved in the bulk homopolymer (15).

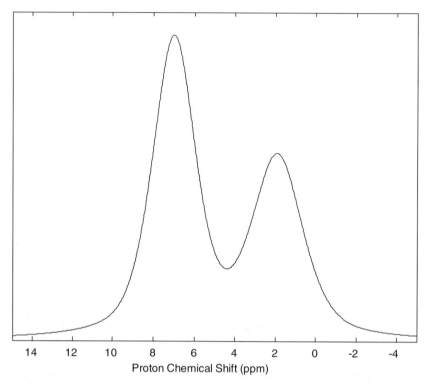

FIGURE 4.15 The 400-MHz solid-state proton NMR spectrum of poly(vinyl phenol) acquired with CRAMPS.

4.2.3 Elastomers

Elastomers were among the earliest polymers studied by solid-state NMR because they often give high-resolution NMR spectra in the absence of magic-angle sample spinning and dipolar decoupling. Many elastomers are well above their T_g at ambient temperature, and the broadening from chemical shift anisotropy and dipolar interactions are nearly averaged by chain motion. This makes it possible to directly observe the signals in the solid state using one-pulse experiments. The degree to which chain motion narrows the signals depends on the polymer of interest and how far above T_g the spectrum is acquired. For swollen polymers or polymers far above T_g, it may be possible to directly observe the high-resolution spectrum. In *cis*-polybutadiene, for example, molecular motions at ambient temperature average the proton dipolar linewidths from 50 kHz to 300 Hz (16), and high-resolution spectra can be observed with 1 kHz magic-angle sample spinning. For polymers closer to T_g, magic-angle sample spinning (2-20 kHz) may be required for line narrowing. In addition, since the dipolar interactions are partially averaged by chain motion above T_g, cross polarization is frequently very inefficient. In some polymers a weak signal can be obtained with long cross polarization times, while in others the dipolar couplings

4.2 CHAIN CONFORMATION IN POLYMERS

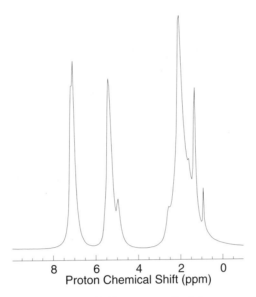

FIGURE 4.16 The solid-state proton NMR spectrum of poly(styrene-*co*-butadiene) obtained with 12-kHz magic-angle sample spinning.

are so efficiently averaged by chain motion that no signals are observed using cross polarization.

—CH$_2$—CH— *co* —CH$_2$—CH=CH—CH$_2$—

Poly(styrene-*co*-butadiene)

Figure 4.16 shows the proton spectrum of a random poly(styrene-*co*-butadiene) copolymer obtained with 12-kHz magic-angle sample spinning. This polymer is well above T_g at ambient temperature and several peaks are resolved, including the polystyrene aromatic signals at 7 ppm and the polybutadiene olefinic peaks near 5.4 ppm. The partially resolved signal centered near 2 ppm contains the remaining signals from the polymer main chain. This spectrum demonstrates that near solution-state spectra can be obtained for polymers far above T_g or with fast magic-angle sample spinning.

Once a high-resolution proton spectrum has been obtained using either high temperature or fast magic-angle sample spinning, some of the solution methods for polymer analysis can be used without modification. This is illustrated in Figure 4.17, which shows the 2D NOESY spectrum for the poly(styrene-*co*-butadiene) copolymer obtained with a 0.1-s mixing time and 12-kHz magic-angle sample spinning. Cross peaks are observed not only within the same monomer unit but also between the styrene

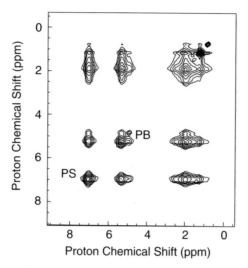

FIGURE 4.17 The 2D NOESY spectrum of poly(styrene-*co*-butadiene) obtained with 12-kHz magic-angle sample spinning.

aromatic signals and the butadiene olefinic peaks. Such peaks are only expected for random copolymers where the immiscible styrene and butadiene monomers are prohibited from phase separating on a longer length scale from the covalent connectivity along the polymer chain.

$$\mathrm{+CH_2-CH_2-O+ \ + CH_2-\underset{CH_3}{CH}-O+ \ + CH_2-CH_2-O+}$$

Poly(ethylene oxide-*b*-propylene oxide-*b*-ethylene oxide)

The NMR properties of elastomers depend not only on the bulk T_g but also on any other materials that may be present. This is illustrated in Figure 4.18, which shows the proton NMR spectrum for the poly(ethylene oxide-*b*-propylene oxide-*b*-ethylene oxide) triblock copolymer in the bulk and as part of a composite with poly(methyl silsesquioxane). Good resolution is observed for both blocks of the triblock copolymer far above T_g. The ethylene oxide block contributes to the signals at 3.5 ppm, while the propylene oxide block has signals at 1 ppm from the methyl group and at 3.5 ppm from the main-chain methine and methylene protons. The signal at 0 ppm in the composite arises from the methyl groups attached to silicon atoms in the methyl silsesquioxanes matrix. The relative intensities of the peaks at 3.5 and 1.0 ppm can be calculated from the molecular weights of the block in the copolymer. For this particular copolymer a ratio of 8:1 is expected. For the bulk triblock copolymer, the ratio is closer to 1:1. This shows that a large part of the intensity is missing from the spectrum acquired with 12-kHz magic-angle sample spinning for line narrowing. The missing intensity

4.2 CHAIN CONFORMATION IN POLYMERS

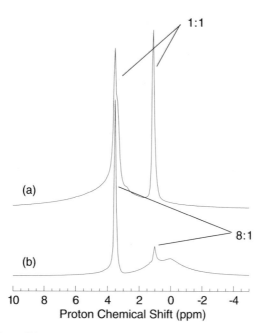

FIGURE 4.18 The solid-state proton NMR spectrum of poly(ethylene oxide-*b*-propylene oxide-*b*-ethylene oxide) triblock copolymer (a) in the bulk and (b) as a composite with poly(methyl silsesquioxane).

is attributed to semicrystalline poly(ethylene oxide) in the bulk triblock copolymer, since 12-kHz magic-angle sample spinning is not fast enough to narrow the signals from crystalline polymers. The 1:1 intensity observed for the bulk triblock copolymer could be due solely to the propylene oxide block, since a 1:1 intensity ratio is expected for the methyl and main-chain signals. The 8:1 main-chain:methyl signal intensity is observed for the triblock copolymer composite. Apparently composite formation suppresses the formation of crystallinity in the poly(ethylene oxide) block, so a high-resolution spectrum is observed for both blocks. This example shows that the local environment (bulk vs. composite) has a large effect of the polymer chain dynamics, and hence on the NMR spectrum.

Kel-F 3700

The resolution in polymers above T_g depends on the temperature above T_g and the rate of magic-angle sample spinning. This is illustrated in Figure 4.19, which shows the one-pulse ^{19}F spectrum of Kel-F 3700, a 69:31 copolymer of vinylidine

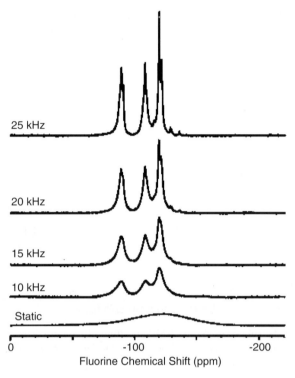

FIGURE 4.19 The solid-state fluorine spectrum of Kel-F 3700 with fast magic-angle sample spinning.

fluoride and chlorotrifluoroethylene (17). Obtaining a spectrum for this polymer is challenging because of the presence of both proton and fluorine dipolar couplings. Figure 4.19 show that a high-resolution spectrum can be obtained with magic-angle spinning alone. The T_g for the copolymer is $-15°C$, and at ambient temperature the static spectrum shows only a broad, unresolved line. The resolution increases with spinning speed and a very high-resolution spectrum can be obtained with spinning at 25 kHz. An even higher resolution can be obtained by increasing the temperature in combination with fast magic-angle sample spinning, and a solution-like spectrum was reported for the copolymer at 150°C (17).

4.2 CHAIN CONFORMATION IN POLYMERS

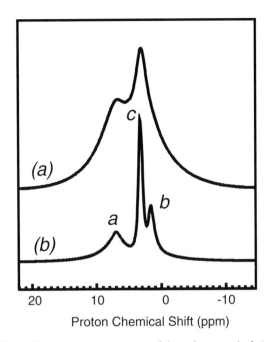

FIGURE 4.20 The solid-state proton spectrum of the polystyrene/poly(vinyl methyl ether) blend at (a) 26°C and (b) 120°C with 12-kHz magic-angle sample spinning.

The combination of temperature and magic-angle sample spinning can be used for line narrowing in the proton spectrum. The temperature relative to T_g as well as the intrinsic chain mobility are both important factors affecting the resolution. This is illustrated in Figure 4.20, which shows the proton NMR spectra of the miscible blend of polystyrene and poly(vinyl methyl ether) at ambient temperature and at 120°C. The chains are well mixed in this blend, and the T_g measured by DSC is 65°C. Polymer blends often have a broad glass transition and a broad distribution in mobility. Furthermore, the dynamics of the chains, even in a miscible blend, can be very different from each other (18,19), so it is possible that the dipolar interactions (and linewidths) may be averaged differently for the two chains. The spectrum at ambient temperature ($T_g - 40°C$) shows more resolution than expected for a typical glassy polymer, presumably because of the chain motion of the more mobile poly(vinyl methyl ether). The resolution at $T_g + 55°C$ (120°C) shows a high-resolution spectrum in which peaks from both polymers (*a* and *c*) are clearly resolved.

Poly(*N*-isopropylacrylamide)

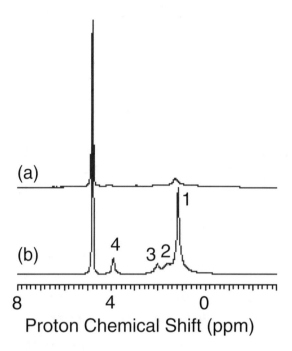

FIGURE 4.21 The proton spectrum of poly(isopropyl acrylamide) in the absence of magic-angle sample spinning.

The methods used to study swollen polymer systems are similar to those used to study elastomers. The experiment to use in such studies depends on the molecular dynamics, which of course depends on the dynamics of the swollen polymer. For polymers with a high degree of solvent uptake, liquid-like spectra are often observed in the absence of magic-angle sample spinning. This is illustrated in Figure 4.21, which shows the proton NMR spectrum of a poly(N-isopropyl acrylamide) gel (20). Poly(N-isopropyl acrylamide) is widely studied in polymer-actuated devices that undergo a volume-phase transition as a function of temperature. Below the phase transition, the swollen polymer gives a liquid-like spectrum and the peaks from the methyl, methylene, main-chain methine, and side-chain methine are resolved at 1.2, 1.6, 2.0, and 3.9 ppm. Above 35°C the chains collapse into a denser structure, and more traditional solid-state NMR methods are required to obtain the spectrum.

Swollen polymer beads are used for the solid-state synthesis of peptides, and have been extensively studied by solid-state NMR methods (21). The goal of many of these studies is the analysis of the peptide product before it is cleaved from the polymer support, often in a drug-discovery effort. The method most often used to observe the swollen gel is fast magic-angle sample spinning. The application of fast magic-angle sample spinning in drug discovery is often referred to as high-resolution magic-angle spinning (HR-MAS).

4.2 CHAIN CONFORMATION IN POLYMERS

The examples just shown illustrate that the NMR properties of polymers above T_g are very different from the glassy and crystalline phases. The chain motion above T_g partially averages the dipolar interactions and it becomes possible to observe the spectra without magic-angle sample spinning and dipolar decoupling. However, the spectra do not cross polarize very efficiently, because the dipolar couplings are averaged by the combination of chain motion and magic-angle sample spinning. In some experiments, especially those utilizing proton magnetization exchange, it is desirable to observe the spectrum using cross polarization. This can often be accomplished using the so-called "ramped" (22) or adiabatic passage Hartman–Hahn (23) cross polarization pulse sequences shown in Figure 2.12. The idea behind these pulse sequences is to vary the power on either the carbon or proton channels during cross polarization so that the Hartman–Hahn condition is matched over a broader range of frequencies. This makes it more likely that polymers with very different molecular dynamics can be cross polarized.

In summary, the increased chain dynamics of polymers above T_g makes it possible to observe the spectrum with a high resolution using simple one-pulse methods, and liquid-like spectra can be observed using the combination of temperature above T_g and fast magic-angle sample spinning. After a high-resolution spectrum has been obtained, many of the methods developed for solution NMR (NOESY, etc.) can be used without modification. If very high-resolution spectra are obtained, it becomes possible to use solution-based methods using though-bond J coupling for the analysis of polymers. It has been reported, for example, that spectral editing methods (DEPT, etc.) can be applied to elastomers (24).

4.2.4 Reactivity and Curing in Polymers

Solid-state NMR is an important method for monitoring changes in solid polymers. Many polymers are used following a curing process that renders them insoluble, so the extent of reaction cannot be measured by solution NMR methods. Although the resolution in solid-state NMR is not generally as good as for solution NMR, curing in polymers leads to large changes in chemical structure that are often obvious from the NMR spectrum. The NMR properties can also be used to study radiation damage and aging in polymers.

The studies of polymer reactivity use all of the methods previously discussed, including cross polarization, magic-angle sample spinning, and dipolar decoupling. The NMR method of choice depends on the molecular dynamics of the polymer of interest. In elastomers, the reactivity can be measured with a high sensitivity using the proton NMR spectrum with fast magic-angle sample spinning. This is illustrated in Figure 4.22 for an acrylate formulation. The polymer was prepared by photochemical polymerization of a mixture of acrylates, so monomers with double bonds are consumed to form the polymer. The protons attached to carbons with double bonds are well resolved from the other protons and can be used to monitor the extent of the reaction. This formulation has a low T_g, so it is possible to obtain a high-resolution spectrum at ambient temperature with fast magic-angle spinning alone. The spectrum in Figure 4.22 was obtained with 15-kHz magic-angle sample spinning

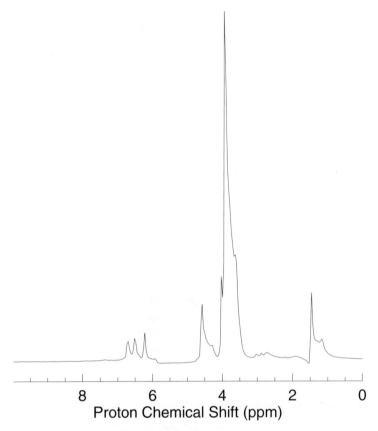

FIGURE 4.22 The solid-state proton NMR spectrum of an elastomeric acrylate formulation after curing. The spectrum was acquired with 15 kHz magic-angle sample spinning.

and the signals from the olefinic protons are well resolved at 6–7 ppm. These data show that 15% of the monomer is unreacted after photopolymerization.

PMDA/*m*-int-*A*

Carbon cross polarization and magic-angle spinning is the method most frequently used to study curing in polymers. Figure 4.23 shows the carbon NMR spectra of a pyromellitic dianhydride (PMDA/*m*-int-*A*) as a function of curing temperature (25).

4.2 CHAIN CONFORMATION IN POLYMERS

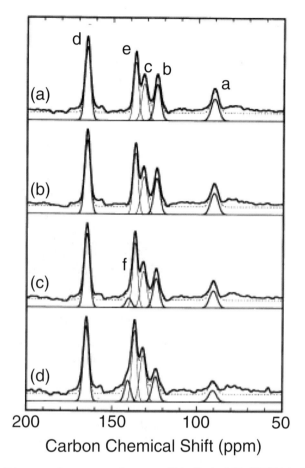

FIGURE 4.23 The solid-state carbon spectra of a pyromellitic dianhydride (PDMA/*m*-int-A) after curing at (a) 200 °C, (b) 300 °C, (c) 350 °C, and (d) 400 °C. The fitted peaks are also shown.

The peaks of interest include the acetylinic carbon and its neighbor (*a* and *b*). The results show that these carbons decrease in intensity for samples cured above 300°C. The decrease in these carbon peaks is due to cross-linking and the reactivity of the acetylinic carbon. The relative intensities of these peaks relative to other signals in the polyimide can be used to measure the reactivity.

Poly[phenylsilylene)ethynylene-1,3-phenyleneethylnylene]

276 THE SOLID-STATE NMR OF POLYMERS

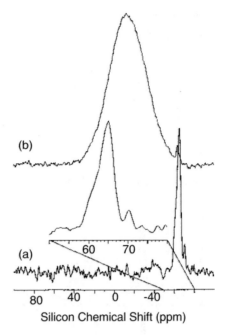

FIGURE 4.24 The solid-state silicon spectra of poly[(phenylsilylene)ethynylene-1,3-pheyleneethynylene] (a) before and (b) after curing at 400 °C.

NMR studies of the reactivity of polymers are particularly important for insoluble materials, such as polyimids and other high-temperature polymers. Figure 4.24 shows how solid-state silicon NMR with cross polarization and magic-angle spinning can be used to monitor the reactivity in poly[(phenylsilylene)ethynylene-1,3-phenyleneethylnylene] (26). The uncured material shows a relatively sharp peak at −60 ppm that is consumed during the cure. The broad lines in the final material result from a variety of chemical products. The reactivity of poly[(phenylsilylene)ethynylene-1,3-phenyleneethylnylene] can also be monitored by ^{13}C NMR.

$$\underset{\text{PFA}}{\overset{-122\text{ ppm}}{+\text{CF}_2-\text{CF}_2+}\overset{-138\text{ ppm}}{\underset{\underset{\underset{-81\text{ ppm}\quad -83\text{ ppm}}{\text{O}-\text{CF}_2-\text{CF}_2-\text{CF}_3}}{|}}{\text{CF}_2-\text{CF}+}}}\;{-131\text{ ppm}}$$

Solid-state ^{19}F NMR with fast magic-angle sample spinning has been used to study polymer reactivity and radiation damage in a perfluoroalkoxy (PFA) resin, a copolymer of tetrafluoroethylene and perfluoroalkyl ether. Figure 4.25 shows the ^{19}F spectra of the copolymer as a function of increasing radiation exposure obtained at ambient temperature with magic-angle sample spinning at 32 kHz (27). Several

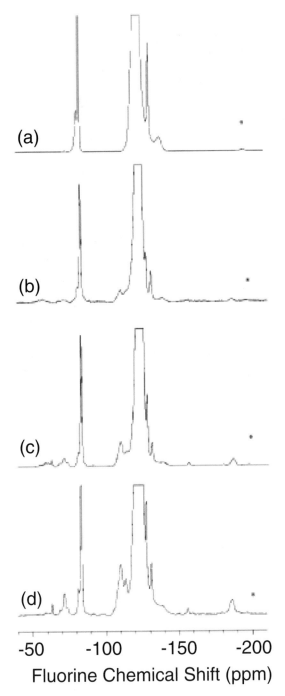

FIGURE 4.25 The ^{19}F spectrum of (a) untreated perfluoroalkoxy resin, and the resin (b) irradiated at 1 MGy at 303 K, (c) irradiated at 1 MGy at 473 K, (d) irradiated at 1 MGy at 573 K. The spectra were acquired with magic-angle sample spinning at 32 kHz.

new peaks are observed in the spectrum as the polymer is exposed at increasingly higher temperatures. The new peaks are assigned by comparison to solution spectra to new saturated chain ends, short and long branches, and new unsaturated chain ends.

In summary, a wide variety of NMR methods can be used to investigate curing and reactivity in solid polymers. The chemical changes give rise to large changes in the spectra that can be observed using traditional solid-state NMR methods, including cross polarization, magic-angle sample spinning, and dipolar decoupling.

4.3 THE STRUCTURE AND MORPHOLOGY OF POLYMERS

4.3.1 Introduction

The chemical shifts in solid polymers provide information about the chain conformation of polymers, but little information about the longer-range structure. Information about the morphology of polymers is obtained by measuring magnetization exchange over longer length scales. The structure on a 2–5-Å length scale can be obtained by monitoring magnetization exchange between two NMR-active nuclei that are in close proximity. Any of a number of spin pairs can be investigated, including ^{1}H–^{13}C, ^{13}C–^{15}N, ^{15}N–^{2}H, and the general strategy is to measure the magnitude of the dipolar couplings, which depends on the internuclear distance.

The measurement of morphology over a longer length scale (5–200 Å) can be accomplished using proton spin exchange. The fundamental process of spin exchange is a mutual spin-flip, as illustrated in Figure 4.26. The ability of two nuclei to undergo such mutual flips depends on their frequencies and the dipolar couplings between them. Mutual spin flips with no net change of energy can only occur between spins that are very close in frequency and have strong dipolar couplings. The dipolar couplings d_{IS} are given by

$$d_{IS} = \frac{\mu_0}{4\pi} \frac{\gamma_I \gamma_S \hbar}{r_{IS}^3} \left(3\cos^2\theta - 1\right) \qquad (4.2)$$

where θ is the orientation of the internuclear vector relative to the magnetic field. The magnitude of the dipolar coupling depends on the inverse third power of the internuclear distances and is too small to promote mutual spin flips when the separation is more than 5 Å. Thus, mutual spin-flips are most often observed between nuclei of the same type (and frequency) that are present at a high density. One exception to this generality is cross polarization, where mutual spin flips are possible because the different nuclei have the same frequency in the spin-locking field.

To observe the magnetization exchange over long distances in polymers, it is necessary to have a large collection of NMR-active nuclei that are strongly dipolar coupled to each other. This makes long-range spin-exchange between carbon atoms very unlikely, because in natural abundance samples ^{13}C is present only at the 1% level, so there is little chance that ^{13}C nuclei will be close enough for spin exchange.

4.3 THE STRUCTURE AND MORPHOLOGY OF POLYMERS

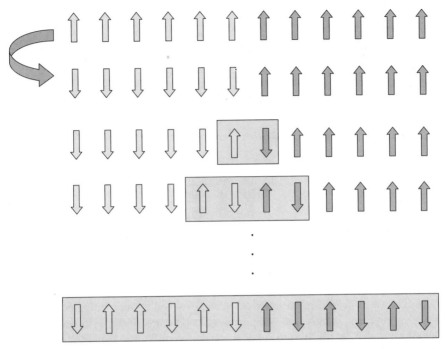

FIGURE 4.26 A diagram showing mutual spin flips and spin diffusion. The top row show the spin system at equilibrium. The second row shows a perturbation to the spin system in which half of the spin are inverted, and the remaining rows show reequilibration of the spin system.

Protons and fluorines are present at a high density and natural abundance and can be used for magnetization exchange over long distances.

The general strategy for measuring the morphology over longer length scales is to create a nonequilibrium distribution of spin states and monitor the return to equilibrium. The key to this experiment is to create this distribution of spin states in such a way that there is a difference in spin populations for different morphologies. This is most often accomplished by taking advantage of differences in the dynamics or proton chemical shifts of different morphologies such that one phase is saturated while the other maintains its equilibrium magnetization. The rate of magnetization exchange between the phases follows the mathematics of diffusion, and for this reason is called spin diffusion. Figure 4.27 schematically shows this process. The experiment begins with a pulse sequence that partially saturates one phase relative to another to create a population difference between the phases. On a short time scale (ms) magnetization is exchanged between the phases through the process of spin diffusion. Eventually the phases equilibrate with each other and slowly (on a time scale of seconds) return to compete equilibrium via spin-lattice relaxation. The rate of spin diffusion can be measured either from the decrease of the mobile signal intensity or the increase in the immobile signal intensity as a function of time after the polarization gradient is applied.

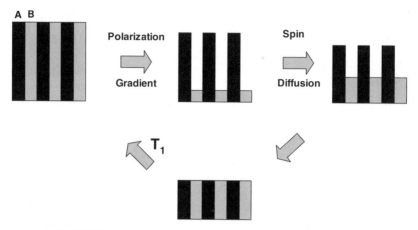

FIGURE 4.27 A diagram showing spin diffusion across an interface.

Spin diffusion has been used to study many different polymer morphologies, including the crystalline, amorphous, and interfacial regions in semicrystalline polymers (28). It has also been used to study other phase-separated morphologies, including block copolymers, blends, and composites (29). The investigations mainly differ in the experimental details used to create a population difference between the different phases. The domain-size measurements depend on the surface area between the phases and cannot be used to independently determine the domain sizes and morphologies. In many studies, where the morphology is known (lamellar, cylindrical, spherical, etc.), the NMR measurements are used to measure the domain sizes. In the event that the morphologies are not known, the NMR domain sizes must be considered estimates. The measurement of domain sizes also depends on the rate of spin diffusion through the spin-diffusion coefficient that must be measured or estimated by comparison to spin-diffusion studies on known polymers.

4.3.2 Spin Diffusion and Polymer Morphology

The general strategy in spin-diffusion studies is to saturate the spins from one part of the material and monitor the return of the spin system to equilibrium. The protons can relax either by spin-lattice relaxation or by mutual spin flips with nearby protons. This process of mutual spin flips (or spin diffusion) is very efficient and can lead to magnetization exchange over long length scales. The transfer of magnetization can be described by the diffusion equation, and is given by

$$\dot{m}(r,t) = D\nabla^2 m(r,t) \tag{4.3}$$

where D is the diffusion constant (in units of m^2/s) and $m(r,t)$ is the local magnetization density. The solutions to this equation are expressed in terms of the response function ΔM_s, which can be obtained from the measured intensities as a function

4.3 THE STRUCTURE AND MORPHOLOGY OF POLYMERS

of time:

$$\Delta M = \frac{M(t) - \bar{M}}{M_0 - \bar{M}} \tag{4.4}$$

where $M(t)$ is the measured intensity as a function of the spin diffusion delay time, M_0 is the intensity after the application of a polarization gradient, but before any time is allowed for spin diffusion or relaxation, and $\bar{M}$ is the intensity after samplewide spin equilibrium has been obtained, but before spin-lattice relaxation.

In a typical spin-diffusion experiment, the intensity of the signal of interest is monitored as a function of the spin-diffusion time, and these experimental data are interpreted with some kind of physical model. Among the earliest models used for spin diffusion are one-, two-, and three-dimensional diffusion, corresponding to lamellar, cylindrical, or spherical morphologies. The time course of spin diffusion for the one-dimensional case is given by

$$\Delta M_s = \phi(t)$$
$$= 1 - \left(\frac{2}{\sqrt{\pi}}\right)\left(\frac{Dt}{\bar{b}^2}\right)^2 \quad \text{for } t \ll \frac{\bar{b}^2}{D} \tag{4.5}$$
$$= \sqrt{\pi}\left(\frac{\bar{b}^2}{Dt}\right) \quad \text{for } t \gg \frac{\bar{b}^2}{D}$$

where $\bar{b}$ is the average domain dimension and D is the diffusion coefficient (30). The time course of magnetization recovery for the two- and three-dimensional models are given by

$$\Delta M_s = \phi(t)^2 \tag{4.6}$$

and

$$\Delta M_s = \phi(t)^3 \tag{4.7}$$

Figure 4.28 shows a plot of the spin-diffusion recovery curves for generic one- and two-dimensional morphologies. The important points to note about Figure 4.28 are that the diffusion is sensitive to both the domain size and the dimensionality. In other words, if the dimensionality is not known, then it may be possible to fit the data with any of the preceding models. If the dimensionality is known, then the spin-diffusion data can be used to make accurate measurements of the domain sizes. In real-world samples it is sometimes observed that the experimental data cannot be fit by any of these simple morphological models. Many samples also contain interfacial material between the two phases that affect the initial part of the spin-diffusion curves.

Since there are few methods able to measure domain sizes in polymer over the 0.2–20-nm length scale, spin diffusion in polymers has been extensively investigated (31–33). There are in general two approaches for extracting domain sizes from

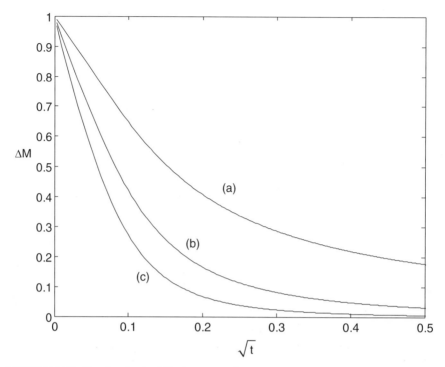

FIGURE 4.28 Simulated spin diffusion curves for (a) 1D, (b) 2D and (c) 3D diffusion using a diffusion coefficients of 0.8×10^{-15} m^2/s and 0.01×10^{-15} m^2/s for the rigid and mobile phases, and a domain size of 2 nm.

spin-diffusion experiments, detailed numerical simulation of the spin-diffusion behavior, or approximate solutions in which the domain sizes are obtained from the linear part of the spin-diffusion curve plotted vs. $\sqrt{t}$. Since there are a number of uncertainties in spin-diffusion experiments, we will focus on the initial rate solution which has been shown to provide accurate estimates of the domain sizes.

Determination of the domain sizes by spin diffusion requires a number of assumptions, including the values for the diffusion coefficients for the two phases and the dimensionality. In addition, most treatments of spin diffusion assume the application of a perfect polarization gradient in which the spins in one phase have been completely saturated, while the spins in the other phase maintain their equilibrium polarization. As we will see later, it is rare that such a perfect polarization gradient can be applied. In addition, if an interface is present, it is very likely that the properties of the interface will be intermediate between the two larger phases, and the protons in this region will be only partially polarized by the polarization gradient.

To understand how the initial rate from a spin-diffusion plot can be related to the domain sizes in phase-separated polymers, let us consider a two-phase system with phases A and B that are characterized by proton densities ρ_A^H and ρ_B^H and diffusion coefficients D_A and D_B. Furthermore, we will assume that the initial polarization gradient is uniform across the domains, and that the polarization gradient between

4.3 THE STRUCTURE AND MORPHOLOGY OF POLYMERS

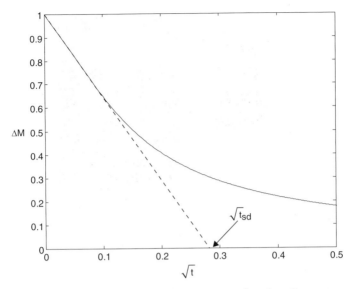

FIGURE 4.29 A simulated spin-diffusion curve for a lamellar system.

phases A and B can be approximated as a step function. If we ignore the relaxation via the spin-lattice relaxation, then the initial slope S of the normalized spin diffusion plot is given by

$$-S = \frac{2S_v \sqrt{D_A D_B} \left(\rho_A^H f_A + \rho_B^H f_B\right)}{f_A f_B \sqrt{\pi} \left(\rho_A^H \sqrt{D_A} + \rho_B^H \sqrt{D_B}\right)} \quad (4.8)$$

where f_A and f_B are the volume fractions of the A and B phases, and S_v is the interface surface-to-volume ratio. The key parameter measured from the initial slope is S_v. This is as much information as can be determined from the normalized spin-diffusion plot. It is important to note that many morphologies may be consistent with a given value for S_v. If other information is available, such as the morphology or dimensionality, then the slope can be interpreted in terms of the domain sizes.

Figure 4.29 shows a spin-diffusion plot calculated for a lamellar system. If we extrapolate the initial slope to the baseline, the intersection point ($\sqrt{t_{sd}}$) is related to the slope by

$$-S = \frac{1}{\sqrt{t_{sd}}} \quad (4.9)$$

Substituting Equation (4.9) into Equation (4.8) gives

$$-S_V = \frac{f_A f_B \sqrt{\pi} \left(\rho_A^H \sqrt{D_A} + \rho_B^H \sqrt{D_B}\right)}{2\sqrt{t_{sd}} \sqrt{D_A D_B} \left(\rho_A^H f_A + \rho_B^H f_B\right)} \quad (4.10)$$

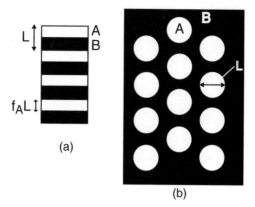

FIGURE 4.30 The characteristic dimensions for the lamellar, cylindrical, and spherical domains.

The surface-to-volume ratio can be related to the characteristic length scales for the lamellar and cylindrical morphologies, as shown in Figure 4.30, where for the lamellar phase L is given by the sum of the thickness of the phases A and B. The relationship between S_v and L is given by

$$S_v = \frac{2}{L} \tag{4.11}$$

Substituting Equation (4.11) into Equation (4.10) gives

$$L = \frac{4\sqrt{t_{sd}}\sqrt{D_A D_B}\left(\rho_A^H f_A + \rho_B^H f_B\right)}{f_A f_B \sqrt{\pi}\left(\rho_A^H \sqrt{D_A} + \rho_B^H \sqrt{D_B}\right)} \tag{4.12}$$

where the characteristic length scale L can be directly related to the morphology, as shown in Figure 4.30. Under certain circumstances, Equation (4.12) can be further simplified. If, for example, $\rho_A^H = \rho_B^H$, Equation (4.12) can be rewritten as

$$L = \frac{4\sqrt{t_{sd}}\sqrt{D_A D_B}}{f_A f_B \sqrt{\pi}\left(\sqrt{D_A} + \sqrt{D_B}\right)} \tag{4.13}$$

and if $\rho_A^H = \rho_B^H$ and $D_A = D_B$, then Equation (4.13) can be further simplified to

$$L = \frac{2\sqrt{t_{sd}}\sqrt{D}}{f_A f_B \sqrt{\pi}} \tag{4.14}$$

From the schematic structures shown in Figure 4.30, we can extract two important features from the spin-diffusion data, the distance across the minor phase (represented as d) and the overall repeat distance, or long period, associated with the domain

4.3 THE STRUCTURE AND MORPHOLOGY OF POLYMERS

TABLE 4.1 The Relationships Between the L, d and d_{1p} in One, Two and Three Dimensions.

Dimensionality	Distance Across Minor Phase (d)	Overall Periodicity (d_{lp})
1D	$f_a L$	L
2D	$2 f_A L$ (= rod diameter)	$1.9046 \, (f_A)^{1/2} L$
3D	$3 f_A L$ (= sphere diameter)	$2.683 \, (f_A)^{2/3} L$

(represented as d_{lp}). For the lamellar structure d is the width of the minor phase and d_{lp} is the sum of the widths of the two phases. For a hexagonal packed cylinders (2D morphology), d is the rod diameter and d_{lp} is the center-to-center distance between the cylinders. For the 3D morphology, d is the sphere diameter and d_{lp} is the spacing between the center of the spheres. The relationship between L, d, and d_{lp} is given in Table 4.1 for the 1D, 2D, and 3D morphologies.

4.3.2.1 Spin Diffusion and Interfaces

The interpretation of the spin-diffusion plots is often complicated by the presence of an interface. In many semicrystalline polymers, for example, the material at the surface of the crystalline region is intermediate in dynamics between the amorphous and crystalline regions. If a difference in chain dynamics is used to create a polarization gradient, then there may not be a sharp polarization gradient across the interface, because this material may have dynamics intermediate between the two phases. The interface may be only partially polarized by the gradient. Frequently the interfacial material is not observed as a well-resolved peak that can be monitored during the spin-diffusion experiment. One consequence of the interface is that it presents a barrier between the two phases, so if we are observing magnetization transfer from the amorphous to the crystalline phase, there will be a time lag before magnetization is observed in the crystalline phase. This will affect the initial part of the spin diffusion curve.

The effect of the interface on the spin diffusion curve is shown schematically in Figure 4.31. The important feature to note is that the linear part of the curve no longer extrapolates back to unity, but instead shows a flat portion at short spin-diffusion times. However, the linear part of the curve is similar to that observed in the absence of an interface. That means that we can use the linear part of the curve to extrapolate to $\sqrt{t_{sd}}$ and use Equation (4.12) and Table 4.1 to calculate the domain sizes and the long periods. The width of the interface can be estimated from the lag between the beginning of the experiment and the point at which the curve is linear. The width of the interface is approximately given by L_{int}

$$L_{int}^2 = \frac{4}{3} Dt \qquad (4.15)$$

where t is the time lag in Figure 4.31.

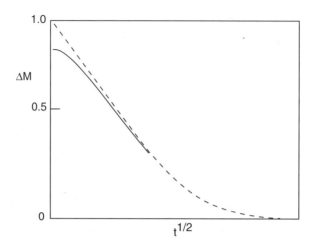

FIGURE 4.31 The simulated spin diffusion curve for a lamellar morphology without (dotted line) and with (solid line) an interface.

4.3.2.2 Spin-Diffusion Coefficients For an accurate estimate of the domain sizes from spin-diffusion experiments, it is important to use proper values for the proton densities and spin-diffusion coefficients. The proton densities can be estimated from the material densities and the mole fraction of protons in the polymer. It can be seen from Equation (4.12) that the rate of spin-diffusion depends strongly on the proton density. This is why protons are most often used for spin-diffusion experiments. Protons have a high natural abundance and density in most polymers, leading to very efficient spin diffusion. Carbons have a much lower natural abundance (1.1%), so the probability of finding two neighboring ^{13}C atoms is very low, so carbon spin diffusion is very inefficient in unlabeled polymers.

A number of approaches have been used to estimate the diffusion constants for proton spin diffusion. The diffusion constants are expected to depend on the dipolar couplings, and therefore on molecular motions that average the dipolar couplings. For this reason, the rates of proton spin diffusion will be very different for rigid polymers and mobile polymers. The values for diffusion coefficients in SI units are on the order of 10^{-15} m^2/s. It has also become common to use the units nm^2/ms (nm^2/ms = 10^{-15} m^2/s) for discussion and comparison of the spin-diffusion constants.

In the earliest studies, the diffusion constant for rigid polymers was estimated from the rates of spin diffusion in solid alkanes as 0.62 nm^2/ms (34). For rigid materials, it was assumed that the rates of spin diffusion in other materials could be calculated by correcting for the proton densities relative to the alkanes.

In the other studies the rate constants for spin diffusion were estimated from T_2 relaxation rates for the more mobile phase, since both the spin-diffusion coefficients and T_2 depend on the dipolar couplings. In some studies the spin-diffusion rate constant has been estimated as

$$D = \frac{2r_0^2}{T_2} \tag{4.16}$$

4.3 THE STRUCTURE AND MORPHOLOGY OF POLYMERS

where r_0 is the radius of the hydrogen atom (35). Using this estimate, good agreement was obtained between the domain sizes in polyurethanes measured by NMR and X-ray scattering. The upper bound for D can also be estimated as

$$D = \frac{13a^2}{T_2} \qquad (4.17)$$

where a is the average distance between adjacent protons in the less mobile domain (30). This estimate was validated in a study of semicrytalline polymers.

More recently, the spin-diffusion coefficients have been calculated by measuring the rate of spin diffusion for a series of symmetric styrene–methyl methacrylate diblock copolymers with a known morphology. Since the morphologies and domain sizes were known from X-ray scattering, the only unknown factor affecting the rate of spin diffusion was the spin-diffusion coefficient. The best results were obtained using a spin-diffusion coefficient of 0.8 nm²/ms for glassy, amorphous polymers below their glass transition temperatures (31).

The situation is more complex with mobile polymers, since the motions that average the dipolar couplings depend strongly on temperature. It has recently been proposed that the diffusion constants can be directly related to the static spin-spin relaxation times for mobile polymers (36). The correlation between spin-diffusion coefficient and spin-spin relaxation time was established by measuring spin diffusion in mobile polymers with known morphologies (determined by scattering and microscopy). For T_2's in the range of 100 to 1000 Hz, the relationship between the diffusion coefficient and the T_2 is given by

$$D = 8.2 \times 10^{-6} T_2^{-1.5} + .007 \text{ nm}^2/\text{s} \qquad (4.18)$$

and for T_2's in the range of 1000 to 3500 Hz it is given by

$$D = 4.4 \times 10^{-4} T_2^{-1} + 0.26 \text{ nm}^2/\text{s} \qquad (4.19)$$

Finally, it should be noted that if the spinning speed is comparable to the magnitude of the dipolar coupling, the rate of spin diffusion may depend on the spinning speed. This is particularly important for mobile polymers like polybutadiene or polyisoprene. For such mobile polymers the most accurate measurements will be those performed without magic-angle sample spinning.

4.3.2.3 Polarization Gradients for Measuring Spin Diffusion

The first step in a spin-diffusion experiment is to establish a polarization gradient. The ideal gradient has a uniform polarization distribution across each phase, but different values for the two phases. The larger the difference in polarization between the phases (i.e., the larger the polarization gradient), the larger the measurable signal will be in the spin-diffusion experiments. Many methods have been used to create the polarization

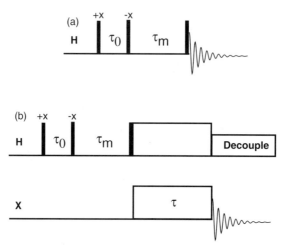

FIGURE 4.32 A pulse-sequence diagram for the Goldman–Shen experiment monitored by (a) the proton spectrum or (b) the cross-polarized spectrum.

gradients, but most take advantage of either differences in the chain dynamics (and relaxation times) or the proton chemical shift of the two phases.

Among the earliest methods used to establish a polarization gradient is the Goldman–Shen experiment, shown schematically in Figure 4.32 (37). This approach uses a difference in the T_2 relaxation time to establish a polarization difference between the two phases. After a $90°_{+x}$ proton pulse, a period of free proton evolution is allowed, during which the signals decay from T_2 relaxation. If we have a rigid phase with a short T_2 and a mobile phase with a longer T_2, then the signals from the rigid phase decay during τ_0, while those from the mobile phase do not. The signals from the mobile phase are then rotated back along the $+z$ axis with the $90°_{-x}$ proton pulse. If there is a large difference in T_2 between the phases, then a population difference between the phases is created with the pulse–delay–pulse part of the pulse sequence. This polarization difference is allowed to evolve during the mixing time τ_m. If the phases are in close proximity, then the polarization between the phases is equilibrated by spin diffusion before the state of the magnetization is measured with the final 90° proton pulse. Data are collected for a range of values for τ_m and the normalized intensity (Equation (4.4)) is plotted against $\sqrt{t}$. For very short spin-diffusion delay times only the more mobile material will be detected. As spin diffusion proceeds, the signals from the rigid phase will build up in intensity, and these signals can be plotted using the standardized spin-diffusion plot give by Equation (4.4). Alternatively, the decay in the mobile phase from proton spin diffusion can also be measured. It is often a good check on the consistency of the data to plot both the growth of the rigid phase and the decay of the mobile phase in the spin-diffusion plots. For polymers with a large difference in chain mobility (and relaxation times), a delay of 20–50 μs is sufficient to create a sharp polarization gradient. Such delays have been used, for example, to create polarization gradients between the crystalline and amorphous (and interfacial)

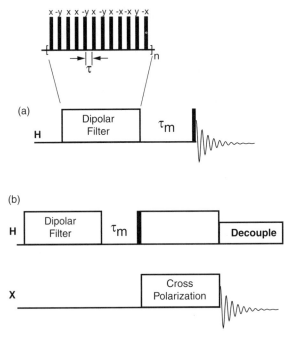

FIGURE 4.33 A pulse-sequence diagram for the dipolar filter experiment monitored by (a) the proton spectrum or (b) the cross-polarized spectrum.

phases of polyethylene at ambient temperature where the amorphous phase is above T_g (30). The difference in relaxation rates for a crystalline phase and an amorphous phase below T_g is smaller and the polarization gradient will not be as large. The Goldman–Shen experiment can be monitored either directly through the proton signals or indirectly through the carbon spectrum using a cross-polarization sequence for detection (Figure 4.32b).

Another effective way to create a polarization gradient based on differences in relaxation times between the phases is to use the dipolar filter pulse sequence shown in Figure 4.33 (38). The polarization gradient is established by applying a 12-pulse decoupling sequence to the protons. When the delay time between the pulses is short, the phases of the 12 pulses are such that they are a very effective decoupling sequence, so the polarization is returned to its equilibrium position after each 12-pulse cycle. As the delays become longer, polarization is lost from T_2 relaxation between the pulses, and the signals are saturated. The idea behind the dipolar filter is that the rigid phase with its strong dipolar couplings will be saturated when the pulse sequence has long delay times, while the mobile phase with its smaller dipolar couplings will not. The strength of the dipolar filter can be adjusted by changing the delay time τ between the pulses and the number of 12-pulse cycles to maximize the polarization gradient. If there is a large difference in dynamics between the phases, as in a crystalline phase and an amorphous phase above T_g, then sharp gradients can be created with several 12-pulse cycles with a delay time of 10–20 μs.

The Goldman–Shen or dipolar filter pulse sequences for studying spin diffusion rely on differences in chain dynamics to generate a sharp polarization gradient. Caution must be used, however, because polymer chains can have a large distribution of correlation times and the tails from these distributions may overlap, making it difficult to create sharp gradients. Furthermore, any interfacial material may be intermediate in dynamics between the rigid and mobile phases, and the mobility may depend on the placement of the chain relative to the rigid or mobile phases. For example, interfacial chains in a semicrystalline polymer may have dynamics more like the crystalline phase for those chains nearest the crystallites, but more like the amorphous phase for the interfacial chains furthest from the crystallites. A gradient in the chain dynamics of the interface that varies with the distance from the crystallites will often result is a more diffuse polarization gradient. The dipolar filter is a little more difficult to implement than the Goldman–Shen experiment, but this experiment can be more finely adjusted to maximize the polarization gradient.

The second class of experiments used to create polarization gradients for spin diffusion measurements rely on a chemical shift differences between protons in the two phases. The proton resolution is not high in most solid polymers, but it is often possible to discriminate between groups of protons, such as the aromatic and aliphatic protons. This makes it possible to selectively excite one group of protons and monitor the spin diffusion as the spin system returns toward equilibrium.

Figure 4.34 shows one of the pulse sequences used to monitor spin diffusion using a chemical shift based polarization gradient. In this sequence the protons evolve under

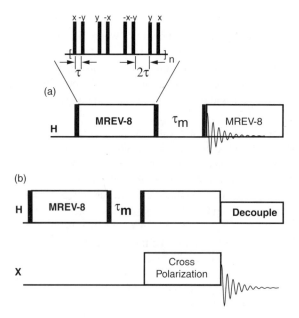

FIGURE 4.34 The pulse-sequence diagram for establishing a polarization gradient based on proton chemical shift differences monitored by (a) the proton spectrum and (b) the cross-polarization spectrum.

4.3 THE STRUCTURE AND MORPHOLOGY OF POLYMERS

the influence of multiple-pulse decoupling, which removes the dipolar couplings and allows the protons to precess under the influence of their chemical shifts. The basic element of the MREV-8 pulse sequence (introduced in Section 2.5.5) is an 8-pulse sequence with the pulses separated by a delay time (either τ or 2τ) (39–42). This same pulse sequence can be used for signal acquisition by acquiring a data point per pulse cycle during one of the 2τ delays.

The effect of the first MREV-8 pulse sequence is to allow the protons to evolve under the influence of the chemical shift Hamiltonian for a set period of time (43). The length of the multiple-pulse sequence is chosen such that a 180° phase difference develops between the two protons of interest (i.e., one along $+y$ and one along $-y$ in the rotating frame). At this point the multiple-pulse decoupling is stopped and a proton pulse is applied to tip one of the signals along the $+z$ axis and one along the $-z$ axis, thereby creating a polarization difference that evolves during the mixing time. The proton signals can be directly monitored using multiple-pulse decoupling during the acquisition time or indirectly monitored by cross polarization after the mixing time (Figure 4.34b).

Another approach is to use selective pulses to invert one of the protons of interest. One pulse sequence that can be used is the so-called delays alternating with nutation for tailored excitation (DANTE) pulse sequence shown in Figure 4.35 (44). The DANTE pulse sequence consists of a series of short pulses (of pulse width α) that are separated by a delay time τ. The frequency excitation profile for such a series of pulses (shown in Figure 4.35b) consists of a centerband and a series of sidebands that are spaced at $1/\tau$. The pulse sequence is used by placing the carrier frequency on one of the peaks of interest while making sure that the other peaks are not excited by one of the sidebands. The pulse tip angle for the on-resonance peak is $n^*\alpha$, where n is the total number of pulses. The largest polarization gradient exists when the peak of

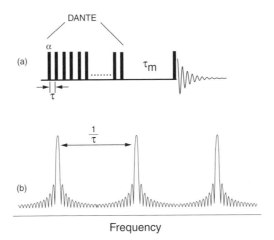

FIGURE 4.35 The (a) pulse sequence diagram for spin diffusion measurements by selective inversion using the DANTE pulse sequence. Part (b) shows the excitation profile.

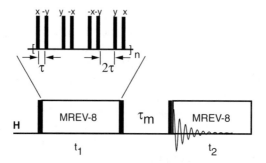

FIGURE 4.36 The pulse-sequence diagram for 2D exchange using multiple-pulse NMR.

interest is inverted with a 180° pulse. This occurs when

$$n^*\alpha = 180° \tag{4.20}$$

The width of the excitation peaks depends on the number of pulses, and a larger number of pulses (and a smaller α) gives a narrower excitation bandwidth. In a typical experiment n may be set to 20 ($\alpha = 9°$) and τ to 0.125 ms. This leads to a sideband separation of 8 kHz, placing the sidebands outside of the normal proton spectrum. The effect of spin diffusion on the spin populations is observed after the mixing time τ_m. As with the dipolar filter, the spectrum can be monitored directly (using protons or fluorines) or indirectly using cross polarization.

Spin diffusion can also be monitored by nD NMR methods. In one early example, the a pulse sequence using multiple-pulse decoupling in both the evolution and detection periods was used to study the length scale of mixing in a polymer blend using solid-state proton NMR (45). The pulse sequence (Figure 4.36) is a modified version of the pulse sequence introduced in Figure 4.34, where the first multiple-pulse evolution period is converted to the t_1 period in the 2D experiment. Spin diffusion can be measured from the cross-peak intensities as a function of mixing time.

Solid-state heteronuclear correlation experiments are another class of experiments that can be modified to monitor spin diffusion. Figure 4.37 shows the pulse-sequence diagrams for two variants of the heteronuclear correlation experiment that include mixing times to allow for spin diffusion. The pulse-sequence diagram in Figure 4.37a shows the heteronuclear correlation experiment often used with medium to slow (2–5 kHz) magic-angle sample spinning (46). A high-resolution spectrum is observed in the proton t_1 dimension because multiple-pulse BLEW-12 pulse sequence is applied on the proton channel and the BB-12 sequence is applied on the carbon channel to remove both proton–proton and proton–carbon dipolar couplings. BLEW-12 is an efficient 12-pulse windowless decoupling sequence that was introduced in Section 2.5.5. Following the mixing time a 24-pulse WIM sequence is applied to both the proton and carbon channels for cross polarization (46). The advantage of the WIM sequence is that proton spin diffusion is quenched during cross polarization, so in the absence of a mixing time the carbon signals are correlated with the closest protons.

4.3 THE STRUCTURE AND MORPHOLOGY OF POLYMERS

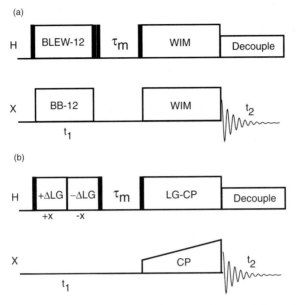

FIGURE 4.37 The pulse-sequence diagrams for heteronuclear correlation using (a) multiple-pulse decoupling and slow or medium speed magic-angle sample spinning and (b) frequency-switched Lee–Goldburg decoupling and fast magic-angle sample spinning.

The strategy behind using heteronuclear correlation to study spin diffusion is that the proton spectrum is more likely resolved in the 2D spectrum compared to the 1D spectrum so the magnetization transfer pathway can be more easily identified.

Another version of the heteronuclear correlation experiment (Figure 4.37b) is used with faster magic-angle sample spinning (47). Multple-pulse experiments frequently fail at high spinning speeds because decoupling becomes inefficient when the cycle time for the multiple-pulse sequence becomes comparable to a rotor period. In this version of the heteronuclear correlation experiment, the proton signals in the t_1 dimension are narrowed by the combination of fast magic-angle sample spinning and FSLG (Section 2.5.5) proton decoupling. After the mixing time cross polarization is achieved using the combination of Lee–Goldburg decoupling on the proton channel and ramped cross polarization on the carbon channel.

For studies of phase-separated polymer mixtures where the chains have very different molecular dynamics, 2D WISE NMR can be used to measure the length scale of mixing using spin diffusion (48). The pulse-sequence diagram for WISE is shown in Figure 4.38. WISE is a heteronuclear correlation experiment without line narrowing in the proton t_1 dimension, and the result is a correlation of the insensitive nuclei chemical shift with the proton lineshape. Since no decoupling is used in the t_1 dimension, the proton linewidth is narrowed only by chain motion. Therefore, polymers above T_g will give narrow lines (1–10 kHz), while semicrystalline or glassy polymers will give much broader (30–50 kHz) lineshapes. These linewidths are often much larger than the dispersion in proton chemical shifts. However, if the mixing

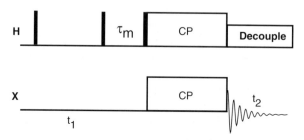

FIGURE 4.38 The pulse-sequence diagram for WISE correlation spectroscopy.

time is long relative to the spin diffusion time, than an averaged lineshape will be observed (48).

4.3.2.4 Proton Relaxation and Morphology The proton relaxation times in solids are strongly affected by spin diffusion and are much more likely to provide information about the morphology than the molecular dynamics in phase-separated polymer samples. These measurements cannot be used to accurately measure the domain sizes, but they can be used to estimate the upper limit on the length scale of mixing.

If we consider a two-phase polymer system in which the domain sizes are less than the length scale of spin diffusion, the relaxation rate k ($1/T_1$ or $1/T_{1\rho}$) is given by

$$k = k_a \frac{N_a \phi_a}{N_a + N_b} + k_b \frac{N_b \phi_b}{N_a + N_b} \tag{4.21}$$

where k_a and k_b are the relaxation rates for the pure phases, and N and ϕ are the number of protons and the mole fractions for the two phases. If the length scale of phase separation is less than the spin diffusion length scale, then a single averaged proton relaxation rate will be measured for both phases. The length scale of mixing L depends both on the relaxation rates and on the spin-diffusion coefficients and is approximately given by

$$L = \sqrt{\frac{6D}{k}} \tag{4.22}$$

Assuming a spin-diffusion coefficient of 10^{-15} m²/s (typical for a glassy polymer), the length scale of spin diffusion will depend on which relaxation time is measured. A typical polymer might have a proton spin-lattice relaxation time of 0.5 s, giving a length scale of spin diffusion of 17 nm. This same polymer may have a $T_{1\rho}$ relaxation time of 0.005 s, giving a length scale of spin diffusion of 1.7 nm.

These approximate measures of the length scale of phase separation in polymers are typically made using cross polarization to measure the proton T_1 and $T_{1\rho}$ (Sections 2.6.2.1 and 2.6.2.2) relaxation times. If the carbon signals for the phases are resolved, then relaxation times are measured for both phases and compared. If the length scale

4.3 THE STRUCTURE AND MORPHOLOGY OF POLYMERS

of phase separation is less than the spin diffusion length scale, then the same, averaged relaxation rate will be measured for both phases. In the event that the length scale of phase separation is slightly larger than the spin-diffusion length scale, nonexponential relaxation will be observed, and if the polymers are phase-separated on a longer length scale, the relaxation times will be the same as those observed for the homopolymers. To use the proton relaxation times to estimate the length scale of phase separation, it is important that the relaxation times for the pure polymers differ by more than a factor of 2. The T_1 and $T_{1\rho}$ relaxation times are often used to estimate the length scale of mixing in polymer blends (49).

4.3.3 Semicrystalline Polymers

The morphology of semicrystalline polymers is critically related to the bulk properties, and has been extensively investigated. The two main parameters investigated in these studies are the fraction of polymer in the crystalline phase and the sizes of the various polymer domains. Semicrystalline polymers are amenable to study by spin diffusion, since there are often large differences in the molecular dynamics between the phases, making it easy to create a polarization gradient using one of the methods outlined in Section 4.3.3.2.

The molecular dynamics of the crystalline and amorphous phases are often very different from each other, and they often have very different spin-spin and spin-lattice relaxation times. In favorable cases, the differences in relaxation times can be used to measure the fraction of polymer in the various phases. This is illustrated in Figure 4.39,

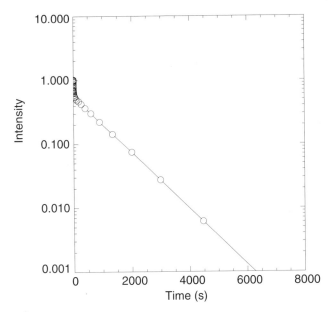

FIGURE 4.39 The semilog plot of the spin-lattice relaxation for semicrystalline polyethylene.

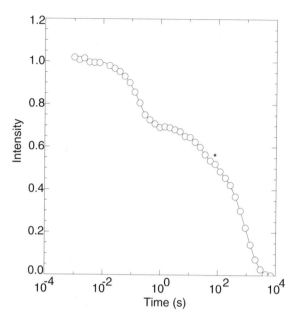

FIGURE 4.40 A semi-log plot of the spin-lattice relaxation in semicrystalline polyethylene.

which shows a semilog plot of the spin-lattice relaxation for a polyethylene sample measured with the cross polarization T_1 experiment (Figure 2.26). In the standard plot where the log of the intensity is plotted vs. time, a curved plot is indicative of multiple relaxation times that are associated with the different phases. The data can be visualized better with a plot of the intensity vs. the log of the time. There are three decay processes in Figure 4.40 that can be assigned to the amorphous, interfacial and crystalline phases. The relaxation is fit to a three exponential fit using the equation

$$y = a_1 e^{-t/T_{1a}} + a_2 e^{-t/T_{1b}} + a_3 e^{-t/T_{1c}} \qquad (4.23)$$

where a represents the amplitude of each decay. The fit to the data gives relaxation times for the amorphous, interfacial, and crystalline phases of 0.17, 25, and 995 s. The fractions of material in each phase can be roughly estimated from the amplitudes as 0.45, 0.09, and 0.46 for the amorphous, interfacial, and crystalline phases.

The crystallinity in polymers can be estimated by a number of methods, and in many cases the numbers are not in exact agreement because the methods are each sensitive to differences in the samples. In the case of NMR relaxation, the crystallinity is measured from the fraction of molecules that have very slow molecular dynamics (and relaxation times). Polymer chains at the edge of the crystallites, for example, may have slightly different dynamics and may be classified as interfacial (as determined by the T_1). These same chains may be considered as crystalline by other methods. In addition, the NMR measurement of crystallinity from the cross-polarization T_1 may differ from other methods because the phases cross polarize at different rates.

4.3 THE STRUCTURE AND MORPHOLOGY OF POLYMERS

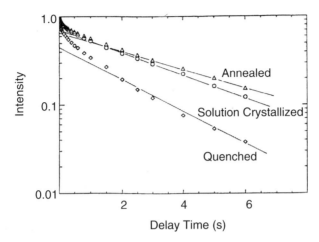

FIGURE 4.41 A semilog plot of the deuterium spin-lattice relaxation in semicrystalline poly(ethylene oxide).

It is important to make the crystallinity measurements at temperatures below the T_g of the amorphous phase where the cross-polarization dynamics of the phases are more likely to be similar. In Section 4.2.1 we showed how the combination of cross polarization and direct polarization can be used to measure the crystallinity in polyethylene samples.

Nuclei other than carbon can also be used to measure the crystallinity in polymers from the T_1 relaxation times. Since the proton relaxation rates can be averaged by spin diffusion, they cannot usually be used to measure the crystallinity. Figure 4.41 shows the deuterium spin-lattice relaxation behavior for a series of poly(ethylene oxide)-d_4 samples that have been subjected to different thermal histories. Poly(ethylene oxide) has three phases, but only two relaxation times can be resolved. The crystallinity in this case is measured by extrapolating the longest relaxation time back to the y axis.

The domain sizes are often of interest in semicrystalline polymers, and in favorable cases they can be measured from proton spin diffusion. The structure of polyethylene was studied in one of the first, classic studies of polymer morphology by spin diffusion (30). The polarization gradient was created using the Goldman–Shen pulse sequence (Figure 4.32) with a 42 μs delay between the first two pulses. The spectrum was not observed in these early experiments, but rather the FIDs were fit to rapidly and slowly relaxing components that were assigned to the rigid and mobile phases. The fits of the amplitudes of the two components as a function of mixing time were used to measure the rate of spin diffusion. It was not possible to distinguish between different morphologies from the spin-diffusion data, and the 1D (lamellar), 2D (cylindrical), and 3D (spheres) morphologies all gave good fits to the data (with different domain sizes). The domain sizes for the different morphologies ranged between 5 and 15 nm.

More recently, both low- and high-density polyethylene have been studied by spin-diffusion methods using more sophisticated pulse sequences and morphological modeling (28). Figure 4.42 shows the spectrum of a low-density polyethylene sample

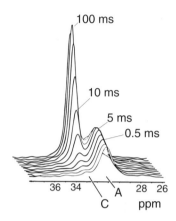

FIGURE 4.42 The carbon spectrum of semicrystalline polyethylene following the application of the dipolar filter pulse sequence.

as a function of mixing time after the application of the dipolar filter. The carbon spectrum was measured by cross polarization following a dipolar filter as a function of the spin-diffusion mixing time, and the relative intensities of the crystalline and amorphous phases were measured from the crystalline and amorphous peaks in the carbon spectra at 32.8 and 31.0 ppm. The data could not be adequately fitted using a simple two-phase model, suggesting that the amorphous and crystalline phases were separated by an interfacial layer that was not detected in the earlier experiments. Direct evidence for the presence of an interfacial layer was obtained by monitoring the intensity of the amorphous component as a function of the spin-diffusion time. Following saturation of the rigid protons with the dipolar filter, the intensity of the amorphous phase reaches a maximum after 5 ms, indicating that there is spin diffusion between the mobile amorphous phase and a less mobile amorphous material, the interface. At a longer delay time (100 ms) there is extensive spin diffusion to the crystalline phase. The diffusion behavior was fitted with a three-phase model containing crystalline, amorphous, and interfacial phases. Despite the large differences in the crystallite thicknesses for the high (40 ± 10 nm) and low (9 ± 2.5 nm) density polyethylene, the thickness of the interfacial layer was similar (2.2 ± 0.5 nm) in both samples.

Poly[bis(3-methylphenoxy)phosphazine]

4.3 THE STRUCTURE AND MORPHOLOGY OF POLYMERS

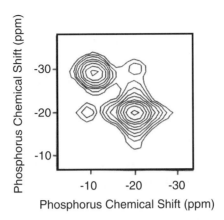

FIGURE 4.43 The ^{31}P 2D exchange spectrum for poly[bis(3-methylphenoxy)phosphazine].

Proton spin diffusion is most frequently used to make domain-size measurements in polymers because the protons have a high natural abundance and the relaxation behavior is easily measured. Other nuclei can also be used, if they are present at a high natural abundance. This is illustrated in Figure 4.43, which shows the ^{31}P 2D exchange spectrum for a semicrystalline sample of poly[bis(3-methylphenoxy)phosphazine] (50). The spin exchange spectrum was measured using a modified experiment where the cross-polarization pulse-sequence element was substituted for the first 90° pulse in the 2D exchange experiment. The cross peaks in Figure 4.43 show the exchange of magnetization between the crystalline and amorphous peaks. There are two possible mechanism by which the magnetization may be exchanged, either by spin diffusion or by chain diffusion between the crystalline and amorphous phases. These possibilities can be distinguished by measurements as a function of temperature, since spin diffusion is not strongly temperature dependent, while chain diffusion is. The small temperature dependence to the exchange spectrum of poly[bis(3-methylphenoxy)phosphazine] shows that the cross peaks are a consequence of spin diffusion.

While the 2D spectra can be used to study spin diffusion, it is often more efficient to use 1D versions of the experiments. The 2D exchange pulse sequence can be used in a 1D experiment with a fixed time between the cross polarization and the first 90° pulse on the insensitive nuclei. This t_1 evolution time is chosen to be

$$t_1 = \frac{1}{2\Delta\omega} \qquad (4.24)$$

where $\Delta\omega$ is the difference in frequency between the crystalline and amorphous peaks. During the t_1 period the signals for the crystalline and amorphous phases precess 180° out of phase so that the next pulse rotates one along the $+z$ axis and one along the $-z$ axis, and the return to equilibrium is monitored after a spin diffusion delay time. Figure 4.44 shows the spectra from this experiment on the poly[bis (3-phenoxy)phosphazine] (51). The ^{31}P spin-diffusion coefficient is assumed to follow

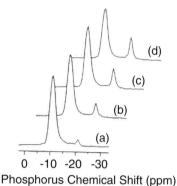

FIGURE 4.44 The phosphorus spin diffusion spectra using a fixed evolution time for poly[bis(phenoxy)phosphazine]. The spectra were obtained for delay times of (a) 0.2, (b) 1, (c) 2 and (d) 4 s.

the same mechanism as for proton spin diffusion, but must be corrected for the differences in the magnetogyric ratio and the density of phosphorus spins vs. protons. This was accomplished by scaling the proton spin-diffusion coefficient of polyethylene ($D_{PE} = 6.2 \times 10^{-15}$ m^2/s) (34) as

$$D(^{31}P) = D(^{1}H)\frac{\rho_P^{1/3}\gamma_P^4}{\rho_H^{1/3}\gamma_H^4} \qquad (4.25)$$

where ρ_P and ρ_H are the phosphorus and proton densities, giving a value of 4.8×10^{-17} m^2/s for the phosphorus spin-diffusion coefficient. This was used to calculate a lamellar domain size of 4.5 nm with a spacing between crystallites of 9.0 nm (50).

$$\underset{\text{Nafion}}{+CF_2-CF_2+CF_2-\underset{\underset{CF_3}{|}}{CF}+}$$
$$O-CF_2-CF-O-CF_2-CF_2-SO_3^-H^+$$

Fluorine is another abundant nucleus that can be used to investigate the morphology of semicrystalline polymers. Fluorine spin diffusion has been used to investigate the structure of Nafion swollen with water and ethanol. A high-resolution ^{19}F spectrum can be observed for Nafion with magic-angle sample spinning, as shown in Figure 4.45. Since the peaks for the side-chain fluorines are well resolved, they can be selectively inverted with a DANTE-type pulse sequence (44) to create a polarization gradient. Figure 4.46 shows the relaxation of the selectively inverted peaks from the amorphous domains swollen with 10% ethanol as a function of spin-diffusion time. The selective pulses excites the side chain CF$_2$ and CF$_3$ peaks at −80 ppm. This magnetization quickly exchanges with other side-chain fluorines and, on a longer time scale, with the crystalline domains that give rise to the peak at −120 ppm.

4.3 THE STRUCTURE AND MORPHOLOGY OF POLYMERS

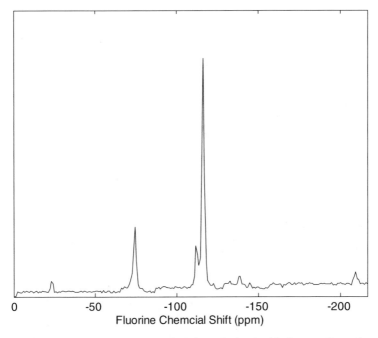

FIGURE 4.45 The fluorine spectrum of Nafion obtained with fast magic-angle sample spinning.

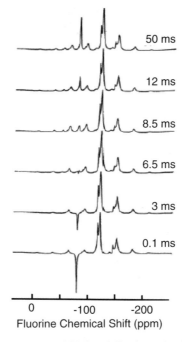

FIGURE 4.46 The fluorine spectra of Nafion following selective inversion of the peak at −80 ppm.

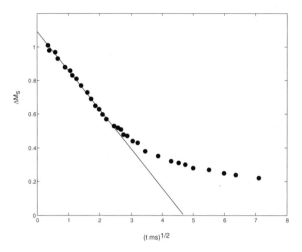

FIGURE 4.47 The spin diffusion plot for Nafion containing 10% ethanol.

Figure 4.47 shows a normalized spin-diffusion plot from which the intercept ($\sqrt{t_{SD}}$) is used to calculate the length scale of spin-diffusion using Equation (4.12). Since the properties and densities of fluorine are similar to protons, the proton spin-diffusion coefficients were used to calculate the domain sizes (12). In the studies with the swollen samples, it was assumed that all of the solvent is absorbed by the pendant groups, which changes the volume fraction of the pendent groups. The spin-diffusion times, volume fractions, and domain sizes for the Nafion swollen by water and ethanol are listed in Table 4.2. The data were interpreted using a 1D morphology. The domain sizes must be considered semiquantitative, since a number of assumptions are used in the calculations.

4.3.4 Block Copolymers

Polymer relaxation and spin diffusion have been used to investigate the morphology of a wide variety of phase-separated polymer systems, ranging from block copolymers to nanocomposites. The NMR methods are of importance because the length scale

TABLE 4.2 The Spin-diffusion Times, Volume Fractions, and Domain Sizes for Nafion Swollen with Water and Ethanol.

Penetrant (%)	$\sqrt{t_{SD}}(ms^{\frac{1}{2}})$	f_a	L(nm)	Pendant Domain Size (nm)
0	2.57	0.33	11.4	3.80
10% (H_2O)	2.73	0.45	10.1	4.59
20% (H_2O)	3.44	0.56	11.8	6.56
10% EtOH	4.68	0.48	16.9	8.12
17% EtOH	5.60	0.56	19.1	10.7
40% EtOH	10.80	0.75	37.7	28.4

4.3 THE STRUCTURE AND MORPHOLOGY OF POLYMERS

of phase separation is difficult to investigate by other means. In addition, NMR has been used to investigate the structure of block copolymers samples that are well characterized by other means to calibrate the spin-diffusion coefficients. These studies will be of increasing importance as synthetic methods advance and more complex block copolymers are prepared.

Block copolymers have contributed greatly to our understanding of spin diffusion, since it is possible to characterize the NMR samples by small-angle X-ray scattering (SAXS) and transmission electron microscopy (TEM). In one study of symmetric diblock poly(styrene-b-methyl methacrylate) copolymers the comparison between the SAX, TEM, and NMR data were studied to calibrate the spin-diffusion coefficients in glassy polymers (31). The morphologies (dimensionalities and domain sizes) were measured by SAX and TEM, and the spin-diffusion coefficients were adjusted to give the best agreement with the known structures. The poly(styrene-b-methyl methacrylate) block copolymers are among the more difficult to investigate because the blocks have similar T_g's and dynamics (and relaxation times). The polarization gradients in these studies were created based on the proton chemical shift differences between the styrene and methyl methacrylate monomers, and both the proton and (cross-polarized) carbon signals were observed. The best fit to the morphological data was obtained with a spin-diffusion coefficient of 0.8×10^{-15} m^2/s (or 0.8 nm^2/s), and this diffusion coefficient is often used for the study of glassy polymers.

Block copolymers of styrene and isoprene and poly(ethylene oxide) and poly(hydroxyethyl-methyl methacrylate) were investigated to calibrate the spin-diffusion coefficients in mobile polymers (36). The results showed that the spin-diffusion coefficients can be correlated with the static spin-spin relaxation times, as given by Equations (4.18) and (4.19). It is important to note that the dipolar couplings in polymers as mobile as polyisoprene are on the same order of magnitude as the spinning speeds, so magic-angle spinning can lead to partial averaging of the dipolar couplings and the spin-diffusion coefficients. For this reason, it is better to perform spin-diffusion measurements in the absence of magic-angle spinning for mobile polymers.

Poly(styrene-b-isoprene-b-styrene)

Proton spin diffusion has been used to investigate the phase structure in poly(styrene-b-isoprene-styrene) triblock copolymers. The domain size in the triblock copolymer is of interest since the film morphology is strongly dependent on the casting solvent. Films cast from CS$_2$ have an ordered hexagonal structure and it

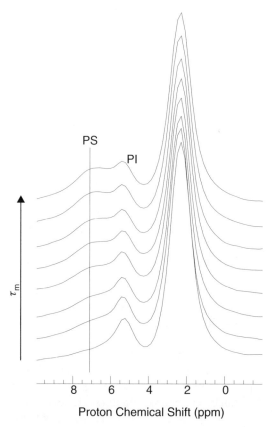

FIGURE 4.48 The proton spectra of the poly(styrene-*b*-isoprene-styrene) triblock copolymer as a function of delay time after the dipolar filter.

is of interest to determine if the gross morphology can be related to differences in the domain structures for the blocks on the nm length scale (52).

There is a large difference in the chain dynamics for the styrene and isoprene blocks, so it is possible to create a polarization gradient from the difference in dynamics between the chains. In this study, the dipolar filter (Figure 4.33) was used to create the polarization gradient and the signal was detected after the spin-diffusion delay time using multiple-pulse MREV-8 homonuclear proton decoupling. Figure 4.48 shows a stacked plot of the spectra as a function of the spin-diffusion delay time. Six cycles of the dipolar filter sequence with 9-μs delays between pulses was applied to saturate the styrene signals while retaining the magnetization from the polyisoprene chains.

Although the resolution is not very high in the solid-state proton spectrum, the aromatic polystyrene and olefinic polyisoprene signals can be resolved at 7 and 5.5 ppm. The peak at 2 ppm contains the overlapped signals from the remaining styrene and isoprene protons. At the shortest delay times the styrene signals are saturated and they recover initially from spin diffusion from the isoprene domains. This leads to a

4.3 THE STRUCTURE AND MORPHOLOGY OF POLYMERS

decrease in the intensity of the isoprene signals at 5.5 ppm and the overlapping peak at 2.0 ppm. As the spin-diffusion time becomes long, it becomes necessary to correct for spin-lattice relaxation using the measured relaxation rate.

Very little difference is observed between the spin-diffusion data for the films cast from carbon disulfide and tetrahydrofuran, showing that changes in the gross morphological features are not associated changes in the domain sizes. The domain sizes can be calculated from intercept ($\sqrt{t_{SD}}$) of the spin-diffusion plots, the volume fractions, and the spin-diffusion coefficients (0.8 nm^2/ms and 0.05 nm^2/ms for polystyrene and polyisoprene) (Equation (4.12)). The domains for the films cast from tetrahydrofuran and carbon disulfide were calculated by assuming a 3D morphology with polyisoprene as the minority phase. The results showed 3.5-nm domains with a spacing between domains of 5 nm for both films.

$$\left[CH_2-CH_2-O \right] \left[CH_2-\underset{\underset{CH_3}{|}}{CH}-O \right]$$

Poly(ethylene oxide-*b*-propylene oxide)

Spin diffusion using the dipolar filter has also been used to investigate the domain sizes in poly(ethylene oxide-*b*-propylene oxide) diblock and poly(ethylene oxide-*b*-propylene oxide-*b*-ethylene oxide) triblock copolymers and organic/inorganic hybrids (53). In this case, both blocks of the copolymer have a low T_g in the amorphous phase, and differences in the dynamics cannot be used to create a polarization gradient. However, the ethylene oxide block crystallizes in the diblock copolymer, so the difference in chain dynamics between the crystalline ethylene oxide and the amorphous propylene oxide can be used to create a polarization gradient.

Figure 4.49 shows the solid-state proton NMR spectrum of the diblock copolymer obtained with 15-kHz magic-angle sample spinning as a function of the delay time in the dipolar filter pulse sequence. Two peaks are observed in the proton spectra that can be assigned to the methyl protons from the propylene oxide at 1 ppm and the main-chain ethylene oxide and propylene oxide methylene and methine protons. Based on the molecular weights of the blocks (5500 g/mol for the ethylene oxide and 1100 g/mol for the propylene oxide) the expected peak intensity ratio for the main-chain and methyl signals is 9.7:1, while the observed ratio is closer to 2.5:1. This shows that some intensity is missing from the spectrum. The missing intensity is assigned to the crystalline ethylene oxide. The dipolar couplings for crystalline ethylene oxide are on the order of 50 kHz, so they will not be averaged by 15-kHz magic-angle sample spinning. The signal at 3.5 ppm is assigned to the main-chain signals from propylene oxide and the amorphous phase of ethylene oxide.

Figure 4.49 shows that intensity is progressively lost from the spectrum as the strength of the dipolar filter is increased as the delay between pulses is changed from 0 to 25 μs. With a strong dipolar filter ($\tau = 25$ μs), the relative ratio of the peaks at 3.5 and 1.0 ppm approaches 1:1, the value expected for the propylene oxide alone. This shows that the ethylene oxide block (both the crystalline and amorphous phases) can be selectively saturated with the dipolar filter.

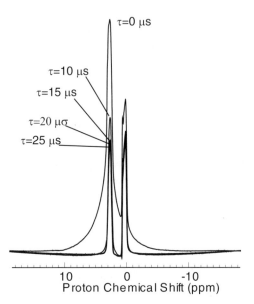

FIGURE 4.49 The solid-state proton spectrum of the poly(ethylene oxide-*b*-propylene oxide) diblock copolymers obtained with 15-kHz magic-angle sample spinning as a function of the strength of the dipolar filter.

It was noted earlier that fast magic-angle sample spinning partially averages the dipolar couplings and can affect the spin diffusion coefficients used to calculate the domain sizes. Therefore, it is desirable to perform the spin-diffusion measurements without spinning, or with spinning at a low speed (~2 kHz in Figure 4.48). The strategy used in the study of the poly(ethylene oxide-*b*-propylene oxide) diblock copolymer was to identify the rigid and mobile phases using the dipolar filter with fast magic-angle sample spinning, but perform the actual spin diffusion measurements without magic-angle sample spinning.

Figure 4.50 shows the static spin-diffusion measurement for the poly(ethylene oxide-*b*-propylene oxide) diblock copolymer. In the absence of magic-angle spinning, only a single broad line is observed with a linewidth of about 3 kHz. Since we know from the fast magic-angle spinning studies that the ethylene oxide block is rigid, the 3-kHz broad line must result from the propylene oxide block alone. With such line widths, the signals from the main-chain methine and methylene near 3.5 ppm are not resolved from the methyl peak at 1 ppm. The ethylene oxide block is saturated with a strong dipolar filter using a 25-μs delay between pulses. The observed signal intensity is at a maximum immediately after the dipolar filter and the intensity decreases as magnetization is exchanged with the ethylene oxide block. The change in intensity is about 30%. Figure 4.51 shows the normalized spin-diffusion plot for the diblock copolymer from which an intercept ($\sqrt{t_{SD}}$) 0.20 s$^{1/2}$ is measured.

Propylene oxide is a mobile polymer, so the spin-diffusion coefficient is determined from correlation between the static T_2 and the diffusion coefficient presented in Equations (4.18) and (4.19). The T_2 was measured using the Carr–Purcell

4.3 THE STRUCTURE AND MORPHOLOGY OF POLYMERS

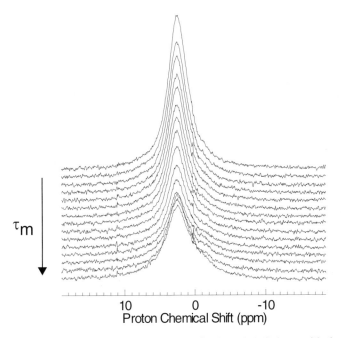

FIGURE 4.50 A static spin-diffusion measurement for the poly(ethylene oxide-*b*-propylene oxide) diblock copolymer.

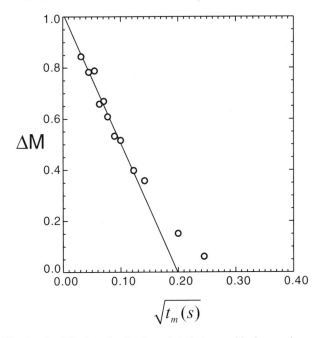

FIGURE 4.51 A spin-diffusion plot for the poly(ethylene oxide-*b*-propylene oxide) diblock copolymer.

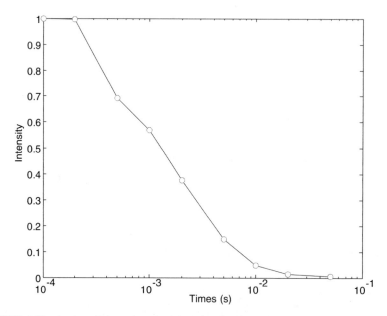

FIGURE 4.52 A plot of the spin-spin relaxation for the poly(ethylene oxide-*b*-propylene oxide) diblock copolymer.

$90°-\tau-180°-\tau$-acquire pulse sequence (Section 2.6.1.2). Figure 4.52 shows a plot of the spin-diffusion decay. Multiexponential behavior is observed, since the sample contains crystalline and amorphous ethylene oxide as well as amorphous propylene oxide. The slowest relaxation rate (116 s^{-1}) is assigned to the mobile propylene oxide block. Using Equation (4.18), this corresponds to a spin-diffusion coefficient of 0.017 nm^2/ms. The diffusion coefficient for the crystalline ethylene oxide block is assumed to be the same as measured for rigid glassy polymer (0.8 nm^2/ms).

The spin-diffusion coefficients and the intercept from the normalized spin-diffusion plots are used to calculate the domain sizes for the diblock copolymer. Table 4.3 lists the other parameters used for the domain-size calculations, including the proton densities and the volume fractions for the ethylene oxide and propylene oxide blocks. These data, along with a dimensionality ε of three, are used with Equation (4.12) to calculate the length scale of spin diffusion L of 11.4 nm. Using Table 4.1, this corresponds to a minor domain size of 5.8 nm and a separation between domains of 9.2 nm.

TABLE 4.3 The Parameters for Domain-size Calculations in the Poly(ethylene Oxide-*b*-propylene Oxide) Diblock Copolymer.

Polymer	D (nm^2/ms)	ρ (g/cm^3)	ρ^H (g/cm^3)	f
Ethylene oxide	0.8	1.15	0.104	0.83
Propylene oxide	0.017	1.04	0.107	0.17

4.3 THE STRUCTURE AND MORPHOLOGY OF POLYMERS

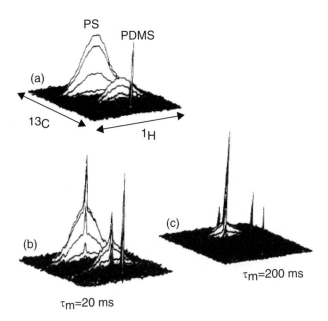

FIGURE 4.53 The 2D WISE spectra of the poly(styrene-b-dimethyl siloxane) copolymer with spin diffusion delays of (a) 0, (b) 0.02 and (c) 0.2 s.

Multidimensional NMR has also been used to investigate the morphology of block copolymers. The NMR experiments used to investigate the morphology usually have a spin-diffusion period incorporated into a correlation or resolved experiment. In Section 1.7.3 the solid-state 2D exchange, heteronuclear correlation, and WISE experiments were introduced with an optional mixing time for spin diffusion. In the WISE experiment, for example, the carbon chemical shift is typically correlated with the proton line width to reveal information about the molecular dynamics (48). In phase-separated block copolymers where the dynamics of the blocks are very different from each other, very different linewidths will be observed in the proton dimension. This is illustrated in Figure 4.53, which shows the WISE spectrum of a 50:50 poly(styrene-b-dimethyl siloxane) copolymer with no mixing time, and with mixing times of 0.02 and 0.2 s. With no spin-diffusion delay time, the styrene carbon signals are correlated with broad lines in the proton dimension, while the poly(dimethyl siloxane) is correlated with a very sharp line, as expected for a rigid and mobile polymer. As the spin-diffusion time is increased, the lineshapes for the rigid and mobile phases begin to mix and both a sharp and broad component are observed for the styrene and dimethyl siloxane signals.

It is of interest to note that relatively homogeneous proton lineshapes are observed for the styrene and dimethyl siloxane signals with no spin diffusion delay. This shows that nearly all of the styrene is immobile and nearly all of the dimethyl siloxane is mobile. Such an observation is only possible if there is a sharp interface between the phases with little or no interfacial material. This behavior can be contrasted with that observed for the 86:14 poly(styrene-*b*-methylphenyl siloxane) copolymer, where two-component lineshapes were observed for each block, demonstrating that a substantial interface exists between the two phases containing mobile polystyrene and immobile poly(methylphenyl siloxane) (48).

The mixing of the lineshapes as a function of the spin-diffusion delay time is a consequence of spin diffusion, and the length scale of spin diffusion can be qualitatively estimated from the time required to average the lineshapes. While such studies provide important qualitative information, they cannot yield the quantitative information available from the dipolar filter, Goldman–Shen, or spin exchange experiments. In addition the 1D spin-diffusion experiment can be performed and analyzed faster.

4.3.5 Multiphase Polymers

Sold-state NMR has been used to study the structure and morphology of a wide variety of multiphase polymer systems, in addition to the phases in semicrystalline polymers and the block copolymers. These systems can be classified based on the length scale of phase separation and on the types of materials. The chains are mixed on a molecular length scale in miscible polymer blends, and special methods are used to measure spin diffusion over the length scale of 0.2–0.5 nm (Section 4.3.6). The methods previously described (dipolar filter, Goldman–Shen, spin exchange, etc.) can be used to study materials phase-separated on the length scale of 1–20 nm, including polyurethanes and organic–inorganic hybrids, that are phase separated on a longer length scale.

Spin diffusion and other methods have been used to study the morphology of polyurethanes containing hard and soft segments. These materials have useful mechanical properties in part because the hard segments are able to cluster into phase-separated islands. NMR provides an insight into the structure of polyurethanes, because we can use spin diffusion to measure the domain sizes and other methods to estimate the fraction of hard and soft segments that are dissolved in the opposite phases.

Polyurethanes were among the first materials studied by spin-diffusion, in part because there is a large difference in the dynamics of the hard and soft phases. The proton NMR free-induction decay of polyurethane has two components that are assigned to the mobile and rigid phases. This allows spin-diffusion in polyurethanes to be studied using the Goldman–Shen experiment with a 75-μs delay between the first two pulses (35). The spin diffusion results were fit to lamellar model with an interface. The domain spacings for different sample (8.3–9.6 nm) were in good agreement with the domain sizes measured by small-angle X-ray diffraction.

The degree of phase separation between the hard and soft segments has been investigated by a number of NMR methods, including 2D WISE NMR and wideline deuterium NMR. Figure 4.54 shows the WISE spectrum for a polyurethane with hard segments made from 4,4′-methylenebis(phenyl isocyanate) (MDI) and butanediol and soft segments of propylene oxide (54). The narrow peaks in the spectrum are assigned to the methyl and main-chain carbons of the propylene oxide at 18 and 75 ppm, while the broad peaks between 110 and 140 ppm are assigned to aromatic carbons in the hard segments. The lower plot shows the cross section through the soft segment

4.3 THE STRUCTURE AND MORPHOLOGY OF POLYMERS

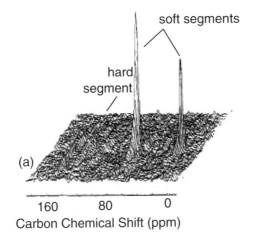

FIGURE 4.54 The (a) 2D WISE spectrum for the MDI-BD urethane polymer and (b) cross section through the soft segment showing the proton line shape.

showing the proton lineshape. Only a narrow line is observed for the soft segment and only broad lines are observed for the hard segment, demonstrating that the system is completely phase separated with very little interface.

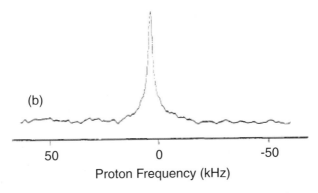

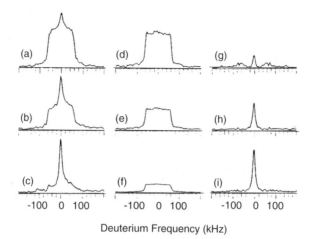

FIGURE 4.55 The solid-state ^{2}H spectra of hard-segment labeled polyurethanes. The three rows correspond to urethanes labeled with 70, 60 and 50% hard segments. The middle column shows the relative intensity of the immobile segments and the right column shows the contribution from the mobile hard segments.

The differences in the dynamics for hard and soft segments of the polyurethanes changes not only the proton lineshapes, but also other lineshapes and relaxation times. This is illustrated in Figure 4.55, which shows the deuterium NMR spectra for a series of polyurethanes in which the central part of the butanediol in the hard segments are labeled with deuterons (55). The soft segments are made of block copolymers of ethylene oxide and propylene oxide. As noted in Section 1.6.5, restricted deuterons give rise to a very broad line (120 kHz) with a "Pake" lineshape. Molecular motion averages these lineshapes, so the lineshapes can be used to measure the molecular mobility. The deuterium spectra for the polymers containing 70, 60, and 50% deuterium-labeled hard segments show a two-component lineshape, with the broad component associated with a restricted environment and the narrow component associated with the more mobile environment. These data show that as the fraction of the soft segment increases, an increasing fractions of the hard segments are dissolved in the mobile phase. Virtually none of the hard segments are mobile in the pure hard-segment polymer, while about half of the hard segments are dissolved in the mobile (soft-segment) phase for the polymer containing a 50:50 mixture of hard and soft segments.

Poly(methyl silsesquioxane)

4.3 THE STRUCTURE AND MORPHOLOGY OF POLYMERS

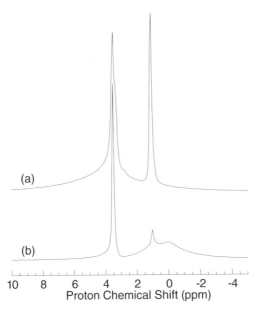

FIGURE 4.56 The solid-state proton NMR spectrum of (a) the poly(ethylene oxide-*b*-propylene oxide-*b*-ethylene oxide) copolymer, and (b) the organic–inorganic composite prepared with 15 wt % polymer. The spectra were acquired with 15-kHz magic-angle sample spinning.

More recently, the spin-diffusion and 2D NMR methods have been used to investigate the morphology in organic–inorganic hybrids for low dielectric-constant applications (56). These hybrid materials are created by mixing sacrificial diblock or triblock copolymers of ethylene oxide and propylene oxide with a poly(methyl silsesquioxane) prepolymer before curing of the inorganic phase. The curing leads to phase separation of the sacrificial polymer, and the performance of the low dielectric-constant devices depends on the length scale of phase separation.

Figure 4.56 compares the solid-state proton NMR spectrum of the bulk poly(ethylene oxide-*b*-propylene oxide-*b*-ethylene oxide) copolymer and the organic–inorganic composite prepared with 15% of a triblock copolymer in a poly(methyl silsesquioxane) matrix. Three groups of signals are resolved in the proton spectrum observed with 15-kHz magic-angle sample spinning, the signals at 3.5 and 1.0 ppm from the main-chain and methyl signals of the triblock copolymer, and the signals near 0 ppm from the poly(methyl silsesquioxane) methyl protons. The dipolar couplings for rigid (crystalline or glassy) protons are near 50 kHz and will not be observed in the spectrum with 15-kHz magic-angle sample spinning. As with the bulk diblock copolymer (Figure 4.49), the expected relative intensities of the polymer peaks at 3.5 and 1 ppm can be calculated from the molecular weights of the ethylene oxide and propylene oxide block (4532 and 2262 g/mol for the ethylene oxide and propylene oxide blocks, respectively). The data for the bulk triblock copolymer show a peak intensity ratio of 2:1, which is much less than the expected value of 8:1. This shows, as with the diblock copolymer, that a large fraction of the ethylene oxide is crystalline and cannot be observed with 15-kHz magic-angle sample spinning. In the composite, however, the peak intensity ratios are closer to the expected value, showing that composite formation suppresses the formation of crystallinity in the ethylene oxide block. Thus, both polymer blocks are mobile in the composite. In fact, the polymer domains are so mobile that it is difficult to observe their carbon spectrum using cross polarization.

The signal at 0 ppm is assigned to the poly(methyl silsesquioxane) methyl signal in the inorganic matrix. At first it might be surprising that such a signal can be observed with 15-kHz magic-angle sample spinning, since the cured silsesquioxanes is expected to be a rigid material. The relatively sharp proton signal is a consequence of two factors: averaging of the dipolar couplings from methyl group rotation, and the low density of protons. Since the protons have a much lower density in the poly(methyl silsesquioxane) compared to most organic materials, smaller dipolar couplings are expected.

Since a high-resolution proton spectrum is observed for the composite, the 2D exchange (or NOESY) experiment can be directly used to probe the structure in the solid state (53). Figure 4.57 shows the 2D exchange spectrum obtained for the composite with 15-kHz magic-angle sample spinning. Three peaks are observed along the diagonal that can be assigned to the polymer main-chain and methyl peaks at 3.5 ppm and 1.0 ppm, and the poly(methyl silsesquioxane) methyl peak at 0 ppm. The off-diagonal intensity shows that there is spin exchange between all of these peaks, but it is difficult to measure the rate of spin diffusion from such plots.

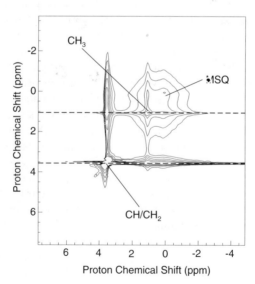

FIGURE 4.57 The 2D exchange spectrum for the triblock copolymer composite with poly(methyl silsesquioxanes).

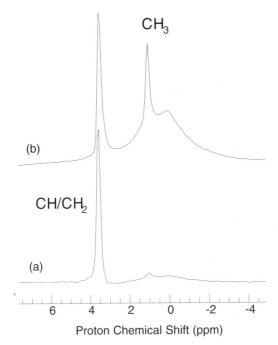

FIGURE 4.58 Cross sections through the 2D exchange spectrum for the triblock copolymer composite at the frequency of (a) the main-chain protons at 3.5 ppm and (b) the methyl protons at 1 ppm.

A better insight into the spin exchange process can be obtained by examining cross sections through the 2D spectra at the frequencies of the main-chain peak at 3.5 ppm and the methyl peak at 1.0 ppm, as shown in Figure 4.58. The largest peak in the cross sections is due the diagonal signal. The cross section through signal at 3.5 ppm (which is due mostly to the ethylene oxide methylene protons) shows only a small amount of exchange to the propylene or poly(methyl silsesquioxane) oxide methyl signals. Cross sections through the propylene oxide methyl signal show very efficient spin exchange to the poly(methyl silsesquioxane) methyl signal. This experiment shows that the polymer domains in the composite adopt a core-shell structure with the ethylene oxide at the center and the propylene oxide at the interface with the poly(methyl silsesquioxane).

The 2D exchange experiments provide a qualitative insight into the structure of the poly(methyl silsesquioxane) composites, but they are difficult to use directly to measure the domain sizes. Since there is a large difference in the dynamics between the polymer and the matrix, the dipolar filter can be used to create a polarization gradient to study spin diffusion. Dipolar filter studies with fast magic-angle sample spinning were used to show that the triblock copolymer in the composite is only slightly affected by a strong dipolar filter (20-μs delay between pulses) while the poly(methyl

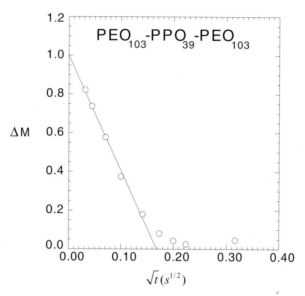

FIGURE 4.59 The spin-diffusion plot for the 15 wt % composite of the poly(ethylene oxide-b-propylene oxide-b-ethylene oxide) copolymer with poly(methyl silsesquioxane).

silsesquioxane) matrix is completely saturated. The strategy for the domain-size measurements is to use the fast magic-angle spinning studies (where the resolution is high) to determine the effect of the dipolar filter. Once it is determined which signals are saturated by the dipolar filter, the actual domain-size measurements are made on samples without spinning.

Figure 4.59 shows the results of a static spin-diffusion experiment on the 15 wt % triblock copolymer obtained using a strong dipolar filter. A value of 0.167 $s^{1/2}$ for $\sqrt{t_{SD}}$ is obtained from the intercept, and this value is used with Equation (4.12) to calculate the length scale of spin-diffusion. The spin-diffusion coefficients for the polymer were estimated from the slowest static T_2 relaxation times (Equation (4.18)) for the composite, and the coefficients for the poly(methyl silsesquioxane) were estimated from static T_2 measurements on the bulk matrix. The volume fraction of the polymer was calculated from the amount of polymer added, and the proton density (ρ^H) for the poly(methyl silsesquioxane) was taken as 0.014 g/cm^3. This leads to a calculated length scale of spin diffusion of 8.6 nm, which corresponds to a domain size of 3.9 nm and a separation between domains of 6.4 nm, assuming a 3D morphology.

These examples have used solid-state proton NMR to probe the morphology of multiphase polymer systems. This approach was chosen because the protons have a high sensitivity, the experiments are easy to perform, and only a small amount of material is required. The experiment in Figure 4.59, for example, was performed on 0.001 g of material from a film spin-cast on a 4-in. silicon wafer. Proton NMR

4.3 THE STRUCTURE AND MORPHOLOGY OF POLYMERS

is also the method of choice to study mobile polymers, because the magic-angle sample spinning required to average the carbon chemical shift anisotropy can partially average the dipolar couplings and the spin-diffusion coefficients. In studies of glassy and crystalline polymers, the dipolar filter and Goldman–Shen experiments can be used with cross polarization and magic-angle sample spinning, since the resolution is generally higher in the carbon spectrum.

Measuring the domain sizes using spin diffusion typically requires a number of assumptions and some estimates of the values for some parameters. In Section 4.3.2, we noted that the rate of spin diffusion depends on the surface-to-volume ratio, and that different morphologies (spheres, cylinders, etc.) can have the same surface-to-volume ratio. This means that spin diffusion alone cannot determine the morphology. NMR is most effectively used in conjunction with other studies (X-ray, TEM, etc.) to provide a more complete understanding of the morphology. This is particularly important for multiphase materials and composites, since many morphologies are possible. In the early studies of polyethylene (30) and polyurethanes (35), for example, the domain sizes were compared with the values from X-ray scattering. In the studies of the ethylene oxide and propylene oxide block copolymer composites, positron annihilation loss spectroscopy was used to determine the dimensionality and for comparison of some of the domain sizes (56).

4.3.6 Polymer Blends

Polymer blends are a special class of materials that can be investigated by NMR. The blends differ from other polymer mixtures in that the length scale of mixing is much smaller, so other NMR methods are needed to investigate the structure of blends. Miscible blends are mixed on the molecular length scale, so methods able to probe mixing on the length scale of 0.2–1.0 nm are often useful. Because of the short length scale of mixing and interactions between chains, blend formation can affect the chemical shifts, relaxation times, and spin diffusion properties. These very local probes of polymer structure and dynamics can in favorable cases provide an insight into the forces driving the molecular-level mixing of polymers.

Blend miscibility is frequently a consequence of strong intermolecular interactions, such as ionic or hydrogen-bonding interactions. In certain cases, these interactions are strong enough to cause shifts in the NMR spectrum. This is the case for the carbonyl peak in the NMR spectrum of the poly(ethylene oxide)/poly(acrylic acid) blend shown in Figure 4.60 (57). A new peak is observed at 185 ppm that can be assigned to intermolecular hydrogen bonding. Chemical shift changes large enough to be observed have also been reported from charge-transfer polymer complexes (58). In cases where the induced shifts are not as large, complex formation is often accompanied by line broadening from a distribution of chemical shifts (59).

Miscibility in many blends is a consequence of weak intermolecular interactions, and these interactions are not strong enough to lead to observable chemical shift

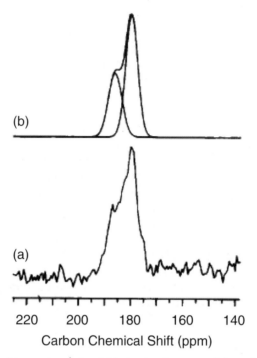

FIGURE 4.60 The (a) experimental and (b) simulated spectra of the carbonyl region of the solid-state NMR spectra of the blend of poly(ethylene oxide) and poly(acrylic acid).

changes in the NMR spectrum. In such cases, the blends can still be investigated using spin diffusion or relaxation methods. The method of choice to investigate these blends often depends on the chemical shift resolution in the carbon and proton spectra. Relaxation measurements are often used to estimate the length scale of mixing in polymer blends even when the spectra are not highly resolved. The relaxation studies can only show the upper limit on the length scale of mixing, so more detailed spin-diffusion and relaxation studies are needed to be more quantitative. In the most favorable cases, it is possible to identify the interacting groups that drive miscibility from the NMR studies.

The simplest way to estimate the length scale of mixing in blends is by measuring the averaging of the proton T_1 and $T_{1\rho}$ relaxation times (60). As shown by Equation (4.21), a volume fraction-averaged relaxation rate is observed in the polymers that are mixed on a length scale shorter than the spin-diffusion length scale as given by Equation (4.22). Since the proton T_1 are longer than the $T_{1\rho}$ they are sensitive to mixing on a longer length scale (17 vs. 1.7 nm) in a typical polymer. If the polymers are mixed on the spin-diffusion length scale, the same averaged value of the proton relaxation rates will be observed for both polymers. This measure of the mixing length scale can only be used if the relaxation rates of the bulk polymers differ from each other by more than a factor of 2.

4.3 THE STRUCTURE AND MORPHOLOGY OF POLYMERS

[-CH₂-CH— co —CH₂-CH-]
         | |
         (phenyl)   CH≡N

Poly(styrene-co-acrylonitrile)

[-CH₂-C(CH₃)-]
         |
         C(=O)-O-CH₃

Poly(methyl methacrylate)

The proton T_1 and $T_{1\rho}$ values have been measured as a function of composition for blends of poly(styrene-co-acrylonitrile) and poly(methyl methacrylate) (60). Figbreak ure 4.61 shows the plots of the T_1 and $T_{1\rho}$ relaxation rates as a function of the volume fraction of poly(methyl methacrylate). Within experimental error, there is a linear relationship between the volume fraction of poly(methyl methacrylate) and the T_1. Thus the polymers appear mixed on the spin-diffusion length scale, which for this sample is estimated from Equations (4.21) and (4.22) as 17 nm. The systematic deviation from linear behavior for the $T_{1\rho}$ relaxation rate shows that the polymers are not well mixed on the shorter (1.7 nm) length scale.

Poly(dimethyl phenylene oxide)

Poly(4-methyl styrene)

The length scale of mixing in polymer blends often depends on both the composition and temperature. The effect of composition on the $T_{1\rho}$ relaxation in blends of poly(2,6-dimethyl phenylene oxide) and poly(4-methyl styrene) are shown in Figure 4.62 (61). The $T_{1\rho}$ relaxation times for bulk poly(2,6-dimethyl phenylene oxide) and poly(4-methyl styrene) are very different from each other and therefore can be used to study the miscibility. The relaxation rates for the 75:25 blends are identical for both chains, so this blend is considered mixed on a molecular length scale. For the 50:50 composition the rates measured for each chain are no longer identical to each other and this blend must be considered homogeneous on the molecular length scale. The nonexponential relaxation is observed for the 25:75 blend is also evidence for molecular-level heterogeneity.

The blends of polystyrene and poly(vinyl methyl ether) have been investigated by a number of NMR methods in an effort to understand the length scale of polymer mixing and the interactions that lead to the formation of miscible blends (45,62–66). Figure 4.63 shows the effect of temperature on the $T_{1\rho}$ relaxation times for the 50:50 blend of polystyrene and poly(vinyl methyl ether) (66). By most criteria (DSC, etc.) the blend is miscible below the lower critical solution temperature, which is measured

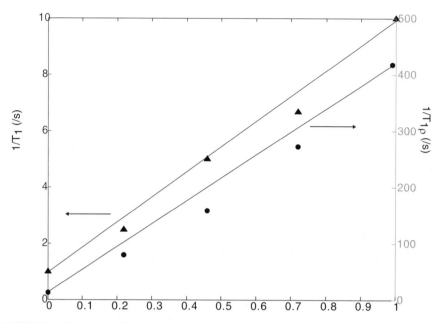

FIGURE 4.61 Plots of the proton (a) T_1 and (b) $T_{1\rho}$ relaxation rates as a function of blend composition.

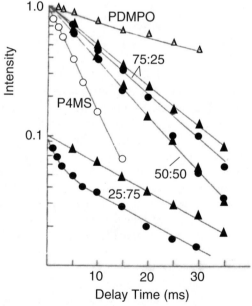

FIGURE 4.62 The $T_{1\rho}$ relaxation data for poly(dimethyl phenylene oxide) (PDMPO) and poly(4-methyl styrene) (P4MS). The open symbols show the relaxation for the pure polymers and the filled symbols show the relaxation for the 75:25, 50:50 and 25:75 blends.

4.3 THE STRUCTURE AND MORPHOLOGY OF POLYMERS

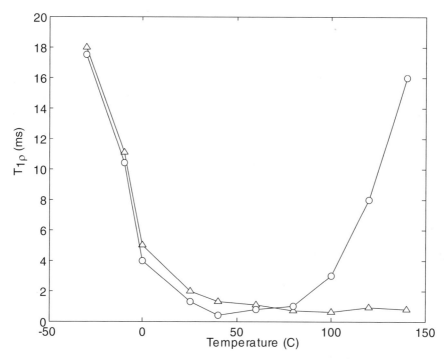

FIGURE 4.63 The proton $T_{1\rho}$ relaxation times for the 50:50 blend of polystyrene and poly(vinyl methyl ether). The data are shown for (○) the polystyrene aromatic and (△) poly(vinyl methyl ether) oxymethine carbons.

at 130°C for the 50:50 blends by cloud point measurements (66,67). Between –50 and 0°C the $T_{1\rho}$ values measured for each chain are identical, suggesting intimate mixing. The $T_{1\rho}$ depends strongly on temperature and has a minimum near 50°C. Above 0°C, the $T_{1\rho}$ times for the two chains begin to diverge from each other, with very strong divergence above 100°C. These data show that as the temperature increases, the blends become increasingly heterogeneous on the NMR length scale, even at well below the lower critical solution temperature measured by other methods.

Spin diffusion in the polystyrene/poly(vinyl methyl ether) blend has also been studied in WISE NMR as a function of the spin-diffusion delay time (48,66). Figure 3.64 shows cross sections through the WISE spectra of the blend with short (100 μs) and long (10 ms) spin-diffusion delay times (66). These cross sections show that the proton lineshape for the aromatic protons is significantly narrowed by spin diffusion to the poly(vinyl methyl ether). The length scale of spin diffusion was estimated from the time required (5 ms) to average the proton lineshapes at 47°C (48). This length scale of spin diffusion is calculated from the spin-diffusion time and the effective spin-diffusion coefficient for a blend as

$$L = \sqrt{\frac{D_{\text{eff}} t_{\text{sd}} \pi}{4}} \quad (4.26)$$

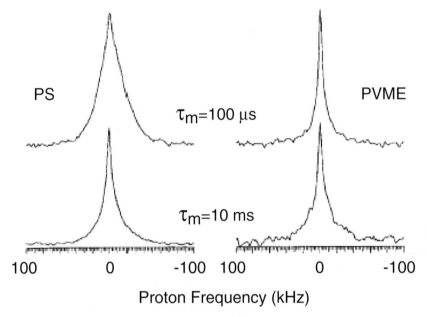

FIGURE 4.64 Cross sections through the 2D WISE spectra of the polystyrene/poly(vinyl methyl ether) blend. The data are shows for the polystyrene aromatic carbon (left) and the poly(vinyl methyl ether) oxymethine carbons (right) for spin diffusion times of 100 μs (top) and 10 ms (bottom).

where the averaged spin-diffusion coefficient for the blend is given by

$$\sqrt{D_{\text{eff}}} = \frac{\sqrt{D_{\text{PS}} D_{\text{PVME}}}}{\frac{1}{2}\sqrt{D_{\text{PS}}} + \frac{1}{2}\sqrt{D_{\text{PVME}}}} \quad (4.27)$$

Using spin-diffusion coefficients of 0.5 and 0.1 nm^2/ms for polystyrene and poly(vinyl methyl ether), a length scale of 0.9 ± 0.3 nm was calculated. For a cylindrical or spherical system, it is estimated that 85% of the spin diffusion has occurred for

$$L \cong \frac{d}{4} \quad (4.28)$$

where d is the domain size. This gives an estimated size for domains of 3.5 ± 1.5 nm. These NMR studies show that the blend is neither phase separated or mixed on the molecular level, but rather that the blend is nanoheterogeneous (48).

The studies of the polystyrene/poly(vinyl methyl ether) blend show that the behavior for blends can be complex. Thus far we have considered the blend as either mixed or phase separated, when in fact it is also possible to have both mixed and unmixed phases in a single blend. This was shown in a proton spin-diffusion study of polystyrene/poly(vinyl methyl ether) blends using selective pulses to invert the signals from either the polystyrene aromatic protons or the poly(vinyl methyl ether) methoxy/methine protons (63). A DANTE pulse sequence (Figure 4.35) was used to invert the signals of interest, and the proton spectrum was measured using magic-angle sample spinning and MREV-8 (Figure 2.15) for homonuclear proton decoupling. In the blend cast from toluene, spin diffusion was observed between polystyrene and the poly(vinyl methyl ether). In a modified version of this experiment several cycles of the DANTE selective inversion separated by a time delay (10 ms) were applied to progressively saturate all of the signals on one chain. This pulse sequence leads to saturation of all protons within the spin-diffusion distance of the saturated protons. Thus, for a completely mixed system, all of the signals would be saturated. Experimental observation showed that there was a fraction of the signals that could not be saturated by this method, and concluded that these signals must arise from chains that are not mixed on the spin-diffusion length scale. They estimated that $79 \pm 3\%$ of the chains were in the mixed phase (63).

In cases where it is possible to obtain a high-resolution proton spectrum, 2D spin exchange can be used to determine if the system is completely mixed. The idea behind this experiment is to compare the 1D line shape with a cross section through the 2D spectrum with a long mixing time. If the system is mixed on a length scale shorter than the spin diffusion length scale, then both line shapes should be identical. This is illustrated in Figure 4.65, which compares the 1D and 2D lineshapes for bulk poly(ethyl acrylate) and a Vycor glass composite (68). For the bulk polymer identical lineshapes are observed for cross sections through the methyl proton signals with a long mixing time (0.25 s). The Vycor glass has 4-nm pores covered with a layer of water at the surface that will be in spin diffusion contact with the polymer if the pores are completely filled. If the some pores are empty, they will not show spin diffusion to the polymer. The comparison of cross sections through the 2D spectra with the 1D spectrum shows that identical lineshapes are not observed, demonstrating that the system is not completely mixed and some of the pores must be empty. It is also possible to estimate the fraction of mixed polymer in a blend using chemical shift-based spin-diffusion methods (41).

Thus far we have discussed using proton spin diffusion to characterize the mixing in polymer blends. Among the advantages of using proton spin diffusion is that the experiments are relatively easy to perform and that protons are present at high concentrations. One limitations of these measurements is that spin diffusion is relatively rapid and proceeds isotropically, both along the chain and between chains. This makes it is difficult to make accurate measurements over a very short length scale. If such measurements can be made, they can provide an insight into the interactions between functional groups that lead to miscibility in blends.

Isotopically labeled samples can also be used to study blend miscibility in experiments where the transfer of magnetization between chains is monitored. This was used in early studies of the miscibility of polystyrene with a block copolymer of styrene and butadiene, where one chain is deuterated and one is protonated (69). The

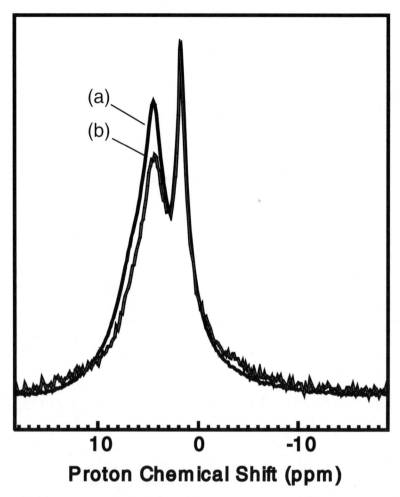

FIGURE 4.65 Comparison of the 1D lineshape and cross sections through the 2D exchange spectrum for poly(ethyl acrylate) and a Vycor glass composite.

idea behind this experiment is that the carbon signals on the deuterated chains cannot be cross polarized because they do not contain protons, and that they can only be observed by cross polarization if they are in close contact (<0.5 nm) with protons from the other chain. This experiment can provide information over a much more limited distance range than proton spin diffusion.

Figure 4.66 shows the cross polarization spectrum for the miscible polystyrene-d_8/poly(vinyl methyl ether) blend cast from toluene (61). The polystyrene in this sample is completely deuterated and cannot be cross polarized. The fact that the polystyrene aromatic carbon signals are observed at 125 ppm in the blend cast from toluene shows that polystyrene is in intimate contact with the poly(vinyl methyl ether). As with all cross-polarization experiments, the intermolecular cross polarization is

4.3 THE STRUCTURE AND MORPHOLOGY OF POLYMERS

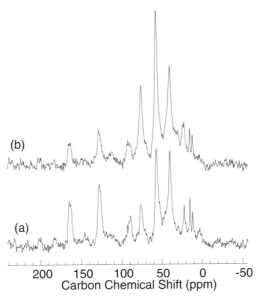

FIGURE 4.66 The cross polarization spectrum of the miscible blend of polystyrene-d_8 and poly(vinyl methyl ether) acquired with cross polarization times of (a) 4 ms and (b) 0.5 ms.

not quantitative and cannot be used to determine the fraction of polymer in the mixed phase.

Intermolecular cross polarization has also been used to study the inclusion complexes formed by cyclodextrin and protonated and deuterated poly(ethylene oxide) (70). Cyclodextrins are cyclic glucose molecules that can assemble to form crystal structures containing long channels that can accommodate a large number of polymers. The diameter of the channel depends on the number of linked glucose units, and α-cyclodextrinm, β-cyclodextrin, and γ-cyclodextrin each form rings with six, seven, and eight glucose units, giving channel diameters between 4.9 Å (α-cyclodextrin) and 7.9 Å (γ-cyclodextrin). The structure and dynamics of the polymer in the cyclodextrin inclusion complex depends both on the polymer structure and the diameter of the channel.

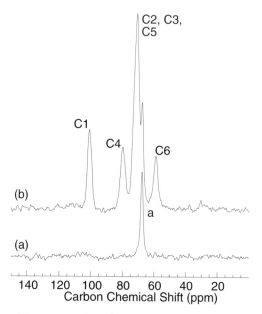

FIGURE 4.67 The carbon spectra of the cyclodextrin inclusion complex with poly(ethylene oxide)-d_4 (a) without and (b) with dipolar dephasing.

Figure 4.67a shows the carbon spectrum obtained with cross polarization and magic-angle sample spinning of the inclusion complex formed by α-cyclodextrin and poly(ethylene oxide)-d_4. The carbon signal for poly(ethylene oxide) appears at 69 ppm, but cannot be clearly resolved because of signal overlap with the glucose carbon signals. However, we can take advantage of the difference in the spin-spin relaxation rates between the glucose and the poly(ethylene oxide)-d_4 to obtain the spectrum of the polymer. The lines in the carbon spectra are broadened by the combination of chemical shift anisotropy and dipolar couplings, which are averaged by the combination of magic-angle sample spinning and dipolar decoupling. However, since the poly(ethylene oxide) contains deuterons instead of protons, the dipolar couplings and the spin-spin relaxation rates are much smaller. We can take advantage of this difference in the dipolar couplings by using the dipolar dephasing experiment (Figure 4.3) (71) to enhance the spectrum of the polymer. The dipolar dephasing experiment differs from the normal cross polarization by the delay time τ between cross polarization and signal observation. During this period the chemical shift anisotropy is averaged by magic-angle sample spinning, but the dipolar couplings are not. Therefore, the signals decay from spin-spin relaxation during this period, and the time constant for this decay depends on the magnitude of the dipolar couplings. Since the dipolar couplings are much larger for the cyclodextrin than the deuterated polymer, we can selectively remove the cyclodextrin signals from the spectrum. This is illustrated in Figure 4.67b, which shows the spectrum of the inclusion complex with a 40-μs delay between the cross polarization and signal acquisition.

4.3 THE STRUCTURE AND MORPHOLOGY OF POLYMERS

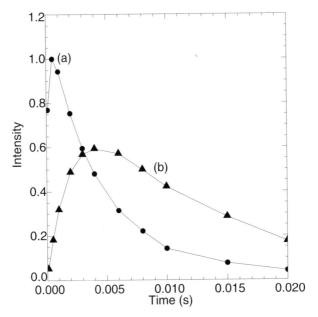

FIGURE 4.68 The time dependence of the cross-polarized signal intensity for the cyclodextrin/poly(ethylene oxide)-d_4 inclusion complex.

The dipolar dephasing can be combined with other pulse sequences to study the structure and dynamics of the complex, even when the carbon signals are overlapped. This is illustrated in Figure 4.68, which shows the time dependence of the cross polarized signal intensity for the α-cyclodextrin and the poly(ethylene oxide)-d_4 in the inclusion complex. The α-cyclodextrin is relatively rigid and rapidly cross polarizes, with the maximum signal observed with a cross-polarization time of 1 ms. The poly(ethylene oxide)-d_4, which cross polarizes more slowly, reaches its maximum intensity after 5 ms of cross polarization. The fact that any cross polarized signal is observed for the poly(ethylene oxide)-d_4 shows that some complex must be formed, since bulk poly(ethylene oxide)-d_4 contains no protons and cannot be cross polarized.

The molecular structure and geometry of interacting groups in polymer blends is in general difficult to determine without introducing isotopic labels into one or both of the polymer chains. In favorable cases the local structure can be measured by taking advantage of molecular motion that may be present on one chain but not on the other through the NOE. The NOE was introduced in Section 1.5.5 as a means to study the structure and dynamics of polymers in solution. The NOE is generally not used for solid-state NMR studies, because there is usually not enough molecular mobility to generate NOEs in most polymers. One exception to this generality is methyl group rotation in polymers, which is typically in the fast motion limit.

Intermolecular NOEs have been used to study the structure of polystyrene/poly (vinyl methyl ether) blends (65). The idea behind this experiment is that at low temperatures (below T_g) the only rapidly moving groups in polymers are the methyl

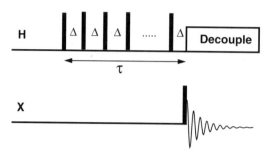

FIGURE 4.69 The pulse sequence diagram for the solid-state NOE experiment.

groups. This methyl group rotation is sufficiently fast to cause NOEs in the vicinity of the methyl groups. This is a very localized, through-space effect, and only carbons within 5 Å of the rotating methyl group will experience a NOE.

Figure 4.69 shows the pulse-sequence diagram for the solid-state NOE experiment. This is a simple experiment where a train of pulses and delays is applied on the proton channel and the carbon signal is monitored by a 90° pulse. The variable in this experiment is the length of the pulse train on the proton channel. The signal is measured as a function of the length of the pulse train, and the NOE enhancement is given by

$$\eta(\tau) = \frac{I(\tau) - I(0)}{I(0)} \qquad (4.29)$$

where $I(\tau)$ and $I(0)$ are the intensities as a function of τ and for $\tau = 0$. The NOE (η) as a function of τ is given by

$$\eta(\tau) = \frac{\sigma}{\rho}(1 - e^{-\rho\tau}) \qquad (4.30)$$

The rates ρ and σ are the relaxation rates for relaxation to the lattice and the cross relaxation rates that depend on the carbon–proton distances and correlation times. It is important to note that the solid-state NOE experiment does not use cross polarization, and therefore has a lower sensitivity, and long delays between scans (or saturation times) may be required. The experiment is typically performed with a long, fixed delay between acquisitions, and the variable τ is the fraction of this delay during which the protons are saturated.

Figure 4.70 shows the solid-state carbon spectrum with and without the NOE, and the difference spectrum for the 1:1 molar blend of deuterated polystyrene and poly(vinyl methyl ether) cast from toluene solution (65). The largest NOE (1.2) is observed for the methoxyl carbon at 57 ppm. This result is expected, since this is the mobile methyl group responsible for creating the NOE. Other NOEs are observed for the poly(vinyl methyl ether) carbons. The signal of most interest is the protonated

4.3 THE STRUCTURE AND MORPHOLOGY OF POLYMERS

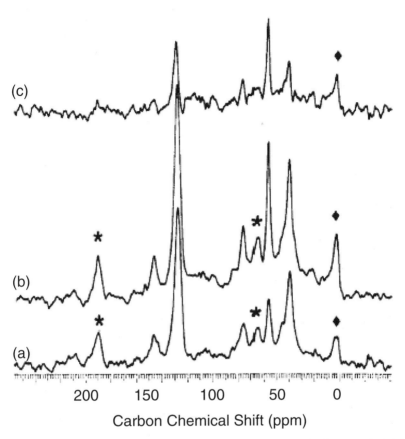

FIGURE 4.70 The solid-state NOE spectra of the miscible blend of polystyrene/poly(vinyl methyl ether). The spectra are shown (a) without proton irradiation, (b) with proton irradiation and (c) the difference spectra. The spinning sidebands and other artifacts are marked (*, ♦).

aromatic carbon signals at 125 ppm from polystyrene, which shows that blend formation places the styrene aromatic carbon in close proximity to the poly(vinyl methyl ether) methoxyl group. Control experiments on phase-separated blends cast from toluene show no intermolecular NOEs.

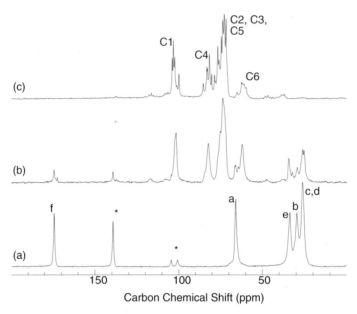

FIGURE 4.71 The carbon spectrum of the inclusion complex of (a) cyclodextrin, (b) the cyclodextrin/polycaprolactone inclusion complex, and (c) polycaprolactone.

As a general rule, it is difficult to determine the chain conformation in polymer blends from the chemical shifts because the lines are broadened by the distribution of conformations through the γ-*gauche* effect. However, for polymers constrained to the channels in inclusion complexes the distribution of chain conformations is limited to those that that are able to occupy the channel (70), and sharper lines are often observed. Figure 4.71 shows the cross-polarization magic-angle sample-spinning carbon spectrum of the polycaprolactone in the bulk and in the inclusion complex with α-cyclodextrin. The signals from both the polymer and the cyclodextrin are observed for the complex. Polycaprolactone adopts an extended all-*trans* conformation in the crystal, and the similarity in chemical shifts between the crystalline polymer and the inclusion complex suggests that it is also in the all-*trans* conformation in the inclusion complex (72). There are several different molecules in the unit cell for pure α-cyclodextrin, giving rise to many different signals for each carbon. A more symmetric structure is formed in the inclusion complex, since only a single peak is now observed for each carbon. The chain conformation can also be investigated by multiple-quantum 2D NMR for polymers with ^{13}C labels (73).

The HETCOR experiment can also be used to probe the mixing length scale in polymer blends. This experiment (Section 1.7.3.3) shows the solid-state correlation between some nucleus (usually carbon) and the protons that cause its relaxation. This is a 2D experiment, and the resolution in the proton dimension is provided either by multiple-pulse (46) or Lee–Goldburg (47) decoupling combined with magic-angle sample spinning. The key to using HETCOR to identify intermolecular interactions

4.3 THE STRUCTURE AND MORPHOLOGY OF POLYMERS

in blends is to identify a carbon signal that is not relaxed by protons on its own chain. The largest peaks in the HETCOR spectrum are due to carbons with directly bonded protons, but it is possible to observe correlations between carbons without directly bonded protons (carbonyl carbons, etc.) and protons on another chain (15). The other strategy is to deuterate one chain in the blend and observe the HETCOR spectrum (74). The HETCOR spectrum then shows correlations between the carbons on the deuterated chain and the protons of the protonated chain.

HETCOR has been used to probe the miscibility and hydrogen bonding in blends of poly(methyl acrylate) and poly(vinyl phenol) (15). Figure 4.72 shows the HETCOR

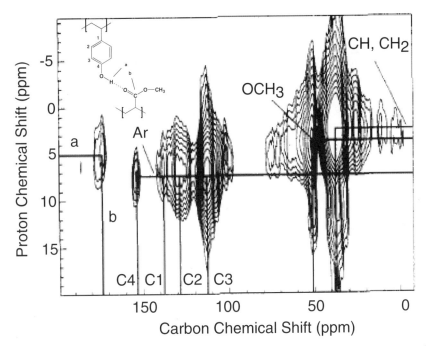

FIGURE 4.72 The carbon–proton HETCOR spectrum for the blend of poly(vinyl phenol) and poly(methyl acrylate).

spectrum for the blend. This is a direct correlation time with no spin diffusion mixing time, so only protons closer than 0.5 nm give rise to correlations. The most intense peaks are from carbons with directly bonded protons, including the aromatic and main-chain carbons. The correlation of most interest is between the poly(methyl acrylate) carbonyl at 176 ppm and a proton signal at 5 ppm. The fact that this correlation is smaller than many of the others suggests that the carbon–proton distance is larger than for the carbons with directly bonded protons. Furthermore, no other carbons are correlated to a proton with a chemical shift of 5 ppm. This suggests that the proton at 5 ppm can be assigned to the hydroxy proton from poly(vinyl phenol) and that the correlation is due to intermolecular hydrogen-bond formation. This assignment is confirmed in experiments in which the poly(vinyl phenol) hydroxyl group is deuterated before casting the blend. The correlation peaks are not observed when the hydroxyl protons are replaced by deuterons. Similar correlation peaks are observed for the blend of poly(methyl methacrylate) and poly(vinyl phenol) (15).

The morphology of blends can also be investigated by inserting a spin-diffusion delay time into the HETCOR pulse sequence (Figure 4.37) to allow for proton spin diffusion. This is illustrated in Figure 4.73, which shows the HETCOR spectra for the inclusion complex of polycaprolactone and γ-cyclodextrin with spin-diffusion delay times of 50 μs and 10 ms. A delay time of 50 μs is too short for significant spin diffusion, so the carbon signals are correlated with their nearest protons. Although the resolution is not very good in the proton dimension, the resolved chemical shifts for the polymer and cyclodextrin peaks can be observed (72). Large changes in the chemical shifts are observed; for the spectra with a 10-ms spin-diffusion delay. Resolved peaks are no longer observed; rather, a single averaged proton chemical shift is observed for both the polymer and the cyclodextrin. This behavior would not be observed if the polymer and cyclodextrin was not mixed on a molecular length scale.

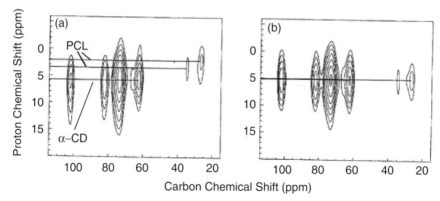

FIGURE 4.73 The HETCOR spectra for the α-cyclodextrin/polycaprolactone inclusion complex with a spin-diffusion delay time of (a) 50 μs and (b) 10 ms.

REFERENCES

1. Bunn, A.; Cudby, E.; Harris, R.; Packer, K.; Say, B. *Chem. Commun.*, 1981, *15*, 15.
2. Hu, W.-G.; Schmidt-Rohr, K. *Polymer*, 2000, *41*, 2979.
3. Torchia, D. A. *J. Magn. Reson.*, 1978, *30*, 613.
4. Frey, M. H.; Opella, S. J.; Rockwell, A. L.; Gierasch, L. M. *J. Am. Chem. Soc.*, 1985, *107*, 1946.
5. Tonelli, A. *NMR Spectroscopy and Polymer Microstructure: The Conformational Connection*, VCH Publishers, New York, 1989.
6. Vander Hart, D. L. *J. Magn. Reson.*, 1981, *44*, 117.
7. Tonelli, A. E.; Gomes, M. A.; Tanaka, H.; Cozine, M. H. In *Solid State NMR of Polymers*, Mathias, L. J., ed., Plenum Press, New York, 1991, 81 pp.
8. Gomez, M. A.; Tanaka, H.; Tonelli, A. E. *Polymer (British)*, 1987, *28*, 2227.
9. Bovey, F. A.; Schilling, F. C. In *Solid State NMR of Polymers*, Mathias, L. J., ed., Plenum Press, New York, 1991, 295 pp.
10. Mathias, L. J.; Davis, R. D.; Jarrett, W. J. *Macromolecles*, 1999, *32*, 7958.
11. Tanaka, H.; Gomez, M.; Tonelli, A.; Chichester-Hicks, S. V.; Haddon, R. C. *Macromolecules*, 1989, *22*, 1031.
12. Meresi, G.; Wang, Y.; Bandis, A.; Inglefield, P. T.; Jones, A. A.; Wen, W.-Y. *Polymer*, 2001, *42*, 6153.
13. Tanaka, H.; Gomez, M. A.; Tonelli, A. E.; Takur, M. *Macromolecules*, 1989, *22*, 1208.
14. Langer, B.; Schnell, I.; Spiess, H. W.; Grimmer, A. *J. Magn. Reson.*, 1999, *138*, 182.
15. White, J. L.; Mirau, P. A. *Macromolecules*, 1994, *27*, 1648.
16. English, A. D.; Debowski, C. *Macromolecules*, 1984, *17*, 446.
17. Isbester, P. K.; Kestner, T. A.; Munson, E. J. *Macromolecules*, 1997, *30*, 2800.
18. Chung, G. C.; Kornfield, J. A.; Smith, S. D. *Macromolecules*, 1994, *27*, 5729.
19. Lutz, T. R.; He, Y.; Ediger, M. D.; Cao, H.; Lin, G.; Jones, A. A. *Macromolecles*, 2003, *36*, 1724.
20. Sun, P.; Li, B.; Wang, Y.; Ma, J.; Ding, D.; He, B. *Eur. Polym. J.*, 2003, *39*, 1045.
21. Shapiro, M. J.; Wareing, M. J. *Curr. Opin. Chem. Biol.*, 1998, *2*, 372.
22. Metz, G.; Wu, X.; Smith, S. O. *J. Magn. Reson. Ser. A*, 1994, *110*, 219.
23. Hediger, S.; Meier, B.; Kurur, N. D.; Bodenhausen, G.; Ernst, R. R. *Chem. Phys. Lett.*, 1994, *223*, 283.
24. Nielsen, N. C.; Bildsoe, H.; Jakobsen, H. J. *Macromolecules*, 1992, *25*, 2847.
25. Nakamura, K.; Ando, I.; Takeichi, T. *Polymer*, 2001, *42*, 4045.
26. Kuroki, S.; Okita, K.; Kakigano, T.; Ishikawa, J.; Itoh, M. *Macromolecules*, 1998, *31*, 2804.
27. Dargaville, T. R.; George, G. A.; Hill, D. J. T.; Scheler, U.; Whittaker, A. *Macromolecules*, 2002, *35*, 5544.
28. Blumich, B.; Hagemeyer, A.; Schaefer, D.; Schmidt-Rohr, K.; Spiess, H. W. *Adv. Mater.*, 1990, *2*, 72.
29. Cai, W. Z.; Schmidt-Rohr, K.; Egger, N.; Gerharz, B.; Spiess, H. W. *Polymer*, 1993, *34*, 267.
30. Cheung, T.; Gerstein, B. *J. Appl. Phys.*, 1981, *52*, 5517.

31. Clauss, J.; Schmidt-Rohr, K.; Spiess, H. W. *Acta Polym.*, 1993, *44*, 1.
32. Schmidt-Rohr, K.; Spiess, H. W. *Multidimensional Solid-State NMR and Polymers*, Academic Press, New York, 1994.
33. VanderHart, D. L.; McFadden, G. B. *Solid State NMR*, 1996, *7*, 45.
34. Douglass, D.; Jones, J. *J. Chem. Phys.*, 1966, *45*, 956.
35. Assink, R. *Macromolecules*, 1978, *11*, 1233.
36. Mellinger, F.; Wilhelm, M.; Spiess, H. *Macromolecules*, 1999, *32*, 4686.
37. Goldman, M.; Shen, L. *Phys. Rev.*, 1966, *144*, 321.
38. Egger, N.; Schmidt-Rohr, K.; Blumich, B.; Domke, W. D.; Stapp, B. *J. Appl. Polym. Sci.*, 1992, *44*, 289.
39. Mansfield, P. *Phys. Lett.*, 1970, *32A*, 485.
40. Mansfield, P. *J. Phys. C: Solid State Phys.*, 1971, *4*, 1444.
41. Mansfield, P.; Orchard, M. J.; Stalker, D. C.; Richards, K. H. B. *Phys. Rev.*, 1973, *B7*, 90.
42. Rhim, W. K.; Elleman, D. D.; Schreiver, L. B.; Vaughn, R. W. *J. Chem. Phys.*, 1973, *60*, 4595.
43. Campbell, G. C.; VanderHart, D. L. *J. Magn. Reson.*, 1992, *96*, 69.
44. Morris, G.; Freeman, R. *J. Magn. Reson.*, 1978, *29*, 433.
45. Caravatti, P.; Neuenschwander, P.; Ernst, R. *Macromolecules*, 1985, *18*, 119.
46. Caravatti, P.; Braunschweiler, L.; Ernst, R. R. *Chem. Phys. Lett.*, 1983, *100*, 305.
47. van Rossum, B. J.; Forster, H.; De Groot, H. J. M. *J. Magn. Reson.*, 1997, *124*, 516.
48. Schmidt-Rohr, K.; Clauss, J.; Spiess, H. *Macromolecules*, 1992, *25*, 3273.
49. Bovey, F. A.; Mirau, P. A. *NMR of Polymers*, Academic Press, New York, 1996.
50. Taylor, S. A.; White, J. L.; Elbaum, N. C.; Crosby, R. C.; Campbell, G. C.; Haw, J. F.; Hatfield, G. R. *Macromolecules*, 1992, *25*, 3369.
51. Takegoshi, K.; Tanaka, I.; Hikichi, K.; Higashida, S. *Macromolecules* 1992, *25*, 3392–3398.
52. Marjanski, M.; Srinivasarao, M.; Mirau, P. A. *Solid State NMR*, 1998, *12*, 113.
53. Mirau, P. A.; Yang, S. *Chem. Mater.*, 2002, *14*, 249.
54. Tao, H. J.; Rice, D. M.; MacKnight, W. J.; Hsu, S. L. *Macromolecules*, 1995, *28*, 4036.
55. Dumais, J. J.; Jelinski, L. W.; Leung, L. M.; Gancarz, I.; Galambos, A.; Koberstein, J. T. *Macromolecules*, 1985, *18*, 116.
56. Yang, S.; Mirau, P. A.; Pai, C.; Nalamasa, O.; Reichmanis, E.; Lin, E. K.; Lee, H.-J.; Gidley, D. W.; Sun, J. *Chem. Mater.*, 2001, *13*, 2762.
57. Miyoshi, T.; Takegoshi, K.; Hikichi, K. *Polymer*, 1995, *37*, 11.
58. Simmons, A.; Natansohn, A. *Macromolecules*, 1991, *24*, 3651.
59. Grobelny, J.; Rice, D.; Karasz, F.; MacKnight, W. *Macromolecules*, 1990, *23*, 2139.
60. McBrierty, V.; Douglass, D.; Kwei, T. *Macromolecules*, 1978, *11*, 1265.
61. Dickinson, L.; Yang, H.; Chu, C.; Stein, R.; Chein, J. *Macromolecules*, 1987, 20, 1757.
62. Kwei, T.; Nishi, T.; Roberts, R. *Macromolecules*, 1974, *7*, 667.
63. Caravatti, P.; Neuenschwander, P.; Ernst, R. *Macromolecules*, 1986, *19*, 1889.
64. Gobbi, G.; Silvestri, R.; Thomas, R.; Lyerla, J.; Flemming, W.; Nishi, T. *J. Polym. Sci. Part C. Polym. Lett.*, 1987, *25*, 61.
65. White, J. L.; Mirau, P. A. *Macromolecules*, 1993, *26*, 3049.

66. Wangler, T.; Rinaldi, P. L.; Han, C. D.; Chun, H. *Macromolecles*, 2000, *33*, 1778.
67. Larbi, F.; Leloup, S.; Halary, J.; Monnerie, L. *Polym. Commun.*, 1986, *27*, 23.
68. Mirau, P. A.; Heffner, S. A. *Macromolecules*, 1999, *32*, 4912.
69. Schaefer, J.; Sefcik, M. D.; Stejskal, E. O.; McKay, R. A. *Macromolecules*, 1981, *14*, 188.
70. Lu, J.; Mirau, P. A.; Tonelli, A. E. *Prog. Polym. Sci.*, 2002, *27*, 357.
71. Opella, S. J.; Frey, M. H. *J. Am. Chem. Soc.*, 1979, *101*, 5854.
72. Lu, J.; Mirau, P. A.; Tonelli, A. E. *Macromolecles*, 2001, *34*, 3276.
73. Schmidt-Rohr, K.; Hu, W.; Zumbulyadis, N. *Science*, 1998, *280*, 714.
74. Li, S.; Rice, D. M.; Karasz, F. E. *Macromolecules*, 1994, *27*, 2211.

5

THE DYNAMICS OF POLYMERS

5.1 INTRODUCTION

NMR has long been used to study the dynamics of polymers, since the relaxation times and lineshapes are extremely sensitive to the chain motions is that occurs over many orders of magnitude. In solution, the relaxation is caused by the combination of rapid librational motions, *gauche–trans* isomerization, and even slower segmental motions. These localized molecular motions are very effective at causing relaxation and, at least for high molecular-weight polymers, the overall reorientation of the entire chain makes only a small contribution to the relaxation. The challenge in solution NMR studies is to describe the correlation function that accounts for the relaxation. These correlation functions can be complex, and may span the time scale from picoseconds to microseconds. Since a single relaxation process does not uniquely characterize the correlation functions, it is important to measure as many relaxation times as possible (T_1, T_2, NOE) as a function of magnetic field strength and temperature to define the correlation function.

The chain dynamics for solid polymers are orders of magnitude slower than for polymers in solution. There are some fast processes that occur in crystalline and glassy polymers, such as methyl group rotation, but for the most part, the chains are restricted. While the fast molecular motions can be characterized using the T_1, T_2, and NOE, other methods are required to study the slow molecular dynamics. The method of choice depends on the time scale of molecular motion. In most studies we are interested in the high-resolution spectrum, and in Section 1.6 we considered the mechanisms that lead to line broadening in solids and how to remove these broadenings. If we observe rather than remove this line broadening, the line shapes can provide important information about the molecular dynamics, since molecular motions lead to averaging

A Practical Guide to Understanding the NMR of Polymers, by Peter A. Mirau
ISBN 0-471-37123-8 Copyright © 2005 John Wiley & Sons, Inc.

of the lineshapes. The line broadening from dipolar interactions in rigid solids is on the order of 50 kHz, and can be used to probe molecular motions on the time scale. The broadening from chemical shift anisotropy and deuterium quadrupolar couplings are on the order of a few kHz and 100 kHz and can be used to study molecular motions on these time scales.

The rates of $T_{1\rho}$ relaxation and the cross-polarization dynamics also can be used to study the molecular motions of solid polymers. These rates are sensitive to the motions on the kHz time scale, and provide information about the slow dynamics. Spin diffusion is very efficient in solids, so the proton relaxation rates must be interpreted with caution. On the slowest time scale, spin exchange can be used to probe the dynamics. The lower limit on the spin exchange time scale is the rate at which the spin system returns to equilibrium via spin-lattice relaxation. The spin exchange experiments are usually performed by monitoring the exchange of peaks within the chemical shift anisotropy or deuterium lineshapes. These same methods can be used to measure the diffusion of chains between the amorphous and crystalline phases in semicrystalline polymers.

A key part of understanding the molecular dynamics of polymers in solution is finding the proper correlation function to describe the molecular motion. As discussed in Section 1.5, relaxation is due to fluctuating fields from nearby NMR-active nuclei. To the extent that these nuclei have components at the frequency of the nuclei of interest, they will cause relaxation. The NMR relaxation is written in terms of the spectral densities (Equation 1.55), which is a mathematical function that describes the density of a relaxation component at a particular frequency. The spectral density is the FT of the correlation function that describes the time dependence of an internuclear vector, such as that between a ^{13}C and its directly attached proton. The observed relaxation rate is the sum of all available relaxation mechanisms (dipolar, chemical shift anisotropy, etc.), so it is necessary to determine which mechanisms are contributing to the relaxation before the relaxation rates can be interpreted in terms of the molecular motions.

5.2 CHAIN MOTION OF POLYMERS IN SOLUTION

NMR has greatly contributed to our understanding of the molecular dynamics of polymers in solution. Information about the relationship between polymer structure and the molecular dynamics was obtained in early studies by comparing the carbon relaxation times for wide varieties of polymers (1). A basic understanding of polymer dynamics and the relationship between structure and dynamics emerged from these studies. The results showed that the polymer dynamics are very sensitive to the size and nature of side chains in vinyl polymers, the presence of heteroatoms in the polymer main chain, as well as double bonds in the main chain and polymer microstructure. The early results were interpreted using the isotropic reorientation correlation function, which does not always give a good fit to the relaxation data. More recently, data have been collected over a wider range of temperatures and magnetic field strengths and fit to more complex forms for the spectral density functions.

Since the initial studies, a wide variety of polymers have been studied in solution and in the melt state. It has been demonstrated that many experiments as a function of temperature and magnetic field strength are needed to distinguish between possible models for the correlation functions and spectral densities (2). The general consensus seems to be that polymers experience molecular motion on a variety of time scales, and bimodal correlation functions often give the best fits to the experimental data (3). The bimodal correlation functions have contributions both from rapid librational motions and slower segmental motions. No correlation function has been presented that fits all NMR relaxation data. Several correlation functions, and several functional forms for the correlation functions, have been proposed (2).

5.2.1 Modeling the Molecular Dynamics of Polymers in Solution

In a typical solution experiment, the relaxation times are measured as a function of magnetic field strength and temperature and the fits to various models are compared. The simplest models (i.e., isotropic motion) usually do not provide a good fit to all of the NMR data, and more complex models are required. Among the successful models are those that incorporate specific polymer motions or distribution of correlation times along with fast librational motions. The data gathered over the widest range of magnetic field strengths and temperatures provide the best test of the motional model.

The basic ideas of relaxation were introduced in Section 1.5. Relaxation times are measured after pulses are applied to perturb the spin system from equilibrium. Fluctuating fields from nearby NMR-active nuclei provide a means for the spin system to return to equilibrium. The relaxation depends on the spectral densities, which describe the power available at a particular frequency to cause relaxation. The spectral density $J(\omega)$ is given by the FT of the second-order autocorrelation funtion $G(t)$ as

$$J(\omega) = \frac{1}{2} \int_{-\infty}^{\infty} G(t) e^{-i\omega t} \, dt \tag{5.1}$$

The correlation function $G(t)$ is given by

$$G(t) = \frac{1}{2} \langle 3 \cos^2 \theta(t) - 1 \rangle \tag{5.2}$$

where $\theta(t)$ is the orientation of an internuclear vector at time, t, relative to its orientation at $t = 0$. The total area under the curve for $G(t)$ is the correlation time τ_c, and is given by

$$\tau_c = \int_0^\infty G(t) \, dt \tag{5.3}$$

A short correlation time is associated with rapid molecular motion and a long correlation time with restricted mobility.

5.2 CHAIN MOTION OF POLYMERS IN SOLUTION

The general procedure for fitting the relaxation data is to find the proper functional form for $G(t)$ that fits all of the relaxation data. This is accomplished by calculating the spectral densities for a model for molecular motion using Equation (5.1), and using the spectral densities to calculate the relaxation times. If the calculated relaxation times are consistent with the observed values, then the correlation function is considered to be a good description of the actual molecular motion. The general procedure is to start with simple models (with fewer parameters) and proceed to more complex models when the simple models fail. The simplest model is isotropic reorientation, which has an exponential correlation function given by

$$G(t) = e^{-t/\tau_c} \tag{5.4}$$

for which the spectral densities is given by

$$J(\omega) = \frac{1}{2} \int_{-\infty}^{\infty} e^{-t/\tau_c} e^{-i\omega t} \, dt = \frac{\tau_c}{1 + \omega^2 \tau_c^2} \tag{5.5}$$

In addition to simple mathematical functions for $G(t)$, it is also possible to write the correlation functions in terms of distributions of motional time constants $F(\tau)$. The correlation function is related to the distribution function by the Laplace transform and is given by

$$G(t) = \int_0^\infty F(\tau) e^{-t/\tau} \, d\tau \tag{5.6}$$

The failure of the isotropic model of polymer mobility to account for the relaxation data in a variety of polymers (1) has led to the search for more complex functions to approximate the autocorrelation functions. The challenge is to find a sufficiently complex function while retaining some physical insight into the molecular motions that give rise to relaxation. In some cases, it is possible to fit the relaxation to generic correlation function using the so-called model-free approach (4,5). In this model the autocorrelation function is approximated as the product of correlation functions for slow reorientation ($G_0(t)$) and for faster reorientation ($G_i(t)$). The correlation function for the model-free can be written as

$$G(t) = G_0(t) G_i(t) \tag{5.7}$$

The assumption implicit in the model-free approach is that fast and slow molecular motions are decoupled from each other and can be treated separately. This may be a good assumption in proteins, where methyl group rotation is decoupled from isotropic tumbling of the entire protein, but it is not always applicable to polymers. Different functional forms of $G_0(t)$ and $G_i(t)$ can be chosen to give the best fit for the data. $G_0(t)$, for example, could be chosen as the correlation function for isotropic or anisotropic rotation, and $G_i(t)$ could be the correlation function for wobbling in a cone or three-site jumps. While the model-free correlation functions can sometimes fit the experimental data, they provide little insight into the motions that cause relaxation.

Models for the correlation functions have been developed for polymers that account for the types of conformational transitions (*gauche–trans* isomerizations, etc.) that might be expected in polymers. One class of models is based on the diamond-lattice, where conformational transitions are modeled as jumps between sites on the tetrahedral lattice (6). Another approach is to model the conformational transitions along the polymer chain as a damped diffusional process. Several types of transitions are classified in Hall–Helfand theory (7), which depend on the displacement of the polymer tails. Some molecular motions, such as crankshaft motions, lead to no change in the polymer tails, while others do. The autocorrelation function for the Hall–Helfand model is given by

$$G(t) = e^{-t/\tau_2} e^{-t/\tau_1} I_0(t/\tau_1) \tag{5.8}$$

where I_0 is the zeroth-order Bessel function, τ_1 is the time constant for correlated jumps responsible for orientational diffusion along the chain, and τ_2 is the time constant for damping from either nonpropagative motions or distortions in the chain from the most stable conformation.

While the Hall–Helfand model gives a good physical picture of the polymer dynamics, fits to the NMR data are generally improved if another term is added to account for fast anisotropic (librational) motion of the chain (the so-called DLM model) (8). With this modification the correlation function is given by

$$G(t) = (1-a) e^{-t/\tau_2} e^{-t/\tau_1} I_0(t/\tau_1) + a e^{-t/\tau_0} e^{-t/\tau_2} e^{-t/\tau_1} I_0(t/\tau_1) \tag{5.9}$$

where

$$1 - a = \left[\frac{\cos\theta - \cos^3\theta}{2(1 - \cos\theta)} \right]^2 \tag{5.10}$$

where the fast anisotropic motion is described as the random reorientation of a vector inside a cone with a half-angle of θ and a correlation time τ_0 (9). If we assume that τ_0 is much faster than τ_1 and τ_2, we can rewrite Equation (5.9) as

$$G(t) = (1-a) e^{-t/\tau_2} e^{-t/\tau_1} I_0(t/\tau_1) + a e^{-t/\tau_0} \tag{5.11}$$

Fourier transformation of Equation (5.11) gives the spectral density

$$J(\omega) = \frac{1-a}{\sqrt{(\alpha + i\beta)}} + \frac{a\tau_0}{1 + \omega^2 \tau_0^w} \tag{5.12}$$

where

$$\alpha = \frac{1}{\tau_2^2} + \frac{2}{\tau_1\tau_2} + \omega^2 \tag{5.13}$$

and

$$\beta = \frac{-2\omega}{\tau_1 + \tau_2} \tag{5.14}$$

We will see examples where this correlation function gives a good fit to the relaxation data over a range of magnetic field strengths and temperatures.

In addition to models based on polymer structure, models with distributions of correlation times have also been used to model the relaxation of polymers. While these interpretations are not as intuitive as models based on the conformational transitions in polymer, they often give good fits to the data. The use of distribution functions is a recognition that the polymer dynamics are sometimes too complex to be described with simple mathematical function.

A number of distribution functions have been used to model the relaxation in polymers, including the Cole–Cole (10), Fuoss–Kirkwood (11), and the log-χ^2 (12) distribution functions. By entering the expression for the distribution function $F(t)$ into Equation (5.6), the spectral densities can be calculated. These distribution functions often have several parameters that can be adjusted to give the best fit to the experimental data.

For the Cole–Cole and Fuoss–Kirkwood distributions, let us consider a distribution of correlation times centered at the correlation time τ_0. We can define

$$\tau = \ln\left(\frac{\tau_c}{\tau_0}\right) \tag{5.15}$$

where the distribution function is normalized as

$$F(\tau) = \int_{-\infty}^{\infty} F(\tau)\,d\tau = 1 \tag{5.16}$$

The Cole–Cole distribution function is given by

$$F(\tau) = \frac{1}{2\pi} \frac{\sin(\varepsilon\pi)}{\cosh(\varepsilon\tau) + \cos(\varepsilon\pi)} \tag{5.17}$$

and Fourier transformation gives

$$J(\omega) = \left(\frac{1}{\omega}\right) \frac{\cos\left[(1-\varepsilon)\frac{\pi}{2}\right]}{\cosh[\varepsilon\ln(\omega\tau_0)] + \sin\left[(1-\varepsilon)\frac{\pi}{2}\right]} \tag{5.18}$$

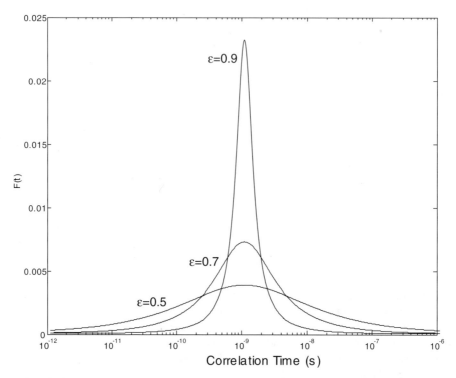

FIGURE 5.1 A plot of the Cole–Cole distribution function for values of ϵ of 0.9, 0.7, and 0.5.

for the spectral density function. The two parameters in this distribution are the correlation time at the center of the distribution τ_0 and the width of distribution ε, which assumes a value between zero and one, with one being a narrow distribution and zero being a broad distribution. Figure 5.1 shows a plot of the Cole–Cole distribution function for $\tau_0 = 1$ ns and ε equal to 0.9, 0.7, and 0.5.

The Fuoss–Kirkwood distribution function is given by

$$F(\tau) = \frac{\beta}{\pi} \frac{\cos\left(\frac{\beta\pi}{2}\right)\cosh(\beta\tau)}{\cos^2\left(\frac{\beta\pi}{2}\right) + \sinh(\beta\tau)} \tag{5.19}$$

and Fourier transformation gives

$$J(\omega) = \frac{2\beta}{\omega} \frac{(\omega\tau_0)^\beta}{1 + (\omega\tau_0)^{2\beta}} \tag{5.20}$$

where again τ_0 is the center of the distribution, and β is an adjustable parameter with values between zero and one. Figure 5.2 shows a plot of the Fuoss–Kirkwood distribution function centered at $\tau_0 = 1$ ns and values for β of 0.9, 0.7, and 0.5.

5.2 CHAIN MOTION OF POLYMERS IN SOLUTION

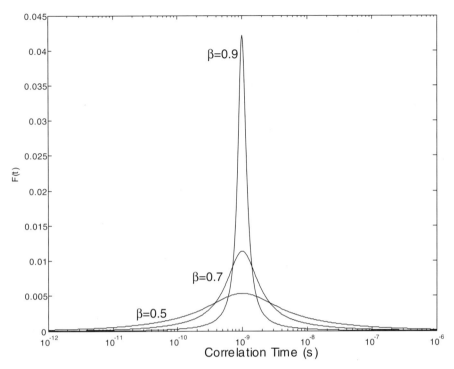

FIGURE 5.2 A plot of the Fuoss–Kirkwood distribution function for $\tau_0 = 1$ ns and values for β of 0.9, 0.7, and 0.5.

One feature of the Cole–Cole and Fuoss–Kirkwood distributions is that they are symmetric with respect to the mean correlation time. In some cases a better fit to the data is obtained with an asymmetric distribution, such as the so-called log-χ^2 distribution (12). By defining

$$s = \log_b \left[1 + (b-1) \frac{\tau}{\tau_0} \right] \tag{5.21}$$

we can obtain the log-χ^2 distribution function as

$$F(s) = \frac{p}{\Gamma(p)} (ps)^{p-1} e^{-ps} \tag{5.22}$$

where $\Gamma(p)$ is the gamma function, and b and p are adjustable parameters. The spectral densities are obtained numerically as

$$J(\omega) = 2 \int_0^\infty \frac{\tau_0 F(s) [b^s - 1]}{(b-1) \left\{ 1 + \omega^2 \tau_0^2 \left[\frac{(b^s - 1)}{(b-1)} \right]^2 \right\}} \tag{5.23}$$

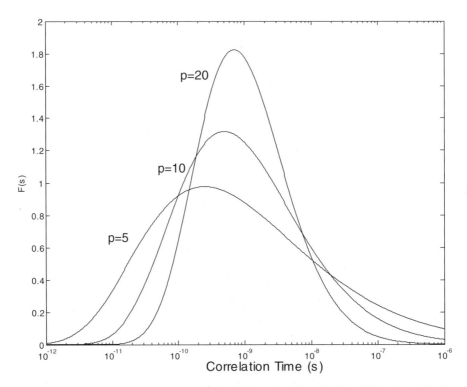

FIGURE 5.3 A plot of the called log-χ^2 distribution function with $b = 1000$ and $p = 5, 10,$ and 20.

The parameters b and p are not independent, and the value of b is typically set to 1000 and the value of p is adjusted to give the best fit to the NMR data. Figure 5.3 shows a plot of the distribution function for p values of 20, 10, and 5. Note that the distribution becomes broader and more asymmetric as the value for p decreases.

The stretched exponential, commonly known as the KWW function, is extensively used in polymer science. This correlation function can also be adapted to NMR measurements using Equation (5.6). The stretched exponential correlation function is given by

$$G(t) = e^{-(t/\tau)^\beta} \tag{5.24}$$

and the spectral densities are given by (3)

$$J(\omega) = \sum_{i=1}^{n} F_{\text{KWW}}(\tau_i) \left[\frac{\tau_i}{1 + (\omega \tau_i)} \right] \tag{5.25}$$

where $F_{\text{KWW}}(\tau_i)$ is evaluated with the series method (13) as

$$F_{\text{KWW}}(\tau_i) = -\frac{\tau_0}{\pi \tau_i} \sum_{k=0}^{\infty} \frac{(-1)^k}{k!} \sin(\pi \beta k) \, \Gamma(\beta k + 1) \left(\frac{\tau_i}{\tau_0} \right)^{\beta k + 1} \tag{5.26}$$

The width of the distribution is given by β and $F_{\text{KWW}}(\tau_i)$ is normalized.

5.2 CHAIN MOTION OF POLYMERS IN SOLUTION

The molecular dynamics of polymers are complex and a variety of correlation functions have been used to fit the relaxation data. As more data are accumulated, these correlation functions are tested over a wider range of frequencies and temperatures, and the fitting criteria becomes more stringent. In a study of polybutadiene, for example, it was reported that the log-χ^2 distribution did not give a good fit to the relaxation data, but a good fit could be obtained if an additional term is added to account for the fast librational motions (14). As a general rule fits to the relaxation data are improved with the bimodal distribution function that simultaneously accounts for the faster localized motion occurring on a picosecond time scale and slower segmental motions.

The general strategy for modifying the correlation functions to account for the faster librational motion is to assume that the faster librational motions are independent of the slower segmental motions described by the distribution functions. The modified correlation functions are written as

$$G(t) = G_{\text{lib}}(t) + G_{\text{seg}}(t) \tag{5.27}$$

where the segmental correlation function is one of those just described. The data fitting does not appear to depend strongly on the functional form of the librational correlation function, so an exponential can be used as

$$G_{\text{lib}}(t) = e^{-t/\tau_{\text{lib}}} \tag{5.28}$$

where τ_{lib} is the correlation time for the librational motion. The modified spectral density function for the KWW function, for example, is given by

$$J(\omega) = \frac{a\tau_{\text{lib}}}{1 + \omega^2 \tau_{\text{lib}}^2} + (1-a) \sum_{i=1}^{n} F_{\text{KWW}}(\tau_i) \left[\frac{\tau_i}{1 + (\omega \tau_i)} \right] \tag{5.29}$$

where the coefficient a describes the fraction of relaxation due to the rapid librational motion. The fitting to the modified correlation functions does not appear to be sensitive to the value of τ_{lib}, as long as the correlation time is much faster than the segmental correlation time.

5.2.2 Relaxation Mechanisms for Polymers in Solution

The first step in the analysis of polymer relaxation is determining the relaxation mechanism. The observed relaxation rate is given by the sum of all of the relaxation rates, and for the spin-lattice relaxation this is given by

$$\frac{1}{T_1} = \frac{1}{T_1^{\text{DD}}} + \frac{1}{T_1^{\text{CSA}}} + \frac{1}{T_1^{Q}} \tag{5.30}$$

where T_1^{DD}, T_1^{CSA}, and T_1^{Q} are the relaxation contributions from dipolar interactions, chemical shift anisotropy, and quadrupolar interactions. The relative contributions of the various mechanisms depend on the nucleus under observation and the chemical structure. The interpretation is generally easiest if the relaxation can be assigned to a single mechanism. There are additional mechanisms (spin rotation, scalar, paramagnetic, etc.) that can give rise to relaxation, but they are generally not important for polymers.

Proton relaxation is often the easiest to observe, but the most difficult to interpret in terms of the correlation times. The most important factor complicating the interpretation of proton relaxation is magnetization exchange via mutual spin flips (spin diffusion). These mutual spin flips exchange magnetization among the protons in such a way that the rate of spin-lattice relaxation depends on all of the dipolar couplings. If the molecular motion is in the slow-motion regime, then the spins will equilibrate internally before relaxing, so all of the protons will have the same relaxation time, regardless of the internuclear distances and correlation times. Another problem with proton relaxation is that the distances between pairs of protons are not fixed, but depend on the conformation. Since most polymers are mobile in solution, the proton–proton distances are fluctuating during the experiment. For these reasons, other nuclei are frequently used to study the molecular dynamics of polymers even though they have a lower sensitivity. Carbon NMR is most commonly used to study the dynamics of polymers because almost all polymers contain carbon. Furthermore, the relaxation for protonated carbons is usually dominated by the carbon–proton dipolar interactions from attached protons, so the rates can often be interpreted using only the dipolar couplings. The rates are more difficult to interpret for nonprotonated carbons because the distances to the nearest protons may depend on the conformation, which is fluctuating for polymers in solution. In addition, the relaxation from aromatic, carbonyl, and olefinic carbons may have contributions from chemical shift anisotropy, so at least two mechanisms must be considered.

The relaxation times of other nuclei also provide information about the molecular dynamics of polymers. The relaxation of silicon, phosphorus, and nitrogen are usually dominated by dipolar couplings, although, depending on the structure, chemical shift anisotropy may also contribute to the relaxation. Fluorine has a good sensitivity, but can suffer from the same problems as protons, and fluorine has a large chemical shift anisotropy. Deuterium NMR can also be used, but isotopically enriched polymers are required. The deuterium relaxation is relatively easy to interpret because the relaxation is dominated by quadrupolar interactions.

Relaxation by dipolar interactions between carbons and their directly attached protons is often the dominant relaxation mechanism in carbon NMR studies. The spin-lattice and spin-spin relaxation rates for dipolar relaxation are given by

$$\frac{1}{T_1^{DD}} = \frac{n}{10}\left(\frac{\mu_0}{4\pi}\right)\frac{\gamma_C^2\gamma_H^2\hbar^2}{r^6}\{J(\omega_H - \omega_C) + 3J(\omega_C) + 6J(\omega_H + \omega_C)\} \quad (5.31)$$

and

$$\frac{1}{T_2^{DD}} = \frac{n}{20}\left(\frac{\mu_0}{4\pi}\right)^2\frac{\gamma_C^2\gamma_H^2\hbar}{r^6}\{4J(0) + J(\omega_H - \omega_C) + 3J(\omega_C)$$
$$+ 6J(\omega_H) + 6J(\omega_H + \omega_C)\} \quad (5.32)$$

5.2 CHAIN MOTION OF POLYMERS IN SOLUTION

where n is the number of protons, μ_0 is the vacuum magnetic permeability, γ_C and γ_H are the carbon and proton magnetogyric ratios, ω_H and ω_C are the proton and carbon frequencies, and r is the internuclear distance. The value of r for carbon–proton relaxation is taken as 1.08 Å for methine and methylene carbons and 1.09 Å for aromatic carbons.

The NOE also depends on the on the molecular dynamics and offers another relaxation parameter for the evaluation of polymer dynamics. The NOE, which is due to dipolar interactions, is given by

$$\text{NOE} = \frac{\gamma_H}{\gamma_C} \frac{6J(\omega_H + \omega_C) - (\omega_H - \omega_C)}{J(\omega_H - \omega_C) + 3J(\omega_C) + 6J(\omega_H + \omega_C)} \tag{5.33}$$

Care must be taken when measuring the relaxation from carbons that have a significant contribution from chemical shift anisotropy, since this is a source of relaxation other than dipolar interactions. The assumption implicit in Equation (5.33) is that the relaxation is due only to dipolar interactions.

Chemical shift anisotropy can make a significant contribution to the relaxation of aromatic, olefinic, and carbonyl carbons, as well as to the relaxation of fluorine, nitrogen, silicon, and phosphorus atoms. The chemical shift anisotropy contribution to the T_1 and T_2 relaxation of axially symmetric atoms is given by

$$\frac{1}{T_1^{CSA}} = \frac{2}{15} \omega_C^2 \Delta\sigma^2 J(\omega_C) \tag{5.34}$$

and

$$\frac{1}{T_2^{CSA}} = \frac{1}{45} \omega_C^2 \Delta\sigma^2 \{4J(0) + 3J(\omega_C)\} \tag{5.35}$$

where

$$\Delta\sigma = \sigma_{33} + \frac{1}{2}(\sigma_{11} + \sigma_{22}) \tag{5.36}$$

Note that the relaxation by chemical shift anisotropy can be identified, since it depends on the square of the magnetic field strength through the $\Delta\sigma$ term.

Polymers that have been isotopically labeled with deuterium are sometimes used to study polymer dynamics. The interpretation of the relaxation mechanism is simplified because quadrupolar interactions dominate the relaxation. The relaxation rates for quadrupolar nuclei are given by

$$\frac{1}{T_1^Q} = \frac{3\pi^2}{10} \left(\frac{e^2 qQ}{h}\right)^2 (1 + \xi)\{J(\omega_D) + 4J(2\omega_D)\} \tag{5.37}$$

and

$$\frac{1}{T_2^Q} = \frac{3\pi^2}{20} \left(\frac{e^2 qQ}{h}\right)^2 (1 + \xi)\{3J(0) + 5J(\omega_D) + 2J(2\omega_D)\} \tag{5.38}$$

where ξ is the asymmetry of the electric field gradient (which is taken as zero for deuterons) and (e^2qQ/h) is the quadrupole coupling constant.

5.2.3 NMR Relaxation in Solution

The NMR relaxation for polymers in solution is relatively easy to measure using the pulse sequences introduced in Section 2.6. Care must be taken to exclude oxygen from the samples with long relaxation times, since paramagnetic oxygen can decrease the relaxation times. The oxygen can be excluded by freeze–thaw cycles under vacuum or by bubbling an inert gas (usually argon) through the sample.

The spin-lattice relaxation times can be measured using the inversion-recovery pulse sequence introduced in Section 2.6.1.1. One potential problem in measuring the relaxation rates for nuclei with large chemical shift dispersions is that nuclei widely separated in frequency will not experience the same tip angle from the initial 180° pulse. This limitation can be partially overcome by using a composite pulse to invert the spins. Composite pulses are a series of pulses given back-to-back with different phases and duration. One of the commonly used composite 180° pulses is the $90°_x$–$180°_y$–$90°_x$ pulse sequence (15), which increases the bandwidth of the 180° pulse by a factor of 6. Figure 5.4 shows the heteronuclear inversion recovery pulse sequence with a composite 180° pulse. In the heteronuclear pulse sequence, the protons are saturated during the relaxation delay so that the relative populations of the proton spin states do not change as a function of the delay time. The failure to saturate the protons can lead to nonexponential relaxation. The power level required to saturate the protons is much lower than required for scalar decoupling, which uses WALTZ decoupling (16) during the acquisition period.

The spin-spin relaxation rates can be estimated from the linewidths using Equation 1.73 or by the Gill–Mieboom modification of the Carr–Purcell pulse sequence introduced in Section 2.6.1.2. The T_2 is rather difficult to measure accurately and is usually not used to measure the correlation times.

The NOE is measured by comparing the intensity from a spectrum with decoupling only during acquisition (no NOE) to a spectrum acquired with decoupling during

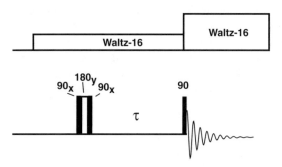

FIGURE 5.4 A pulse-sequence diagram for heteronuclear inversion-recovery using a composite 180° pulse for inversion and proton saturation during the entire experiment. Higher decoupler power levels are used during acquisition.

5.2 CHAIN MOTION OF POLYMERS IN SOLUTION

the entire pulse sequence, as shown in Section 2.6.1.3. As with the heteronuclear T_1 experiment, lower power decoupling can be used during the period before acquisition. The delay between acquisitions must be longer than five times the T_1.

5.2.4 The Relaxation of Polymers in Solution

The solution relaxation of polymers has been extensively studied in order to understand polymer structure–property relationships. In the early days it was somewhat of a surprise that polymers in the melt or in solution gave such high-resolution spectra. The expectation was that very broad lines would be observed for high molecular-weight materials. The correlation time for a random coil molecule is in general given by

$$\tau_r = \frac{2M\,[\eta]\,\eta_0}{3RT} \qquad (5.39)$$

where M is the molecular weight, $[\eta]$ is the intrinsic viscosity for the polymer, and η_0 is the solvent viscosity. In the absence of molecular motions that are rapid compared to τ_r, this equation predicts that τ_r will be very long and that broad lines would be observed for even moderate molecular-weight polymers. The observation of narrow lines demonstrates that some mechanism other than reorientation of the entire chain leads to averaging of the dipolar couplings. This additional motion has been identified as librational and segmental motions that typically occur over a time scale of picoseconds to nanoseconds. It is sometimes observed that the lines for side-chain peaks are sharper than the main-chain signals. This shows that the side chains can experience additional motion relative to the main-chain atoms.

The hypothesis that segmental motion dominates the relaxation for high molecular-weight polymers can be tested by measuring the relaxation times as a function of polymer molecular weight. One such result is shown in Figure 5.5, which shows the relaxation times for the methine and methylene carbons of polystyrene as a function of polymer molecular weight (17). The relaxation times are not very sensitive to molecular weight for polymers with molecular weights above 10,000 g/mol. This result shows that in high molecular-weight polymers, the relaxation is due only to segmental relaxation.

Over the years the polymer relaxation data has been interpreted with increasingly sophisticated models. The initial studies fit the relaxation data to the isotropic model for the correlation time, which more recent studies have shown is not a good model. However, the T_1 and correlation times calculated from the simple model can be used to characterize gross changes in the molecular dynamics as a function of polymer structure.

Table 5.1 lists the relaxation times and correlation times calculated from the isotropic correlation time model for a variety of polymers (1). While the correlation times are by no means quantitative, they do show gross trends relating polymer structure and dynamics. It is clear, for example, from the comparison of polyethylene, polypropylene, and polystyrene that the presence and mass of the side chain

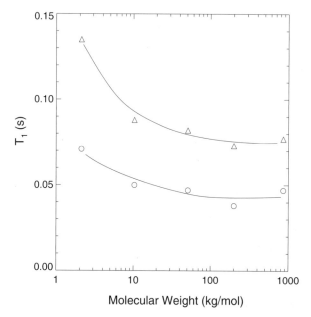

FIGURE 5.5 The carbon spin-lattice relaxation times for the (△) methine and (○) methylene carbons for polystyrene as a function of polymer molecular weight.

grossly affects the chain dynamics of atoms in the main chain. The correlation times for polyethylene at 30°C are a factor of 3 faster than those for polypropylene. The correlation times for polypropylene are approximately the same as for poly(vinyl chloride), and both are a factor of 4 shorter than for polystyrene with its larger side chain. The chain dynamics in many polymers involves *gauche–trans* isomerizations along the main chain, and increasing the size of the side chain increases the steric interference between the groups, leading to a slowing of the chain dynamics.

Heteroatoms in the polymer main chain can also influence the correlation times and lead to either an increase or decrease in the mobility. This is shown by the comparison of polypropylene (0.13 ns) and poly(propylene oxide) (0.049 ns), where the introduction of an oxygen into the main chain leads to a greater flexibility. The addition of sulfone groups leads to extremely long correlation times (23 ns) and broad lines in the solution spectrum.

The introduction of double bonds into polymers often leads to an increase in the mobility. The double bonds have a high barrier to rotation, but the barriers are reduced at the allyic bonds. The correlation times for *cis* and *trans* polybutadiene are similar to those observed for polyethylene. Polymer stereochemistry can in some cases influence the chain dynamics, as shown by the comparison of the correlation times for atactic, isotactic, and syndiotactic poly(methyl methacrylate). This is often a rather subtle effect.

As a general rule, the isotropic correlation time model does not provide a good fit to polymer relaxation data over a range of magnetic field strengths and temperatures.

5.2 CHAIN MOTION OF POLYMERS IN SOLUTION

TABLE 5.1 The Carbon Spin-lattice Relaxation Times for a Variety of Polymers and Correlation Times Calculated from the Isotropic Model.

Polymer	Solvent[a]	Temperature (°C)	Concentration (wt %)	τ_c (ns)
Polyethylene	TCB	110	33	0.019
	ODCB	100	25	0.018
	ODCB	30	25	0.040
Polypropylene	ODCB	100	25	0.044
		30	25	0.13
Polyisobutylene	CDCl$_3$	30	5	0.16
Polybutene-1	CCl$_3$CHCl$_2$	100	10	0.14
1,4-Polybutadiene				
cis	CDCl$_3$	54	20	0.016
trans	CDCl$_3$	54	20	0.021
Poly(vinyl chloride)	TCB	107	10	0.15
Poly(vinylidine chloride)	HMPA-d$_{18}$	40	15	0.63
		89	15	0.20
Poly(vinylidine fluoride)	DMF	41	20	0.079
Poly(vinyl alcohol)	DMSO	30	20	0.55
Polyacrylonitrile	DMSO	50	20	0.33
Polystyrene				
atactic	Toluene-d$_8$	30	15	0.49
isotactic	Toluene-d$_8$	30	15	0.55
Poly(methyl methacrylate)				
isotactic	CDCl$_3$	38	10	0.39
syndiotactic	CDCl$_3$	38	10	0.62
atactic	DMF	41	20	0.81
Poly(α-methyl styrene)	CDCl$_3$	30	10	0.49
Polyoxymethylene	HFIP	30	3	0.082
Poly(ethylene oxide)	C$_6$D$_6$	30	5	0.018
Poly(propylene oxide)	CDCl$_3$		5	0.049
Poly(methyl thiirane)	CDCl$_3$	30	10	0.057
Poly(phenyl thiirane)	CDCl$_3$	25	15	0.25
Poly(styrene peroxide)	CHCl$_3$	23	22	0.21
Poly(butene-1-sulfone)	CDCl$_3$	40	25	23

[a] Solvent abbreviations: TCB—1,2,4-trichlorobenzene; ODCB—o-dichlorbenzene, HMPA—hexa(methyl-d$_3$)phosphoramide; DMF—dimethylformamide; DMSO—dimethyl sulfoxide, HFIP—hexafluoro-2-propanol.

Perhaps the best way to investigate the relaxation in polymers is to measure the relaxation times over as wide a range in field strengths as possible. This is often not practical, because most NMR spectrometers are designed to work at specific field strengths and it is very difficult to change the field on a superconducting magnet. The range of magnetic field strengths under which the relaxation rates can be measured is usually limited by the maximum and minimum field strengths available. Since the resolution is increased at a higher field, low-field magnets are not commonly available.

It is often the case when studying the relaxation in polymers that only fields between 200 and 600 MHz (for protons) are available. While this makes a large difference in the resolution, this corresponds only to a factor of 4 in field strength, and it would be desirable to study the relaxation over a wider range of frequencies.

The second variable in NMR relaxation studies is the temperature. Since the dynamics are thermally activated, changing the temperature can lead to changes in the correlation times. This is illustrated in Table 5.1 for polyethylene at 30°C and 100°C, where the correlation time changes by more than a factor of 2. Thus, by combining changes in magnetic field with changes in temperature it becomes possible to probe the widest possible variation in correlation times.

When temperature is used as a variable in the relaxation studies care must be taken to ensure that the temperature dependence is the same for all dynamics that contribute to the relaxation. A number of studies have shown that both fast librational motions and slower segmental motions contribute to the relaxation. To use temperature as a variable, it must be demonstrated that these motions all have the same temperature dependence. It is possible to determine if this criterion is fulfilled with a plot of $\omega_0/T_1(T)$ vs. $\log(\omega_0\tau_c(T))$ (18). In the event that all of the correlation times that contribute to the relaxation have the same temperature dependence, all of the relaxation data will fall on a single curve. This is illustrated in Figure 5.6 for proton relaxation in polybutadiene relaxation data acquired at 32, 60, and 100 MHz (18).

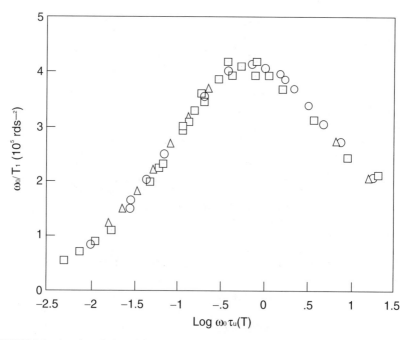

FIGURE 5.6 A reduced plot of the proton relaxation data for poly butadiene acquired at 32 (□), 60 (○) and 100 MHz (△).

5.2 CHAIN MOTION OF POLYMERS IN SOLUTION

Polyisoprene

The relaxation behavior for a number of polymers has been studied as a function of temperature and magnetic field strength. Many different models for the correlation functions have been used, but the best fits to the relaxation data often have fast and slow components, corresponding to librational and segmental motions. The relaxation of the C3 and C4 atoms of polyisoprene have been studied and fitted to several of the possible correlation functions (19). The relaxation analysis of polyisoprene illustrates the general approach for the interpretation of polymer relaxation data.

The carbon relaxation rates for polyisoprene has been measured over a range of temperatures at carbon field strengths corresponding to 125.8, 90.6, and 25.2 MHz (19). The spin-lattice relaxation for C4 was assumed to arise solely from dipolar interactions, while the C3 relaxation was due to the combination of dipolar interactions (Equation (5.31)) and relaxation by chemical shift anisotropy (Equation (5.34)). Figure 5.7 shows a plot of the log of the spin-lattice relaxation times as a function of

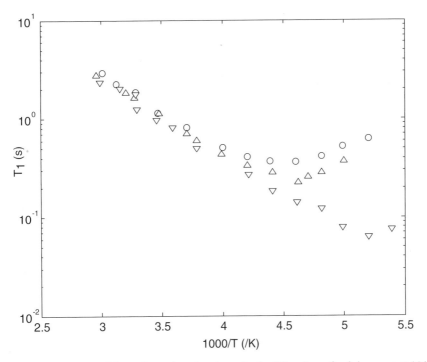

FIGURE 5.7 A plot of the carbon relaxation times for the C4 carbon of polyisoprene at (O) 25.2 MHz, (△) 90.6 MHz and (▽) 125.8 MHz.

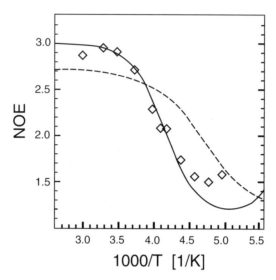

FIGURE 5.8 A plot of the temperature dependence of the NOE for the C4 carbon of polyisoprene at 125 MHz. The solid line shows the fit to the biexponential model and the dotted line shows the fit to the Cole-Cole model derived from the T_1 relaxation data.

inverse temperature for the C4 carbons at the three different magnetic field strengths. At high temperature the relaxation rate is strongly dependent on temperature and not very sensitive to the magnetic field strength. Near 220 K ($1/T = 0.0045$ K^{-1}), a T_1 is a minimum is observed. Such a T_1 minimum is characteristic of molecular motion on a time scale near the inverse of the spectrometer frequency. This makes the T_1 minimum very sensitive to the spectrometer frequency. The value and shape of the relaxation curve in the vicinity of the T_1 minimum very sensitive to the correlation function, and can be used to discriminate between different models for the molecular motion. Figure 5.8 shows a plot of the NOE as a function of inverse temperature.

The assumption implicit in using the data as a function of temperature is that the molecular motions that cause relaxation are thermally activated, and that the conformational transitions are characterized by the activation energy E_a. According to Kramer's theory the relationship between the correlation time and the temperature is given by

$$\tau_c = A\eta e^{E_a/RT} \tag{5.40}$$

where A is the prefactor and η is the viscosity. Poly(ethylene oxide) and polyisoprene have been studied as a function of solvent viscosity and it has been observed that the data are fitted better by

$$\tau_c = A\eta^\alpha e^{E_a/RT} \tag{5.41}$$

where the coefficient α is determined empirically (20). For polyisoprene dissolved in ten solvents covering a factor of 70 in viscosity, a value of 0.41 ± 0.02 was measured for α. This coefficient is used in Equation (5.41) to calculate the activation energy from the measurement of the correlation time as a function of temperature.

5.2 CHAIN MOTION OF POLYMERS IN SOLUTION

TABLE 5.2 Fits to the Carbon Spin-lattice Relaxation for Polyisoprene using Several Models for the Correlation Functions.

Model	Width[a]	τ_1/τ_2	f	χ_R^2
Exponential				5.9
Cole–Cole	0.79			4.3
Fuross–Kirkwood	0.79			4.6
log χ^2	25			3.4
Biexponential		10	0.6	3.4
		100	0.52	1.24
		1,000	0.49	1.08
		10,000	0.49	1.08
Hall–Helfand	20			3.5
DLM	1	1,000	0.48	1.1
	5	1,000	0.43	1.5
	10	1,000	0.39	1.9
Modified Cole–Cole	0.95	1,000	0.47	0.99

[a] Width corresponds to ε, β, p, or τ_1/τ_2.

The relaxation data in Figure 5.7 was fitted to several types of unimodal and bimodal correlation functions (19), and some of the results are listed in Table 5.2. The parameters of interest in this table are the width of the distribution (ε, β, or p), the ratios of the fast to slow correlation times (τ_1/τ_0), the fraction of the relaxation attributed to the rapid motion (f), and the goodness-of-fit parameter χ_R^2. The lower the value of χ_R^2, the better the fit of the relaxation data.

The fits to the relaxation data using unimodal models, including the isotropic motion model and the Cole–Cole, Fuoss–Kirkwood, and log-χ^2 distributions, generally resulted in poor fits and a large value for χ_R^2. A typical fit to the data from a unimodal model is shown in Figure 5.9 for the Cole–Cole distribution function. The unimodal models do a poor job of predicting the T_1 minimum and the frequency dependence. The models with a distribution of correlation times do a better job of fitting the data than does the single exponential function.

The biexponential correlation function gives a better fit than the unimodal function and distribution functions, although a ratio larger than 1000 for τ_1/τ_0 is required. The fit to the biexponential function does not seem to be sensitive for ratios larger than 1000. In the best fits, about 50% of the relaxation, can be attributed to the faster librational motion.

The DLM model, which combines the Hall–Helfand correlation function with faster librational motion, also gives a reasonable fit to the data, although the best fit is obtained with a width (corresponding to the τ_1/τ_2 ratio) of one.

The best fit to the polyisoprene T_1 data is obtained using a Cole–Cole distribution function modified to include a term for the faster librational motion. Figure 5.10 shows the fit to the T_1 data using the modified Cole–Cole function with the width set to 0.95, a ratio for τ_1/τ_2 of 1000 where 47% of the relaxation attributed to the faster librational motion. Although only the T_1 data have been used to model the molecular dynamics, the best fits to the T_1 data also give the best fits to the NOE data.

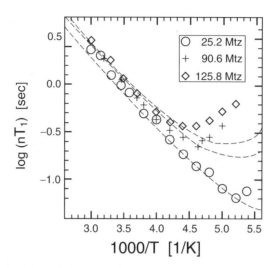

FIGURE 5.9 A fit of the polyisoprene relaxation data for the C4 carbon to the Cole–Cole distribution function.

In most studies the authors report fits to the data using only one of several possible models. The dynamics of poly(vinyl methyl ether) in solution and the bulk above T_g, for example, have been studied at 25.15 and 62.5 MHz (8). Poor fits were observed to unimodal functions, including the single exponential model. Good fits were obtained using the DML model (Equation (5.11)) and some of the results are shown in Figure 5.11. The best fit to the experimental data was obtained with 40% of the relaxation attributed to fast librational motions, τ_1/τ_0 set to 200, and τ_1/τ_2 set to 2. The same parameters were used to fit poly(vinyl methyl ether) in solution, except that τ_1/τ_0 was set to 400.

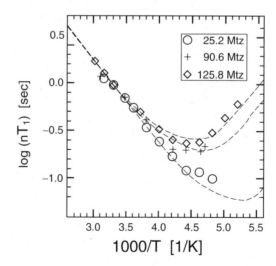

FIGURE 5.10 A fit of the spin-lattice relaxation data for the C4 carbon of polyisoprene to the modified Cole–Cole distribution function.

5.2 CHAIN MOTION OF POLYMERS IN SOLUTION

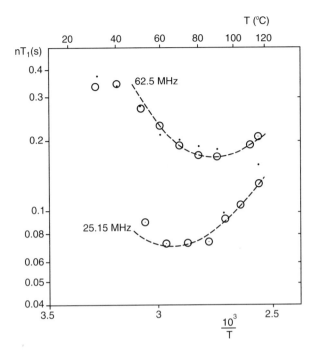

FIGURE 5.11 Fits of the carbon relaxation data for the (O) methine and (•) methylene carbons of poly(vinyl methyl ether) as a function of temperature at 62.5 and 25.15 MHz.

We noted several times that data must be gathered over a range of frequencies and temperatures to discriminate between different models for the correlation functions, and that the data are acquired over the largest range provides the best test of the correlation functions. In a study of atactic polypropylene the carbon relaxation times were measured over the frequency range of 5 to 125 MHz, and this large range in observation frequency provides the means to discriminate between some of the models (2). For the relaxation times for the data acquired at 25, 75, and 125 MHz a modified KWW distribution function provided a better fit to the data than either the modified log-χ^2 or the DML model, and the T_1 and NOE data are shown in Figure 5.12 (3). The modified KWW and log-χ^2 distribution functions include a term to account for fast molecular motion.

Another feature of the relaxation becomes observable when the relaxation of atactic polypropylene melts is measured at 5 MHz. The 5-MHz data were not well fit by the modified KWW distribution function that gave good fits to the data at higher frequencies. The fit was improved by including an additional term for a normal mode described by Rouse dynamics modified for entanglement effects, as shown in Figure 5.13 (3). The inclusion of the normal mode term in the correlation function was only required to fit the relaxation data at 5 MHz and did not improve the fit at higher observation frequencies.

A number of polymers in dilute solution and in the bulk have been studied by NMR and fitted to one of the bimodal models for the correlation function. While it is

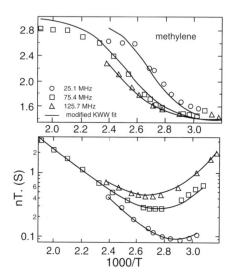

FIGURE 5.12 The NOE (top) and spin-lattice relaxation times (bottom) for the methylene carbons in atactic polypropylene at 25.1, 75.4 and 125.7 MHz. The solid line shows the fit to the modified KWW model.

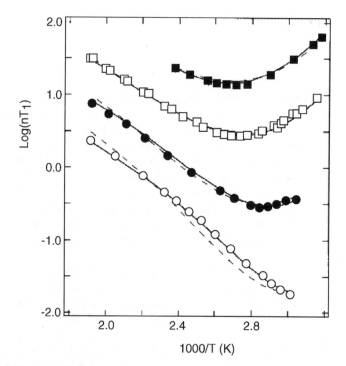

FIGURE 5.13 A semilog plot vs. inverse temperature of the methylene carbon spin-lattice relaxation times at field strengths of (■) 125.7, (□) 75.4, (●) 25.1 and (○) 5 MHz. The dotted lines show the fits to the modified KWW model and the solid line shows the fit including Rouse dynamics.

5.3 NMR RELAXATION IN THE SOLID STATE

TABLE 5.3 The Contribution of Fast Librational Motion to the Relaxation in Several Polymers.

Polymer	Bond	%
Polyisoprene	CH	41
Polyisoprene	CH_2	48
Polyisobutylene	CH_2	21
1,2-Polybutadiene	CD/CD_2	42
Poly(methyl vinyl ether)	CH/CH_2	40
Poly(1-napthylacrylate)	CH	22
Poly(1-napthyl methyl acrylate)	CH	15
Poly(1-napthyl ethyl acrylate)	CH	16
Bulk Polyisobutylene	CH_2	21
Bulk Poly(methyl vinyl ether)	CH/CH_2	40

often difficult to compare the correlation times and distribution widths in the different models, it is instructive to compare the amplitude of the correlation function attributed to fast molecular motions as a function of polymer structure. Table 5.3 compares the amplitude of the term attributed to fast molecular motions for a variety of polymers (14). Polyisobutylene and polyisoprene have been studied both in the bulk above T_g and in dilute solution, and identical contributions to the correlation function from fast molecular motions are reported. Note that in polyisoprene that the librational motions contribute differently to the relaxation of the CH and CH_2 groups. This shows that the correlation functions are different for the two carbons and explains why the value of nT_1 is different for the methine and methylene carbons. This result is most likely a consequence of the double bond. The data also show that as the size of the polymer side chain increases (i.e., poly(napthyl acryaltes)), the faster librational motions are restricted.

5.3 NMR RELAXATION IN THE SOLID STATE

5.3.1 Introduction

The observation of NMR lineshapes, particularly those due to quadrupolar couplings in deuterated polymers or the chemical shift anisotropy in the carbon, silicon, phosphorus, or nitrogen spectra, are a rich source of information about the molecular dynamics of polymers. The lineshapes have a characteristic appearance in the absence of molecular motion and the changes due to the molecular dynamics depend on the geometry, amplitude, and rate of molecular motions. Solution-like spectra are obtained above T_g, since large-amplitude motions with a nearly isotropic distribution effectively average the lineshape.

NMR is a particularly good method to study the dynamics of polymers, because the relaxation times and lineshapes are sensitive to the dynamics over a wide range of time scales, from picoseconds to seconds. No single method is able to probe

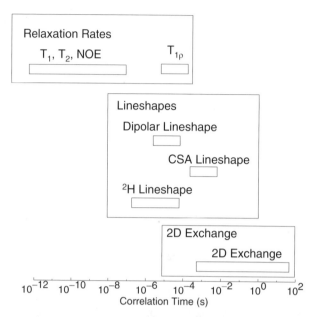

FIGURE 5.14 A plot showing the experiments used to study the dynamics of polymers and their time scales.

the dynamics over such a wide time scale, but rather several methods are used in combination. These methods used to probe the dynamics on the different time scales shown schematically in Figure 5.14. The fastest motions in solid polymers ($\tau_c < 10^{-7}$ s) can be measured using the T_1, T_2, and NOE experiments used to measure the dynamics in solution. Several different experiments can be used to measure the dynamic in the intermediate time scale (10^{-7} s $> \tau_c > 10^{-4}$ s), by measuring the chemical shift anisotropy, dipolar and deuterium lineshapes, the deuterium echo intensity, and the $T_{1\rho}$ relaxation times. nD exchange experiments can probe the dynamics on the longer time scale ($10^{-4} > \tau_c > 10^2$), including jumps between sites in a crystalline lattice and chain diffusion between the amorphous and crystalline regions (21).

The dynamics of polymers are of particular importance because they can often be directly related to the bulk properties of polymers. The mechanical properties are often related to the mechanism by which polymers dissipate energy, and these same mechanism may contribute to the NMR relaxation. The very slow dynamics probed by exchange NMR can lead to the molecular-level assignments of the transitions observed by dielectric and dynamic–mechanical spectroscopy (21).

5.3.2 NMR Relaxation in Solid Polymers

The NMR relaxation times for the carbon, silicon, nitrogen, and phosphorous atoms in polymer provide information about the fast molecular dynamics. As noted previously (Section 4.3.2), the relaxation rates of protons (and fluorines) in solids are dominated

5.3 NMR RELAXATION IN THE SOLID STATE

by spin diffusion, so the relaxation times are more likely to provide information about the morphology than the molecular dynamics. For nuclei other than protons and fluorines, the relaxation times are very sensitive to the molecular dynamics, and large differences in the relaxation rates are observed for crystalline, amorphous, and rubbery polymers.

The spin-lattice relaxation rates are often used to measure the molecular dynamics of polymers. As with polymers in solution, the relaxation times can have contributions from dipolar interactions, chemical shift anisotropy, or quadrupolar interactions (Equation (5.30)), and the relaxation times are easiest to interpret if the relaxation can be attributed to a single relaxation mechanism. For carbons with directly bonded protons the relaxation is dominated by the carbon–proton dipolar interactions. Aromatic and carbonyl carbons can also have contributions from chemical shift anisotropy in addition to the dipolar interactions, while the relaxation for deuterons is due almost exclusively to quadrupolar interactions. For dipolar interactions the relaxation times depend on the number of nearby protons, their distances, and correlation times.

We noted earlier that the T_1 has a parabolic dependence on the correlation time (Figure 1.42). The relaxation times are very long for fast and slow motions, and at a minimum when the correlation time is near the inverse of the spectrometer frequency. The dynamics for many polymers in solution and in the bulk above T_g have correlation times near the inverse of the spectrometer frequency, and a T_1 minimum is often observed. The correlation times in crystalline and glassy polymers are far into the slow-motion regime, and long spin-lattice relaxation times are observed. In extreme cases the relaxation times can be on the order of thousands of seconds (22).

The long relaxation times of many crystalline polymers make it difficult to measure the relaxation times using the inversion-recovery pulse sequence (Figure 2.19), because a relaxation delay between scans in excess of five times the longest T_1 is required. The relaxation times in solids are more commonly measured using the cross polarization T_1 (CPT_1) pulse sequence (Figure 2.26). In the cross-polarization experiment, it is necessary to wait only five times the proton T_1 between scans. This is an advantage because the proton relaxation times are usually on the order of a few seconds. For those cases where it is not desirable to use cross polarization for signal enhancement, the relaxation times can be measured using the heteronuclear saturation recovery pulse sequence shown in Figure 5.15. Lower decoupling power is used during the relaxation delay, while high-powered decoupling is used during acquisition. As

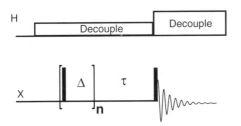

FIGURE 5.15 The pulse-sequence diagram for heteronuclear saturation recovery.

with other solid-state experiments, the signals are recorded with magic-angle sample spinning and dipolar decoupling.

The spin-lattice relaxation behavior for many solid polymer samples does not show the single-exponential relaxation observed for polymers in solution. We noted earlier (Figure 4.39) that the spin-lattice relaxation for polyethylene is multiexponential because different relaxation times are associated with the crystalline, interfacial, and amorphous phases. Nonexponential relaxation times are sometimes observed for glassy polymers because the local environments for individual chains can vary.

The spin-lattice relaxation measurements for polymers in solution provide a direct link between the relaxation behavior and the segmental and librational chain motions. This link is not as clear for crystalline and glassy polymers in part because the relaxation is so slow in crystalline and glassy polymers that other factors can affect the relaxation. This is illustrated in Figure 5.16, which shows a plot of the relaxation times for the methine and methylene carbons of isotactic polypropylene as a function of inverse temperature (23). The relaxation is strongly temperature dependent and shows a minimum near 170 K. Other studies have shown that there is little molecular

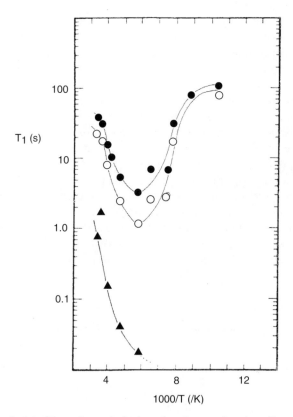

FIGURE 5.16 A plot of the carbon spin-lattice relaxation as a function of inverse temperature for the (O) methine, (•) methylene and (▼) methyl carbons of isotactic polypropylene.

5.3 NMR RELAXATION IN THE SOLID STATE

motion at this temperature for isotactic polypropylene, so the relaxation must be attributed to something other than chain motion. In this case, the relaxation of the main-chain methine and methylene is due to methyl group rotation, since the methyl group shows the same T_1 minimum. The conclusion from these studies is that there is insufficient main-chain motion at low temperatures to cause efficient relaxation, so molecular motion several atoms away dominates the relaxation. This makes it difficult to determine the correlation times from the spin-lattice relaxation of carbons in the solid state.

The spin-spin relaxation times for solid polymers are usually not measured directly, but they can have a large effect on the appearance of the spectrum. We typically use magic-angle sample spinning and dipolar decoupling to remove the broadening from chemical shift anisotropy and dipolar couplings. As noted in Sections 2.5.1 and 2.5.3, the line broadenings from chemical shift anisotropy and dipolar couplings are on the order of a few kHz and 50 kHz, and we average the lineshapes by rotating the sample at a speed that is fast compared to the chemical shift anisotropy broadening or by irradiating the spectrum with sufficient proton power to cause rapid transitions between the proton spin states. If molecular motion occurs at the same frequency as the averaging (a few kHz or 50 kHz) then the averaging becomes ineffective and broadened lines are observed. The interference of molecular motion and proton decoupling is known as Rothwell–Waugh broadening, and the spin-spin relaxation time is given by (24)

$$\frac{1}{T_2} = \frac{\gamma_C^2 \gamma_H^2 \hbar^2}{5r^6} \left(\frac{\tau_c}{1 + \omega_2^2 \tau_c^2} \right) \tag{5.42}$$

where ω_2 is the decoupling field strength. When $\omega_2^2 \tau_c^2 \cong 1$, the linewidth will be at a maximum and the signals will be broadened beyond detection.

The interference of molecular motion and decoupling is illustrated in Figure 5.17, which shows the carbon spectra of isotactic polypropylene observed with cross polarization and magic-angle sample spinning as a function of temperature (23). Three well-resolved peaks are observed at higher temperature from the methylene, methine,

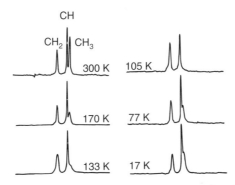

FIGURE 5.17 The spectrum of isotactic polypropylene as a function of temperature.

and methyl carbons. At temperatures near 170 K, the methyl signal begins to broaden. The methyl peak completely disappears from the spectrum at 105 K, but appears again at even lower temperatures (17 K). These data are consistent with the interference of methyl group rotation and the proton decoupling. The methyl signals in isotactic polypropylene broaden when the correlation time for methyl group rotation is near the inverse of the decoupling field strength. Such Rothwell–Waugh broadening can be identified by its dependence on the decoupler field strength. The studies with isotactic polypropylene show that at low temperature where the methyl signals are broadened that the linewidth shows the expected dependence (Equation 5.42) on the decoupler field strength (23).

The NOEs are sensitive to the faster motions in polymers, but they are not very useful for studying the dynamics of solid polymers. In the crystalline and amorphous phases below T_g, the chain dynamics are restricted and only small NOEs from chain motion are expected. The NOEs in solids are due mostly to methyl group rotation rather than chain motion, and they are very difficult to interpret in terms of the correlation times. The solid-state NOEs are more useful for studying the structure of polymers (25).

The cross-polarization contact time and the rotating-frame relaxation times $T_{1\rho}$ can in favorable circumstances be used to study the molecular dynamics of solid polymers. As shown in Equation (1.93), the buildup of magnetization during cross polarization is due to T_{CH}, while the decay is due to $T_{1\rho}(H)$ and $T_{1\rho}(C)$. The $T_{1\rho}$ relaxation rates are of interest because the rates depend on the strength of the spin-locking field, and provide valuable information about the molecular dynamics on a kHz frequency time scale. As noted previously, the proton $T_{1\rho}$ is dominated by spin diffusion, and in the presence of spin diffusion only an averaged relaxation rate for the entire chain is observed. Thus, the $T_{1\rho}(H)$ can only provide information about the relaxation of the entire system rather than the dynamics of segments along the chain. The $T_{1\rho}(H)$ relaxation data are most useful when a minimum in the relaxation times is observed, because this clearly demonstrates the presence of a dynamic process with a time constant on the order of the inverse of the spin-locking field.

The T_{CH} relaxation rate is most commonly measured from the initial buildup of the cross-polarization intensity. The cross-polarization time dependence can be fitted to a growing and decaying exponential (Equation (1.93)), and the time constant for the growing exponential is T_{CH}. It is in general not possible to interpret T_{CH} in terms of the correlation times, but rather the T_{CH} can be used to distinguish between more rigid and less rigid materials.

The proton relaxation times T_1 and $T_{1\rho}$ are dominated by spin diffusion and provide information about the average state of polymer chains. Spin diffusion makes the relaxation times very difficult to interpret in terms of the correlation times. The exception to this generality is, when a polymer undergoes large-amplitude molecular motion on the relevant time scale, either the inverse of the spectrometer frequency for T_1 relaxation or on the inverse of the spin-locking field for $T_{1\rho}$ relaxation. This is illustrated in Figure 5.18, which shows the temperature dependence of the proton $T_{1\rho}$ relaxation for polycarbonate (26). The $T_{1\rho}$ relaxation has a minimum near 270 K when the relaxation time is measured using a 43-kHz spin-locking field. This shows that

5.3 NMR RELAXATION IN THE SOLID STATE

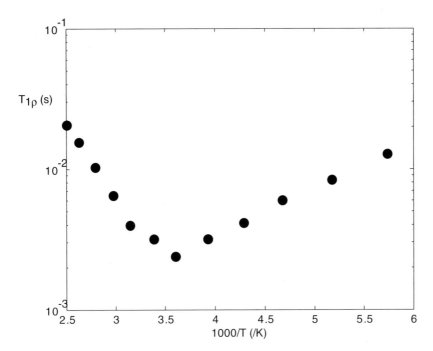

FIGURE 5.18 The proton $T_{1\rho}$ relaxation in polycarbonate as a function of temperature.

polycarbonate has a large-amplitude molecular motion with a correlation time near 43 kHz. Molecular motions on the kHz time scale have been related to the mechanical and impact properties of polycarbonates (27).

In favorable cases, the carbon $T_{1\rho}$ can provide important information about the polymer dynamics on the kHz time scale. The carbon $T_{1\rho}$ is measured using the modified version of the cross-polarization pulse sequence (Figure 2.28). The variable in this experiment is the length of time that the carbon spin-locking field is turned on after the end of cross polarization, and the time constant for the exponential decay is the carbon $T_{1\rho}$.

The carbon $T_{1\rho}$ is of interest because it may provide information about the molecular motions of polymers on a kHz time scale, but the relaxation times must be carefully interpreted. One complicating factor is that there are two relaxation mechanisms that must be distinguished before the carbon $T_{1\rho}$ relaxation rates can be interpreted in terms of the polymer dynamics. The observed carbon relaxation rate $T_{1\rho}^*$ is given by

$$\frac{1}{T_{1\rho}^*} = \frac{1}{T_{1\rho}^C} + \frac{1}{T_{1CH}^D} \tag{5.43}$$

where $T_{1\rho}^C$ is the desired relaxation rate, and T_{1CH}^D is the time constant for cross relaxation between the spin-locked carbons and the unlocked protons. The relaxation times are given by

$$\frac{1}{T_{1\rho}^C} = \frac{n}{10}\left(\frac{\mu_0}{4\pi}\right)^2 \frac{\gamma_H^2 \gamma_C^2 \hbar^2}{r^6} \{4J(\omega_1) + 4(\omega_H - \omega_C) + 3J(\omega_C) + 6J(\omega_H)$$
$$+ 6J(\omega_H + \omega_C)\} \tag{5.44}$$

and

$$\frac{1}{T_{CH}^D} = \frac{1}{2}\sin^2\theta M_{CH}^2 \{\pi\tau_D e^{-\omega_1\tau_D}\} \tag{5.45}$$

where M_{CH}^2 is the second moment of the heteronuclear dipolar coupling interaction, θ is the angle between the internuclear vector and the magnetic field, and τ_D is the correlation time for spin fluctuations. Note that both mechanisms depend on the magnitude of ω_1, but the $T_{1\rho}^C$ relaxation rate depends quadratically on the spin-lock power ω_1, while the T_{CH}^D rate depends exponentially on ω_1. The relative contributions of the two pathways can be evaluated by measuring the dependence of the relaxation rate on ω_1. The studies generally show that the relaxation in glassy polymers is due predominantly to $T_{1\rho}^C$ relaxation, while crystalline polymers can have contributions from the cross-relaxation term (23,28).

The preceding discussion illustrates the difficulty in quantitatively interpreting the carbon $T_{1\rho}$ values in terms of the molecular dynamics and rotational correlation times. The carbon $T_{1\rho}$ values are more commonly used to qualitatively compare the dynamics of polymers in different environments. This is illustrated in Table 5.4, which compares the carbon spin-lattice and rotating-frame spin-lattice relaxation times for semicrystalline polycaprolactone and the inclusion complex with cyclodextrin (29). There are large differences between the local environment for polycaprolactone in the semicrystalline environment and the inclusion complex. The impediments to chain motion are much larger in the crystalline environment than in the inclusion complex, where the polymer chain is constrained to a pore with a diameter of 5–7 Å. This is expected to give rise to a large difference in the chain dynamics.

The carbon spin-lattice relaxation times are very long for the semicrystalline polymer, where relaxation times between 169 and 223 s are observed. The increased

5.3 NMR RELAXATION IN THE SOLID STATE

TABLE 5.4 The Carbon Spin-lattice and Rotating-frame Spin-lattice Relaxation Times for Polycaprolactone and the Cyclodextrin Inclusion Complex.

Carbon	T_1 (s)		$T_{1\rho}$ (ms)	
	PCL	PCL/αCD	PCL	PCL/αCD
1	169	0.24	23	17
2	160	0.26	17	15
3,4	161	0.31	15	15
5	223	0.29	18	12
6			169	24

chain motion for the inclusion complexes leads to a shortening of the T_1 to less than a second. By comparison, the rotating frame relaxation times show a much smaller change. The relaxation times in the semicrystalline material are on the order of 15–23 ms for the methylene carbons and 169 ms for the carbonyl. The methylene carbon relaxation times are slightly shortened to 12–17 ms for the inclusion complex, while the carbonyl relaxation time is reduced to 24 ms. The T_1 relaxation is due to molecular motion on the MHz frequency scale, and this motion changes quite significantly as the molecule moves from the crystal to the inclusion complex. The rotating-frame relaxation times have contributions from the same motions, but also to relaxation on the frequency scale of the spin-locking field (30–60 kHz). The small differences in relaxation times show that the motions on this time scale do not change drastically between the polycaprolactone in the crystalline conformation and the inclusion complex.

5.3.3 Spin Exchange in Solid Polymers

Spin exchange experiments, usually monitored using 2D NMR, have become an important method to monitor the molecular dynamics of polymers. These studies can be used to quantitatively measure the dynamics of polymers, both in the amorphous and crystalline phases. The high-resolution studies typically utilize cross polarization and magic-angle sample spinning and can only be used when there is a difference in chemical shift for the conformations under investigation. Spin exchange is sensitive to the slow molecular dynamics that occur on a time scale longer than milliseconds, but shorter than the spin-lattice relaxation times (hundreds of seconds for some crystalline polymers). The spin exchange experiments in static (wideline) experiments are sensitive to these same features, but also to the geometry of molecular motion (Section 5.3.4).

Figure 5.19 shows the pulse-sequence diagram for a heteronuclear experiment used to measure carbon, silicon, phosphorus, or nitrogen spin exchange. The sequence is analogous to the three-pulse proton spin exchange (or NOESY) sequence, except the first 90° pulse is replaced by the cross-polarization sequence. This generates magnetization from the cross-polarized nucleus that evolves during the t_1 period. The 90° pulse rotates this magnetization along the $-z$ axis for the mixing time, and the final

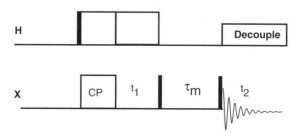

FIGURE 5.19 The pulse-sequence diagram of the 2D exchange experiment using cross polarization.

90° pulse converts the signals into observable magnetization. The most critical part of the pulse sequence is the mixing time τ_m, during which magnetization exchange occurs. If the polymer undergoes a conformational transition that results in a change in frequency for the peak of interest, then the signal will have a different frequency in the t_1 and t_2 time domains and will show up as an off-diagonal peak in the 2D exchange spectrum. The intensity of the off-diagonal peak depends on the relative ratio of the exchange lifetime to the mixing time.

Figure 5.20 shows the 2D exchange spectrum for atactic polypropylene at 250 K (30). The spectrum of atactic polypropylene shows three peaks for the methylene

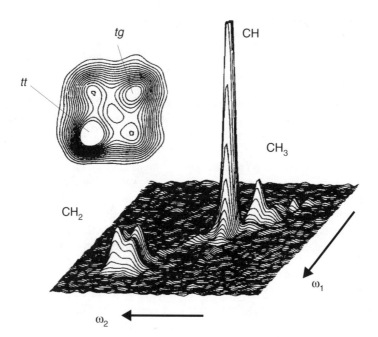

FIGURE 5.20 The 2D exchange spectrum for atactic polypropylene at 250 K. The spectrum was obtained with a mixing time of 0.5 s.

5.3 NMR RELAXATION IN THE SOLID STATE

carbons near 47 ppm that are assigned to the *tt*, *tg*, and *gg* chain conformations. At low temperature, conformational exchange is slow enough that the three peaks are resolved, while at higher temperatures (above 260 K), only a single broad line is observed from the rapid exchange of the conformations. The 2D exchange spectrum acquired with a mixing time of 0.5 s shows off-diagonal intensity connecting the signals assigned to the *tt* and *tg* conformation, indicating that these conformations are experiencing exchange during the mixing time. The relative intensities of the diagonal and cross peaks are given by the 2D exchange equations given in Section 1.7.1.2. By measuring the peak volumes as a function of mixing time and temperature, it is possible to determine both the exchange rates and their activation energies.

2D exchange experiments can be used to study the slow molecular dynamics in polymers as long as the lifetimes of the states are less than the spin-lattice relaxation times, which can be extremely long in crystalline polymers. One important use of this method is to study the exchange of chains between the crystalline and amorphous phases in semicrystalline polymers. Figure 5.21 shows the 2D exchange spectrum for semicrystalline polyethylene obtained at 363 K with a mixing time of 1 s (31). The spectrum of polyethylene shows two peaks, one from the crystalline phase (33 ppm) and the amorphous phase (31 ppm). The sample under study is a high-density polyethylene semicrystalline. The exchange spectrum shows cross peaks between the crystalline and amorphous phases that are labeled *ac* and *ca* in Figure 5.21. The exchange rates and activation energies (105 kJ/mol) are consistent with α-transition observed by other spectroscopies. This shows that the 2D exchange experiments can provide molecular-level assignments of the peaks observed by dielectric or dynamic mechanical spectroscopy. It is estimated that such ultraslow exchange experiments can be used to study chain diffusion rates in the range of 10^{-17} to 10^{-21} m²/s (31).

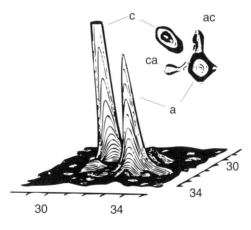

FIGURE 5.21 The 2D exchange spectrum for semicrystalline polyethylene showing magnetization exchange between the crystalline and amorphous phases.

It is important to note one important difference between the spin exchange experiments for protons and for other nuclei. Protons are an abundant nuclei, and exchange cross peaks can be observed both from chemical exchange, which results in a change in peak frequency, and from spin diffusion (Section 4.3.2). It is possible to distinguish between exchange and spin diffusion because they have very different temperature dependences. Spin diffusion has a very small temperature dependence, while conformational exchange and chain diffusion typically have high activation energies. In the cases of conformational exchange and chain diffusion, a change of 10 K often changes the exchange rate by a factor of 2, while little change is observed in the rate of spin diffusion.

The cross peaks of very dilute nuclei (such as carbon at natural abundance) in 2D exchange experiments are due almost exclusively to conformational exchange. Since carbon is present at a natural abundance of 1.1%, the probability of having two neighboring ^{13}C's that can exchange magnetization is on the order of 1/10,000. It may be possible to observe these effects at very long mixing times in samples with long relaxation times. Spin diffusion becomes important for isotopically labeled samples or for nuclei with a high natural abundance. Phosphorus, for example, has a natural abundance of 100% and exchange peaks could arise either from chemical exchange or spin diffusion. The temperature dependence of the exchange cross peaks has been used to distinguish between chemical exchange and spin diffusion in ^{31}P exchange studies of polyphosphazines (32,33).

5.3.4 Polymer Dynamics and Lineshapes

The wideline spectra of polymers often contain information about the molecular dynamics. The lines are broadened in solids from chemical shift anisotropy and dipolar interactions, and, for quadrupolar nuclei, by quadrupolar interactions. In high-resolution solid-state NMR studies, we remove the broadening using dipolar decoupling or magic-angle sample spinning because the broad lines cause all of the signals to overlap. The direct observation of the polymer lineshapes is a powerful method for the study of the dynamics because the averaged lineshapes depend not only on the rate of molecular motion, but also the geometry.

The frequency of molecular motion probed by lineshape studies is related to the inverse of the width of the lineshape. The broadening from chemical shift anisotropy is on the order of a few kHz, while the dipolar lineshape is on the order of 30–50 kHz and the quadrupolar lineshape has a width on the order of 100 kHz. The lineshapes are a better tool for studying motions in the intermediate range than are relaxation rates that are most sensitive to molecular motions near the Larmour frequency (MHz) through the spectral density terms. This makes lineshapes a valuable tool to study the intermediate time scale dynamics that can be related to the mechanical properties of polymers.

The effects of molecular motion on polymer lineshapes can be easily understood by considering the effect of changing orientation on the spectra. As noted in the discussion of solid-state NMR in Section 1.6, the broadening from chemical shift anisotropy, dipolar couplings, and quadrupolar interactions is inhomogeneous. This means that the observed lineshape is the sum of the lineshapes for each of the

NMR-active nuclei in the sample. The frequency for each nucleus depends on its orientation relative to the magnetic field, and summing over all possible orientations gives the observed lineshape. Molecular motion causes a change in orientation, and therefore a change in frequency, and perhaps a change in lineshape.

The effect on the lineshape of this change in orientation (and frequency) depends both on the amplitude and rate of molecular motion. We can understand this by considering a conformational transition between two equally energetic conformations. If we consider the chemical shift anisotropy, for example, of just these two conformations, we would observe two peaks in the spectra and the chemical shifts would be determined by the orientations relative to the static magnetic field. By considering only these two orientations, we can see that the conformation exchange is reduced to the chemical exchange problem considered in Section 1.3.1.4. If the exchange is slow relative to the inverse of the frequency separation, then there will be little or no effect on the line positions, the linewidths, or intensities of the peaks (Figure 1.17; slow exchange limit). As the rate of molecular motion approaches the inverse of the frequency separation, the lines broaden (intermediate exchange), and in the fast exchange limit a single sharp line is observed halfway between the two frequencies (fast exchange).

The strategy in most lineshape studies is to measure the lineshape as a function of temperature. At very low temperatures the molecular motions are frozen out and the static lineshape is obtained. It is necessary to measure (or calculate) the static lineshape in order to calculate the effects of chemical exchange on the observed lineshape at higher temperature. The lineshape is then measured as a function of temperature and compared with a calculated lineshape that takes into account the geometry and rate of molecular motion.

Fast and intermediate molecular motion in polymers can be studied by measuring the lineshape as a function of temperature. Molecular motion in the slow exchange regime (10^{-3}–10^2 s) can be studied using the spin exchange experiment on the wideline spectrum. Spin exchange of the wideline spectrum does not generate simple cross peaks as in the high-resolution spectrum, but rather ridges or patterns in the wideline 2D spectrum. The jump angle in simple cases can be determined by inspection of the wideline spin exchange spectrum. The intensity of the ridges will depend on the rate of molecular motion and the mixing time.

5.3.4.1 Wideline Deuterium NMR Deuterium NMR is often used to study polymer dynamics because the relaxation is dominated by the quadrupolar coupling, so the relaxation mechanism is well characterized (34). Deuterium has a low sensitivity and extremely broad lines, so polymers with site-specific labels are required. This is a disadvantage in that new polymers must be synthesized, but an advantage in that the labels are incorporated into the polymer at well-defined sites, such as the main-chain or the side-chain atoms. In the absence of molecular motion the frequency of a given deuteron is given by

$$\omega = \omega_0 \pm \delta(3\cos^2\theta - 1 - \eta \sin^2\theta \cos 2\phi) \qquad (5.46)$$

where ω_0 is the resonance frequency; $\delta = 3e^2qQ/8\hbar$, $e^2qQ/\hbar$ is the quadrupolar coupling constant; η is the asymmetry parameter; and the orientation of the magnetic field in the principle axis of the electric field gradient tensor is specified by the angles θ and ϕ. For $C-D$ bonds in rigid solids, $\delta/2\pi = 62.5$ kHz and $\eta \approx 0$. Thus, two lines are observed for each deuteron, and the separation between the lines depends on the orientation of the $C-D$ bond vector relative to the magnetic field. In isotropic samples, averaging over all possible orientations gives rise to the well-known "Pake" spectrum shown in Figure 1.54.

Molecular motion in the polymer will cause the lineshapes to change in a way that depends on the geometry and time scale of the molecular motion. If the motion is rapid on the time scale defined by the coupling constant, $1/\delta$, it is said to be in the fast motion limit. For deuterium this corresponds to $\tau_c < 10^{-7}$ s, and leads to a characteristic change in lineshape given by

$$\omega = \omega_0 \pm \bar{\delta}(3\cos^2\theta - 1 - \bar{\eta}\sin^2\theta \cos 2\phi) \qquad (5.47)$$

where $\bar{\delta}$ and $\bar{\eta}$ are the coupling constant and asymmetry parameter, respectively, for the *averaged* electric field gradient tensor. It is important to note that $\bar{\eta}$ may be different from zero even though $\eta = 0$.

The strategy in many deuterium NMR studies is to measure the lineshape as a function of temperature and compare the observed spectrum with that calculated assuming some model for the molecular motion. It is generally advantageous to acquire a spectrum at a low enough temperature to freeze out all large-amplitude molecular motion so that the values for δ and η can be determined experimentally. In the absence of motion, a good fit should be obtained with the experimental spectrum by adjusting the values for δ and η.

Finding the proper model for the deuterium lineshape can be either a simple or complex process, depending on the polymer under study and the site of the deuterium label. The spectrum for a polymer with a main chain deuteron may be difficult to model, for example, while the spectrum for a ring-deuterated sample undergoing 180° ring flips may be very easy to correctly model. Amorphous polymers are inherently inhomogeneous, and models that incorporate a distribution of correlation times are often required.

A variety of models have been used to model the wideline deuterium lineshapes for a variety of polymers (35). While a detailed discussion of the many models for the molecular motion is beyond the scope of this text, it is useful to consider the deuterium lineshape expected from some of the simple models (36). One of the simplest and most successful approaches is to simulate the wideline spectrum using the motion in a cone model, as shown in Figure 5.22. To use the cone model to simulate methyl group rotation we take the carbon–carbon bond vector as the rotation axis and assume that the methyl deuterons undergo a three-site hop in the plane of the cone. The parameters needed for the lineshape simulations are the cone angle α, the jump angle β, the populations of the three sites, and the hopping rate. For the methyl group we assume that the three sites are equally populated and the motion is in the fast-motion limit. The angle α is determined by the bond angle geometry ($\alpha = 180° - 109.5° = 70.5°$), and the hops occur between three sites on a circle, so $\beta = 120°$. Figure 5.23 shows the calculated spectrum for a deuterated methyl group in a polymer. The three-site hop averages the coupling constant by one-third, but leaves the asymmetry unchanged

5.3 NMR RELAXATION IN THE SOLID STATE

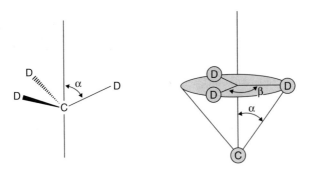

FIGURE 5.22 A diagram showing the geometry of motion in a cone.

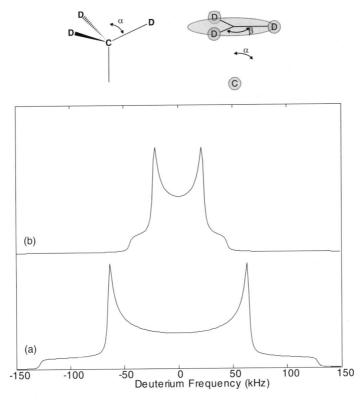

FIGURE 5.23 The calculated deuterium spectrum assuming methyl group rotation.

($\eta = 0$), so the spectrum appears the same as the static spectrum except on a much smaller scale. Molecular motion in addition to methyl group rotation would lead to additional changes in the lineshape, provided they were on the fast or intermediate time scale.

Another simple molecular motion encountered in polymers is ring flips for deuterated aromatic rings. Ring flips for the *ortho* and *meta* deuterons lead to a reorientation (Figure 5.24) that can be modeled by the motion in a cone model with values for α and β of 60° and 180°. It can be seen from this simulation that the lineshape for a ring deuteron differs substantially from the static lineshape and can be easily identified. It should be noted that rotation about the 1–4 axis of an aromatic ring does not lead to a change in lineshape for the *para* deuteron, since this deuteron is on the rotation axis. The spectrum for a fully deuterated aromatic ring with *ortho*, *meta*, and *para* deuterons undergoing rapid ring flips would show a complex lineshape resulting from the motionally averaged lineshapes from the *ortho* and *meta* deuterons and the static lineshape for the *para* deuterons.

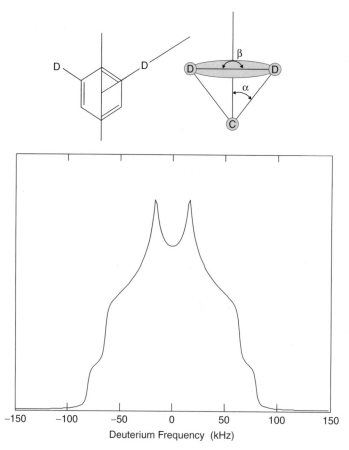

FIGURE 5.24 The model for ring flips and the calculated deuterium wideline spectra.

5.3 NMR RELAXATION IN THE SOLID STATE

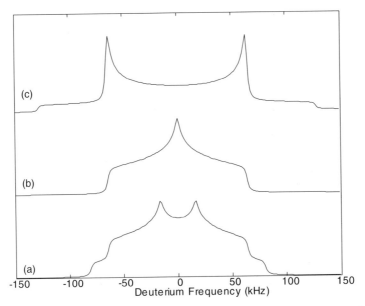

FIGURE 5.25 The effect of cone angle on the deuterium spectra for cone angles of (a) 30°, (b) 54.7°, and (c) 90°.

The molecular motion of main-chain deuterons is more complex to model because a large number of possible motions that could lead to chain reorientation. In favorable cases, the spectra can be compared with those calculated from simple jump models. Figure 5.25 compares the effect of a change in cone angle on the appearance of the spectra for a two-site jump model. Comparison of the spectra shows that the lineshape is extremely sensitive to the cone angle. However, it should be noted that several different models can lead to the same values for $\bar{\delta}$ and $\bar{\eta}$, and specification of the cone angle and a good fit to the spectra are not sufficient to establish the motional mechanism. Note, for example, that the spectra for the two-site jump with a cone angle of 30° gives the same averaged values for the coupling constant and asymmetry ($\bar{\delta}/\delta = 0.625$ and $\bar{\eta} = 0.6$) as for the ring flip with a cone angle of 60°. Thus, either model would give an equivalent fit to the data. In such cases, it may be possible to distinguish between the models by measuring the so-called T_{1Q} lineshape (37).

The lineshapes can also be affected by a distribution in jump angles or correlation times. Such distributions might be expected for amorphous polymers where the chains can experience a range of local environments. Figure 5.26 compares the simulated deuterium spectra for the two-site jump with a cone angle of 54.7° as the jump angle distribution is increased from zero of 60°, assuming a Gaussian distribution in jump angles. This comparison shows that the scaled coupling constant is extremely sensitive to the distribution width, and the scaled coupling constant ($\bar{\delta}/\delta$) decreases from 0.5 to 0.28 as the distribution width increases from zero to 60°.

As with all chemical exchange processes, the deuterium lineshape is very sensitive to the site populations. Figure 5.27 compares the lineshapes expected from a two-site

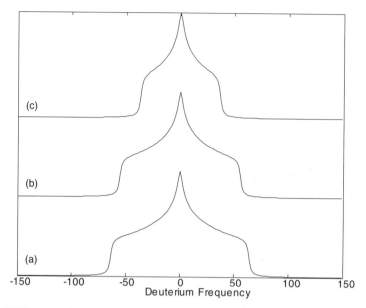

FIGURE 5.26 The effect of jump-angle distributions on the deuterium lineshape for a two-site jump with $\alpha = 54.7°$ for a distribution width of (a) 0°, (b) 30°, and (c) 60°.

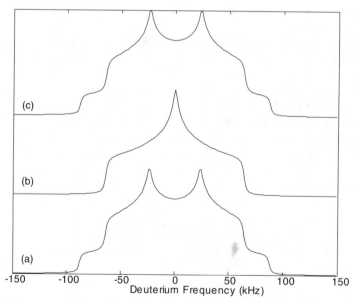

FIGURE 5.27 The effect of site population on the deuterium lineshape for a two-site jump with $\alpha = 54.7°$ for a populations of (a) 25/75, (b) 50/50 and (c) 75/25.

5.3 NMR RELAXATION IN THE SOLID STATE

jump with a cone angle of 54.7° for the spectrum with equally populated sites with the spectrum with a 25/75 population of the two sites. This shows that the lineshape is extremely sensitive to the site populations. In the model for methyl group rotation or ring flips, it can generally be assumed that the populations are equal, but this is not necessarily true for main-chain motion.

The preceding discussion shows that while the deuterium lineshapes are sometimes complex, they are extremely sensitive to the details of molecular motions. It is generally not possible to extract such detailed information from relaxation-rate measurements. It should also be noted that the deuterium lineshapes are most informative for molecular motion that partially averages the quadrupolar coupling. As with other couplings, large-amplitude near-isotropic molecular motion almost completely averages the quadrupolar couplings, and narrow Lorentzian lineshapes are observed. These narrow lineshapes (such as those for a polymer above T_g) are not very sensitive to the details of the molecular motion.

Nylon 6

Deuterium NMR has been used to study a wide variety of crystalline and amorphous polymers. Figure 5.28 shows the deuterium spectra for a series of nylons in which various positions in the diamine and adopyl moiety have been selectively deuterated (38). These spectra show that the deuterium spectra are extremely sensitive to the details of the molecular dynamics, and that the molecular dynamics of nearby deuterons are not identical. These data can be fitted to a model incorporating librational motion that increases in amplitude as the temperature is increased (39). Analysis of the lineshapes and relaxation times for nylons deuterated at the amine groups shows that the ND groups act as pinning points that restrict the mobility of the chains (40).

Amorphous polymers, including polystryene and polycarbonate, have also been studied by deuterium NMR following site-specific labeling. As might be expected for amorphous chains, the variations in the local environments lead to distributions in mobility. Figure 5.29 compares the experimental and simulated deuterium spectra for ring-deuterated polystyrene at 373 K with different values for the delay time between pulses (41). The differences in the dynamics of the amorphous chains give rise to both differences in the relaxation times and lineshapes, and by acquiring the spectrum with a short delay between pulses, it is possible to enhance the spectra in the more rapidly relaxing component. The spectra acquired with a longer delay time (3 s) shows the spectrum for all (rapidly and slowly relaxing) of the deuterons. This spectrum is a composite from those rings that are immobile on the deuterium time scale and give rise to a static spectrum, and those rings that are moving rapidly and give the motionally average spectrum expected from rapidly flipping rings. The spectrum with the short delay time (0.5 s) is mostly from the more mobile rings, as these rings have a shorter

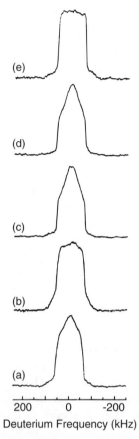

FIGURE 5.28 The fully relaxed deuterium spectra at 97°C for nylon-66 polymers specifically labeled with deuterium at the (a) C3/C4 and (b) C2/C5 carbons of the adipoyl moiety, and polymers labeled at the (c) C3/C4, (d) C2/C4 and (e) C1/C6 carbons of the diamine moiety.

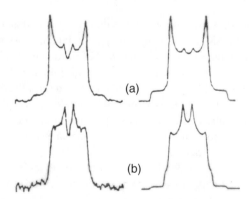

FIGURE 5.29 The experimental (left) and simulated (right) spectra for ring-deuterated polystyrene-d_5 acquired with relaxation delays of (a) 3 s and (b) 0.05 s.

5.3 NMR RELAXATION IN THE SOLID STATE

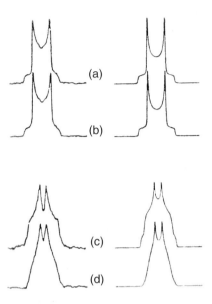

FIGURE 5.30 The wideline experimental (left) and simulated (right) deuterium spectra for methyl-deuterated polycarbonate and ring-deuterated polycarbonate. The spectra are shown for (a) partially and (b) fully relaxed methyl-deuterated and (c) partially and (d) fully relaxed ring-deuterated polycarbonate.

spin-lattice relaxation time. Once the lineshape for the rapidly flipping rings has been identified, the contribution to the total spectrum can be calculated. The results show that at 373 K, 20% of the styrene aromatic rings are rapidly flipping. Ring flips in deuterated polymers have also been studied in crystalline polymers (42,43).

Figure 5.30 shows the experimental and simulated spectra for methyl- and ring-deuterated polycarbonate (41). A good simulation of the experimental spectra for the methyl-deuterated polycarbonate can be obtained by assuming that the only motions that average the quadrupolar couplings are methyl group rotations. Since the methyl group is rigidly attached to the polymer backbone, motions that average the quadrupolar coupling would be indicative of main-chain motion for polycarbonate. The spectrum for ring-deuterated polycarbonate shows the features expected for aromatic rings undergoing 180° flips. The best fit to the experimental spectrum is one in which the rings undergo small-amplitude oscillations in addition to the ring flips. The amplitude of the oscillations increases from $\pm 15°$ at ambient temperature to $\pm 35°$ at 383 K.

Deuterium NMR can also be used to provide information about the structure and dynamics of semicrystalline polymers. In most cases the dynamics are expected to be restricted in the crystalline phase. The dynamics of the amorphous and interfacial phases will depend strongly on temperature, and complex lineshapes may be observed due to the way that the quadrupolar couplings are averaged differently in the three phases. Typical behavior for a semicrystalline polymer is illustrated in Figure 5.31,

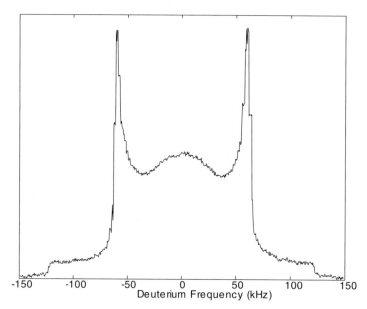

FIGURE 5.31 The deuterium NMR spectrum of linear polyethylene-d_4.

which shows the deuterium NMR spectra for linear polyethylene-d_4 crystallized from the melt (34). The deuterium lineshape has two-components, a broad lineshape from immobile deuterons in the crystalline phase and a narrower component from the more mobile chains in the amorphous phase. The lineshape for the amorphous polyethylene is very sensitive to temperature and is much narrower than the signals from the crystalline component. This shows that fast, large-amplitude motions nearly average the quadrupolar coupling in the amorphous phase. As the line narrows, the signal from the amorphous phase appears to grow in intensity, but this is because it is much easier to detect sharp lines.

Thus far we have considered the effects of rapid molecular motion on the deuterium lineshape, but molecular motions in the intermediate time scale ($\tau_c = 10^{-7}$–10^{-5} s) can also affect the spectrum and the signal intensity. The lineshapes are best understood by considering the effects of chemical exchange as the individual deuterons undergo motion and change orientation with respect to the magnetic field. We noted in our original discussion of chemical exchange (Section 1.3.1.4) that broadened lineshapes are observed in the intermediate region. The deuterium lineshapes are also broadened in the intermediate exchange regime. It is possible to simulate the lineshapes for molecular motion in the intermediate regime, but a more sophisticated approach is required (36).

Molecular motion in the intermediate regime can also affect the deuterium intensity. In Section 2.5.4 we noted that different experimental methods are required to acquire a wideline deuterium spectrum. The main problem is that the deuterium spectrum is so broad and decays so quickly (μs) that data have to be gathered immediately after the rf pulse. This is not possible, as the probe and rf system needs a few μs to recover after the application of an intense rf pulse. For this reason, deuterium spectra are acquired

5.3 NMR RELAXATION IN THE SOLID STATE

as an echo using the 90°–τ–90°–τ quadrupolar echo pulse sequence (Figure 2.13). The delay times for the quadrupolar echo are on the order of 10 to 30 μs, and if any molecular motion occurs during the delay time, τ, then the signals will not be refocused and a lower signal intensity will be observed. Motion in the intermediate time regime can be identified in a study of the echo intensity as a function of the delay time, τ. If the intensity is insensitive to the echo delay time, then molecular motion on the intermediate time regime is not a problem. If there is motion on the intermediate time scale, then care must be taken in fitting the spectra, because the lineshape as well as the intensity could depend on the quadrupolar echo delay time.

The effect of echo delay time on the lineshape is shown in Figure 5.32 for methyl-deuterated poly(methyl acrylate) in the bulk at 37°C (44). Note that for this sample the μs delay times in the quadrupolar echo led to large differences in the lineshape, particularly in the center and at the edges. The lineshape for simulation is obtained by extrapolating the lineshape back to $\tau = 0$. As a general rule, changes in intensity and lineshape as a function of the echo delay time are a signature of molecular motion on the intermediate time scale.

Molecular motion in the slow time scale ($\tau_c < 10^{-4}$ s) cannot be identified directly from the lineshape, but it can be identified from nD spin exchange experiments. These spin exchange experiments are a powerful means for studying the molecular dynamics of polymers, because the ridge pattern in the 2D exchange spectrum is extremely sensitive to molecular geometry, and it is some times possible to determine the jump angle between sites to a precision of a few degrees.

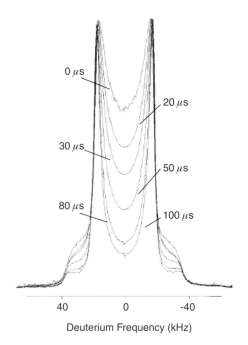

FIGURE 5.32 The effect of the quadrupolar echo delay time on the spectrum of for methyl-deuterated poly(methyl acrylate) in the bulk at 37°C.

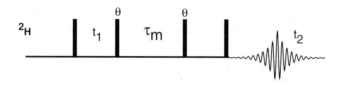

FIGURE 5.33 The pulse-sequence diagram for the 2D wideline deuterium exchange experiment.

The pulse-sequence diagram for measuring the deuterium exchange spectrum is similar to that for other nuclei, except the sequence is modified to take into account that an echo spectrum must be observed. Figure 5.33 shows the pulse sequence diagram for the deuterium 2D exchange experiment (45). The key part of the experiment is the mixing time, during which spin exchange occurs. The format for the 2D exchange spectrum is a correlation plot for the deuterium lineshape, where the lineshape appears along the diagonal and an off-diagonal pattern arises from exchange. It is the shape of this pattern that contains information about the reorientation angle.

Figure 5.34 shows a simulated 2D exchange pattern for a deuterium undergoing a 60° jump between two sites with equal populations (36). The most apparent features in the spectrum are the ridge patterns perpendicular to the diagonal that result from the exchange. The shape of the pattern is extremely sensitive to the jump angle. From simple geometric considerations, the jump angle can be determined from the ratio of the principal axes of the ellipse a and b as

$$|\tan \theta| = \frac{a}{b} \tag{5.48}$$

In this way, the jump angle can be calculated without reference to a particular model for the molecular motion.

Deuterium spin exchange experiments have been used to characterize the dynamics in a variety of semicrystalline polymers, including isotactic polypropylene (46) and poly(vinylidine fluoride) (47). Figure 5.35 shows the 2D spin exchange spectrum for poly(vinylidine fluoride) acquired with a 0.2-s mixing time at 370 K (47). In addition to the "Pake"-type spectrum along the diagonal, an off-diagonal ridge is observed that is the result of jumps between sites in the crystalline lattice. The best fit between the experimental and simulated spectra is a model in which the deuterons are undergoing jumps of either 67° or 113° between sites on the crystalline lattice. These data are consistent with only one of the proposed crystal structures for poly(vinylidine fluoride), and it can be concluded from these data that the chain motion corresponds to a transition between the $tgtg^-$ and g^-tgt conformations. In a similar way, it was shown that the deuterons in isotactic polypropylene undergo jumps of 113° in the crystalline lattice as monomers are translated from one site on the crystalline lattice

5.3 NMR RELAXATION IN THE SOLID STATE

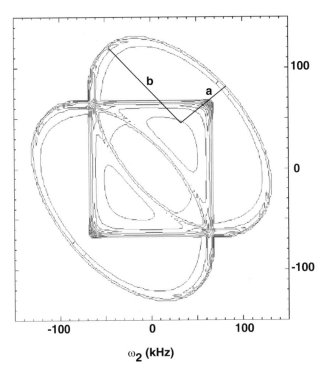

FIGURE 5.34 The simulated deuterium exchange spectrum for a deuteron undergoing a 60° jump between two sites.

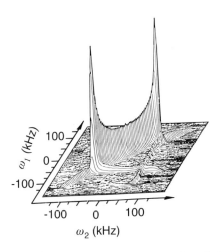

FIGURE 5.35 The 2D deuterium spin exchange spectrum for poly(vinylidine fluoride) acquired with a 0.2-s mixing time at 370 K.

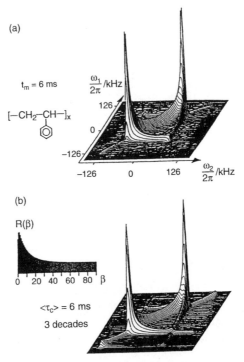

FIGURE 5.36 The experimental (top) and simulated (bottom) 2D exchange spectra for main-chain deuterated polystyrene at 391 K with a mixing time of 6 ms. The inset plot shows the distribution in jump angles after the mixing time.

to another. Such 2D NMR experiments are an unambiguous way to provide molecular-level assignments to the transitions observed by dielectric and dynamic mechanical spectroscopy.

Deuterium spin exchange spectroscopy can also be used to investigate the dynamics of amorphous polymers. Amorphous polymers are by their nature heterogeneous and do not undergo the well-defined jumps observed in crystalline polymers. Instead, there is a wide distribution in the jump angles and correlation times. Figure 5.36 shows the 2D exchange spectrum for main-chain deuterated polystyrene acquired with a 0.012-s mixing time at 391 K (48,49). Among the things to note in this spectrum are the extended ridges in the 2D plane at the frequency of the singularities and the broadening and loss of signal for the center part of the spectrum along the diagonal. The data are best fitted by a model using isotropic rotational diffusion with a broad distribution of correlations times. The mean correlation time is 0.006 s, and the width of the distribution in correlation times is three decades. Similar methods have been used to study the molecular dynamics of methyl-deuterated polyisoprene (50), atactic polypropylene (46), and blends of polyisoprene and poly(vinyl ethylene) (51).

The various experiments using deuterium NMR to probe the dynamics of polymers are each sensitive to a particular frequency of molecular motion. The various methods

5.3 NMR RELAXATION IN THE SOLID STATE

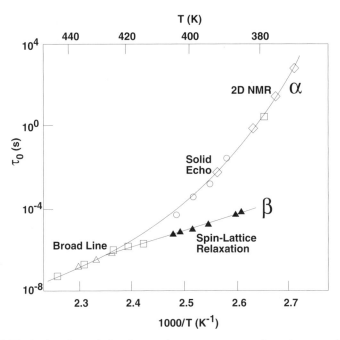

FIGURE 5.37 A plot of correlation time vs. inverse temperature for polystyrene for several different NMR methods.

can be combined to give a more global picture of the chain dynamics. This is illustrated in Figure 5.37, which shows a plot of correlation times measured from different experiments vs. inverse temperature (52). The data points are the correlation times calculated from 2D exchange, solid echo, broad-line lineshapes, and deuterium spin-lattice relaxation time measurements, and the experimental data spans 12 orders of magnitude in correlation time. The data lie along two curves. The data for the 2D exchange, solid echo, and broad-line lineshapes lie along a curve that can be fitted to the Williams–Landel–Ferry equation (53) describing the temperature dependence of polymers above T_g. These dynamics are associated with the so-called α transition. The T_1 data lie along another curve and are associated with the β transition.

5.3.4.2 Chemical Shift Anisotropy and Polymer Dynamics

The dynamics of solid polymers can also be investigated using the line broadening that arises from the chemical shift anisotropy. The lineshape from the chemical shift anisotropy is given by

$$\omega = \omega_{\text{iso}} + \frac{\delta}{2}(3\cos^2\theta - 1 - \eta\sin^2\theta\cos\phi) \tag{5.49}$$

where ω_{iso} is the isotropic chemical shift, η is the asymmetry parameter, δ is the coupling constant, and the angles θ and ϕ relate the principal axis of the chemical shift to the magnetic field direction. If there is motion that is fast compared to the width of the anisotropic lineshape that reorients the principal axis of the chemical shift, then an averaged anisotropic lineshape is observed that is given by

$$\omega = \omega_{\text{iso}} + \frac{\bar{\delta}}{2}(3\cos^2\theta - 1 - \bar{\eta}\sin^2\theta\cos\phi) \tag{5.50}$$

where $\bar{\delta}$ is the averaged coupling constant and $\bar{\eta}$ is the averaged asymmetry parameter.

The strategy for studying polymer dynamics via the chemical shift anisotropy is similar to that used for deuterium NMR. The spectra are acquired as a function of temperature and fit to a motional model that gives the best agreement between the experimental and simulated lineshapes. The chemical shift anisotropy lineshape has a smaller width than the deuterium lineshape (a few kHz vs. 100 kHz), so the chemical shift anisotropy lineshape probes the molecular motion on a slower time scale. As with deuterium NMR, isotopic labeling is frequently usually required for lineshape studies.

The effect of molecular motion on the chemical shift anisotropy lineshape can be understood by considering the effects of chemical exchange. As noted in Section 1.6.1, the chemical shift anisotropy lineshape arises because chemical shift for a particular carbon atom depends on its orientation relative to the magnetic field, and if molecular motion changes its orientation, then it will experience a change in frequency. If we consider a jump between two orientations, then the problem becomes a chemical exchange problem, as considered in Section 1.3.1.4, and the effect on the lineshape will depend on whether the exchange is in the fast-, intermediate-, or slow-motion limit. The lineshape can be calculated by considering the exchange between sites for all possible orientations relative to the magnetic field.

The effect of fast molecular motion on the chemical shift anisotropy lineshape can be calculated from the motion in a cone model introduced in Section 5.3.4.1. The input for the lineshape simulation is the principal values of the chemical shift tensor, the cone angles, and the jump angles. For an accurate simulation it is important to obtain the principal values of the chemical shift tensor by acquiring the spectrum at a low enough temperature that that all motions are frozen out. If such a spectrum cannot be observed, it is sometimes permissible to use the principal values for the chemical shift from the most closely related chemical structure reported in the literature.

Figure 5.38 shows the simulated lineshapes for a carbon chemical shift anisotropy lineshape for a static carbon, a carbon undergoing methyl group rotation, and for an *ortho* carbon in an aromatic ring undergoing ring flips (36). For the purposes of this comparison, the principal values for the carbon chemical shift are 260, 200, and −10 ppm, and the value of the asymmetry is 0.375. The spectrum for the static carbon shows a broad lineshape with easily observable values for the principal values of the chemical shift. Methyl group rotation can be modeled as a three-site jump using the motion in a cone model with a cone angle of 70.5° and a jump angle of 120°. The averaged chemical shift anisotropy lineshape is axially symmetric ($\eta = 0$) and appears reversed relative to the static pattern. Ring flips for the *ortho* carbons in aromatic rings can be modeled using a cone angle of 60° and a jump angle of 180°. The resulting spectrum is almost axially symmetric ($\eta = 0.08$) and is reversed relative

5.3 NMR RELAXATION IN THE SOLID STATE

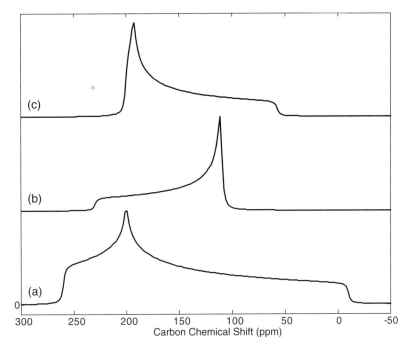

FIGURE 5.38 The chemical shift anisotropy lineshape for (a) a static carbon, (b) a rotating methyl group, and (c) a flipping ring.

to the pattern for methyl group rotation. These simulations show that the chemical shift anisotropy lineshape is extremely sensitive to the molecular motion. However, the lineshapes do not uniquely identify the motion, since several motional models can give similar values of the averaged coupling constant and asymmetry parameter. It should also be noted that the large anisotropies in the simulations are more typical of carbonyl or aromatic carbons rather than aliphatic carbons.

The chemical shift anisotropy lineshape for the carbonyl carbons has been used to investigate the so-called β transition in poly(methyl methacrylate) observed by dielectric and dynamic–mechanical spectroscopy (54). The frequency of the β transition at ambient temperature is near 10 Hz, and this motion is believed to be related to the favorable mechanical properties of poly(methyl methacrylate). NMR studies of the carbonyl can provide molecular-level information about the β transition because the chemical shift tensors have been studied for many carbonyl carbons. As shown in Figure 5.39, the σ_{33} axis is perpendicular to the OCO plane, σ_{22} lies nearly along the CO double bond, and the σ_{11} direction bisects the OCO angle. If the side chains are roughly perpendicular to the chain axis, this would lead to a situation where the σ_{33} axis lies along the chain axis.

It is possible to investigate the chemical shift anisotropy lineshapes for carbonyl groups in aliphatic polymers at natural abundance because the carbonyls are shifted away from the other carbon signals. This is illustrated in Figure 5.40, which shows

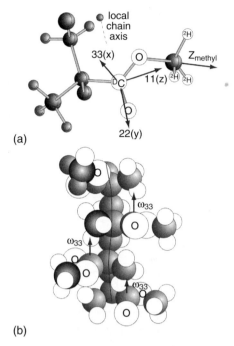

FIGURE 5.39 A drawing showing the chemical shift anisotropy axis in the carbonyl of poly(methyl methacrylate).

the static natural abundance of carbon spectra for poly(methyl methacrylate) as a function of temperature (55). The carbonyl is well resolved from the other peaks, and at 300 K the chemical shifts can be assigned to ω_{11} at 268 ppm, ω_{22} at 150 ppm, and ω_{33} at 112 ppm. Changes in the lineshape are observed as the temperature is increased to 373 K. Note that while the ω_{11} and ω_{22} signals broaden and shift, the ω_{33} signal appears to be unaffected. This finding is consistent with a molecular motion for the β transition in which the side chain undergoes a 180° flip. Such a flip would reorient the direction of the σ_{11} and σ_{22} axes while inverting the direction of σ_{33} axis. Since σ_{33} is insensitive to whether it is parallel or antiparallel to the chain direction, ω_{33} is unchanged by a side-group flip.

The molecular dynamics of the β transition in poly(methyl methacrylate) can be investigated in more detail using 2D exchange NMR. As with the deuterium lineshape, the chemical shift anisotropy lineshape is a consequence of the different orientations of atoms in the magnetic field. If molecular motion changes the orientation of an atom, it will also change the frequency, and in the slow-motion regime, such frequency changes can be identified by spin exchange experiments. The pulse sequence is the same as that used to investigate spin exchange in magic-angle spinning experiments (Figure 5.19). Although it is sometimes possible to investigate the dynamics without isotopic labeling, the SNR can be greatly improved with isotopic labeling.

5.3 NMR RELAXATION IN THE SOLID STATE

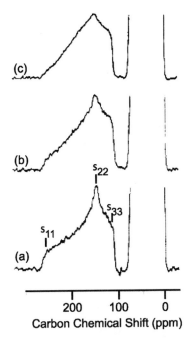

FIGURE 5.40 The carbon spectrum of poly(methyl methacrylate) showing the chemical shift anisotropy lineshape for the carbonyl as a function of temperature.

Figure 5.41 compares the experimental 2D spin exchange spectrum for poly(methyl methacrylate) enriched with ^{13}C at a level of 20% for the carbonyl carbon with simulated spectra calculated for various models (55). The spectrum calculated by assuming only 180° flips of the carbonyl has pronounced elliptical ridges that are not observed in the experimental spectra. However, if these 180° flips are combined with main-chain rotations with an amplitude 20°, an excellent fit to the experimental spectrum is obtained. The main-chain rotations without the 180° flips also gives a poor fit to the experimental data. The correlation times determined from the 1D and 2D NMR studies have the same value and temperature dependence as those reported from the dielectric and dynamic mechanical studies, establishing a direct link between the β transition and the dynamics of the poly(methyl methacrylate) side group.

2D spin exchange experiments can also be used to investigate the molecular dynamics of crystalline polymer for those cases where the chemical shift anisotropy lineshapes are well resolved, or with isotopic labeling. Figure 5.42 shows the 2D spin exchange spectrum for poly(oxymethylene) at 360 K with a mixing time of 2 s (56). In addition to the chemical shift anisotropy lineshape along the diagonal, an off-diagonal ridge is observed that is indicative of a jump between equivalent sites in the crystal. Poly(oxymethylene) has nine monomer units in five turns (9_5) in the crystal structure and a 200° change in orientation between methylene carbons in neighboring sites. The best fit between the experimental and simulated spectra is for a change in orientation

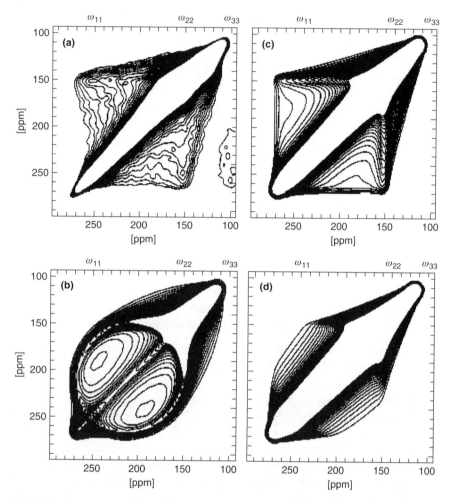

FIGURE 5.41 Comparison of (a) the experimental 2D exchange spectrum with the simulated spectra from models, assuming (b) side-chain flip with main-chain rotation, (c) side-chain flips only, and (d) main-chain rotation only.

of 200°, demonstrating that the dynamics observed by spin exchange NMR can be assigned to the movement of chains between neighboring sites in the crystal.

The use of the chemical shift anisotropy lineshape for investigating polymer dynamics was illustrated for the carbonyl carbon, which has a large and often well-resolved lineshape, and for poly(oxymethylene), which has a single carbon signal. Other atoms can also be used with isotopic labeling. Other nuclei can be used as well, since many of these, including ^{15}N, ^{31}P, and ^{29}Si, have relatively large anisotropies.

5.3.4.3 Dipolar Lineshapes and Polymer Dynamics The lineshapes arising from dipolar couplings can also be used to investigate the molecular dynamics of polymers.

5.3 NMR RELAXATION IN THE SOLID STATE

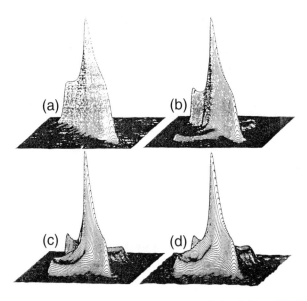

FIGURE 5.42 The 2D exchange spectra of poly(oxymethylene) (a) at 252 K with a mixing time of 1 s, (b) at 330 K with a mixing time of 1 s, (c) at 360 K with a mixing time of 2 s and (d) at 370 K with a mixing time of 4 s.

These experiments are extremely simple to run, but the lineshapes tend to be broad and featureless and cannot be interpreted with the atomic level of detail as for the chemical shift anisotropy and deuterium lineshapes.

The broadening from dipolar interactions is a consequence of the close proximity of NMR-active nuclei (Section 1.6.3). For most nuclei, the broadening is mainly due to interactions with protons, which are abundant in the lattice. For a carbon nucleus, each interaction with a proton results in a coupling given by

$$D = \frac{\hbar \gamma_C \gamma_H}{2\pi r^3}(3\cos^2\theta - 1) \qquad (5.51)$$

where r is the internuclear distance and θ is the angle between the C–H vector and the magnetic field direction. As with the deuterium and chemical shift anisotropy lineshapes, the dipolar lineshape is obtained by summing over all possible orientations of the C–H vector. The broadness and lack of easily observable features in the dipolar lineshape is a consequence of the fact that there are many carbon–proton and proton–proton couplings, since the density of protons is high in most organic solids. The many interactions give rise to a multitude of coupling constants that, when added together, give rise to a broad featureless line. It is the average width of this line that provides information about the molecular dynamics of polymers.

The dipolar linewidths are difficult to directly measure from the NMR spectrum because the linewidths are typically greater than the chemical shift resolution. For this reason, the dipolar linewidths are typically measured using a heteronuclear 2D

NMR experiment where the resolution is provided by the insensitive nuclei. The most common experiment used to measure the dipolar linewidths is the so-called WISE experiment shown in Figure 1.70 (57). The WISE experiment is a resolved experiment where, the carbon chemical shifts appear along one dimension and the proton linewidths appear along the other. The proton signals evolve after the initial 90° pulse in the absence of multiple-pulse decoupling or fast magic-angle sample spinning, so the broad line reflects the dipolar coupling in the proton dimension. After the evolution period, the proton magnetization is transferred to the carbons by cross polarization and the signals are recorded with proton decoupling. There is an optional mixing time between the evolution period and the cross polarization to allow for proton spin diffusion.

Figure 5.43 shows a contour plot of the WISE spectrum for the miscible blend of polystyrene and poly(vinyl methyl ether). While the polymers are miscible, there is a large difference in the chain dynamics for polystyrene and poly(vinyl methyl ether) that gives rise to large differences in the proton linewidths. The polystyrene is more rigid and the full width at half maximum for the aromatic protons is 35 kHz, while the poly(vinyl methyl ether) is more mobile and has a line of 12 kHz for the methoxyl protons.

The WISE experiment is most commonly used to measure the average chain dynamics because of spin diffusion during the mixing time and during the cross-polarization time. Even when the mixing time is set to zero, spin diffusion during the cross-polarization period can average the lineshapes for protons that are close in space. Spin diffusion is slower by a factor of 2 during cross polarization, but nevertheless sufficiently rapid that the lineshapes tend to be characteristic of nearby groups.

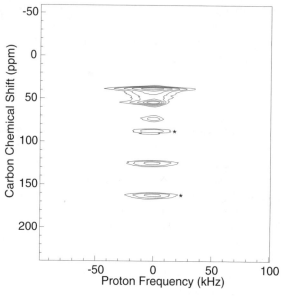

FIGURE 5.43 The 2D WISE spectrum for the polystyrene/poly(vinyl methyl ether) blend.

5.3 NMR RELAXATION IN THE SOLID STATE

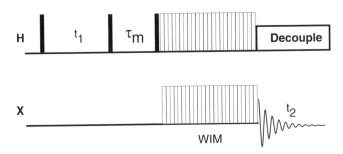

FIGURE 5.44 The pulse-sequence diagram for 2D WISE NMR using WIM for cross polarization.

Methyl group rotation, for example, partially averages the dipolar couplings and leads to smaller linewidths. Therefore, any atoms close to a methyl group will have a smaller linewidth from spin diffusion than expected from the dipolar couplings alone. The longer the cross polarization time (or mixing time), the larger the area for which the linewidths will be averaged by spin diffusion.

The averaging by spin diffusion of the proton lineshapes limits the use of the WISE experiment for measuring the individual chain dynamics, but provides important information about the length scale of chain mixing. In the polystyrene/poly(vinyl methyl ether) blend, for example, it was observed that a mixing time of 5 ms was required to nearly average the linewidths for the two chains (57). This shows that the blend is "nanoheterogeneous" with domain sizes on the order of 3.5 nm.

Slight modifications of the WISE sequence can be introduced to eliminate the spin diffusion during the cross polarization, so that the proton linewidths can be used to measure the dynamics of atoms along both the main chain and side chain. Figure 5.44 shows a variant of the WISE experiment where the cross polarization has been replaced by a WIM sequence (58). The 24-pulse WIM sequence can be used for cross polarization while suppressing spin diffusion (59). Thus, if the mixing time is very short (<50 μs), then the proton linewidths will be a direct measure of the extent to which molecular motion averages the dipolar couplings.

Figure 5.45 shows cross sections through the WIM/WISE spectra of poly(n-butyl methacrylate). The cross sections show that, as expected, the linewidths for the methyl side chain are much smaller than for the neighboring methylene, and that even broader lines are observed for the main-chain methylene. The WIM/WISE experiments as a function of temperature show that the main-chain methylene decreases in linewidth from 38 kHz to 32 kHz as the temperature is increased from 25°C to 80°C, while the methyl linewidth decreases from 16 kHz to 12 kHz over this same temperature range (60).

The WIM/WISE experiment suppresses both intra- and interchain spin diffusion, so it can be used to study the molecular dynamics of polymer mixtures and blends. This is illustrated in Figure 5.46, which shows that polystyrene undergoes a much larger change in dynamics than does the poly(vinyl methyl ether) as the temperature is increased from 27°C to 80°C.

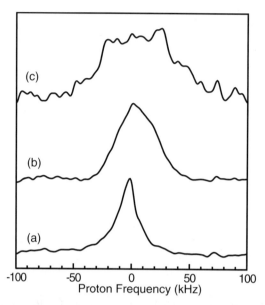

FIGURE 5.45 Cross sections through the WIM/WISE spectrum of poly(n-butyl acrylate) for the (a) methyle, (b) oxymethylene, and (c) main-chain methylene carbons.

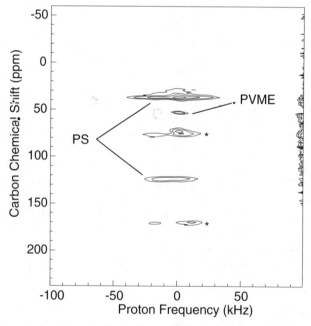

FIGURE 5.46 The 2D WIM/WISE spectrum for the miscible polystyrene/poly(vinyl methyl ether) blend. The spinning sidebands are marked (*).

In summary, the linewidths measured by the WISE experiments cannot provide the atomic-level detail of the dynamics available from the deuterium and chemical shift anisotropy lineshapes, but the experiments are extremely simple to perform and are applicable to a wide range of samples since isotopic labeling is not required.

REFERENCES

1. Heatley, F. *Prog. NMR Spectrosc.*, 1979, *13*, 47.
2. Qiu, X.; Moe, N. E.; Ediger, M. D.; Fetters, L. J. *J. Chem. Phys.*, 2000, *113*, 2918.
3. Moe, N. E.; Qiu, X.; Ediger, M. D. *Macromolecles*, 2000, *33*, 2415.
4. Lipari, G.; Szabo, A. *J. Am. Chem. Soc.*, 1982, *104*, 4546.
5. Lipari, G.; Szabo, A. *J. Am. Chem. Soc.*, 1982, *104*, 4559.
6. Jones, A. A.; Stockmayer, W. H. *J. Polym. Sci. Polym. Phys. Ed.*, 1977, *15*, 847.
7. Hall, C. K.; Helfand, E. *J. Chem. Phys.*, 1982, *77*, 3275.
8. Dejean de la Batie, R.; Laupretre, F.; Monnerie, L. *Macromolecules*, 1988, *21*, 2045.
9. Howarth, W. O. *Faraday Trans. 2*, 1979, *75*, 863.
10. Cole, K. S.; Cole, R. H. *J. Chem. Phys.*, 1941, *9*, 341.
11. Fuoss, R. M.; Kirkwood, J. G. *J. Am. Chem. Soc.*, 1941, *63*, 385.
12. Schaefer, J. *Macromolecules*, 1973, *6*, 882.
13. Lindsey, C. P.; Patterson, G. D. *J. Chem. Phys.*, 1980, *73*, 3348.
14. Zhu, W.; Ediger, M. D. *Macromolecules*, 1995, *28*, 7549.
15. Levitt, M. H. *Prog. NMR Spectrosc.*, 1986, *18*, 61.
16. Shaka, A.; Keeler, J.; Freeman, R. *J. Magn. Reson.*, 1983, *53*, 313.
17. Allerhand, A.; Hailstone, R. K. *J. Chem. Phys.*, 1972, *56*, 3718.
18. Guillermo, A.; Dupeyre, R.; Cohen-Addad, J. P. *Macromolecules*, 1990, *23*, 1291.
19. Gisser, D. J.; Glowinkowski, S.; Ediger, M. D. *Macromolecules*, 1991, *24*, 4270.
20. Glowinkowski, S.; Gisser, D. J.; Ediger, M. D. *Macromolecules*, 1990, *23*, 3520.
21. Spiess, H. W. *Chem. Rev.*, 1991, *91*, 1321.
22. Axelson, D. E. *Carbon-13 Solid State NMR of Semicrystalline Polymers*; VCH Publishers, Deerfiel Beach, FL, 1986.
23. Lyerla, J. R.; Yannoni, C. S. *IBM J. Res. Dev.*, 1983, *27*, 302.
24. Rothwell, W. P.; Waugh, J. S. *J. Chem. Phys.*, 1981, *74*, 2721.
25. White, J. L.; Mirau, P. A. *Macromolecules*, 1993, *26*, 3049.
26. Jones, A. A.; O'Gara, J. F.; Inglefield, P. T.; Bendler, J. T.; Yee, A. F.; Ngai, K. *Macromolecules*, 1983, *16*, 658.
27. Schaefer, J.; Sefcik, M. D.; Stejskal, E. O.; McKay, R. A.; Dixon, W. T.; Cais, R. E. *Macromolecules*, 1984, *17*, 1107.
28. Schaefer, J.; Stejskal, E. O.; Buchdahl, R. *Macromolecules*, 1977, *10*, 384.
29. Lu, J.; Mirau, P. A.; Tonelli, A. E. *Prog. Polym. Sci.*, 2002, *27*, 357.
30. Zemke, K.; Chmelka, B. F.; Schmidt-Rohr, K.; Spiess, H. *Macromolecules*, 1991, *24*, 6874.
31. Schmidt-Rohr, K.; Spiess, H. *Macromolecules*, 1991, *24*, 5288.

32. Taylor, S. A.; White, J. L.; Elbaum, N. C.; Crosby, R. C.; Campbell, G. C.; Haw, J. F.; Hatfield, G. R. *Macromolecules*, 1992, *25*, 3369.
33. Takegoshi, K.; Tanaka, I.; Hikichi, K.; Higashida, S. *Macromolecules*, 1992, *25*, 3392.
34. Spiess, H. *Adv. Polym. Sci.*, 1985, *66*, 23.
35. Bovey, F. A.; Mirau, P. A. *NMR of Polymers*, Academic Press, New York, 1996.
36. Macho, V.; Brombacher, L.; Spiess, H. *Appl. Magn. Reson.*, 2001, *20*, 405.
37. Hirschinger, J.; English, A. D. *J. Magn. Reson.*, 1989, *85*, 542.
38. Miura, H.; English, A. *Macromolecules*, 1988, *21*, 1543.
39. Wendoloski, J.; Gardner, K.; Hirschinger, J.; Miura, H.; English, A. *Science*, 1990, *247*, 431.
40. Hirschinger, J.; Miura, H.; Gardner, K. H.; English, A. D. *Macromolecules*, 1990, *23*, 2153.
41. Spiess, H. *Colloid Polym. Sci.*, 1983, *261*, 193.
42. Schadt, R. J.; Cain, E. J.; Gardner, K. H.; Gabara, V.; Allen, S. R.; English, A. D. *Macromolecules*, 1993, *26*, 6503.
43. Simpson, J. H.; Rice, D. M.; Karasz, F. E. *J. Polym. Sci., Part B, Polym. Phys.*, 1992, *30*, 11.
44. Lin, W.; Blum, F. D. *Macromolecules*, 1997, *30*, 5331.
45. Schmidt, C.; Blumich, B.; Spiess, H. W. *J. Magn. Reson.*, 1988, *79*, 269.
46. Schaefer, D.; Spiess, H.; Suter, U.; Flemming, W. *Macromolecules*, 1990, *23*, 3431.
47. Hirschinger, J.; Schaefer, D.; Spiess, H.; Lovinger, A. *Macromolecules*, 1991, *24*, 2428.
48. Wefing, S.; Spiess, H. *J. Chem. Phys.*, 1988, *89*, 1219.
49. Kaufmann, S.; Wefing, S.; Schaefer, D.; Speiss, H. W. *J. Chem. Phys.*, 1990, *93*, 197.
50. Schaefer, D.; Spiess, H. W. *J. Chem. Phys.*, 1992, *97*, 7944.
51. Chung, G. C.; Kornfield, J. A.; Smith, S. D. *Macromolecules*, 1994, *27*, 964.
52. Pschorn, U.; Rossler, E.; Sillescu, H.; Kaufmann, S.; Schaefer, D.; Speiss, H. *Macromolecules*, 1991, *24*, 398.
53. Williams, M. L.; Landel, R. F.; Ferry, J. D. *J. Am. Chem. Soc.*, 1955, *77*, 3701.
54. McCrum, N.; Read, B.; Williams, G. *Anelastic and Dielectric Effects in Polymeric Solids*, Wiley, New York, 1967.
55. Schmidt-Rohr, K.; Kulik, A. S.; Beckham, H. W.; Ohlemacher, A. O.; Pawelzik, U.; Boeffel, C.; Spiess, H. W. *Macromolecules*, 1994, *27*, 4733.
56. Hagemeyer, A.; Schmidt-Rohr, K.; Spiess, H. *Adv. Magn. Reson.*, 1989, *13*, 85.
57. Schmidt-Rohr, K.; Clauss, J.; Spiess, H. *Macromolecules*, 1992, *25*, 3273.
58. Qiu, X.; Mirau, P. A. *J. Magn. Reson.*, 1999, *142*, 183.
59. Caravatti, P.; Braunschweiler, L.; Ernst, R. R. *Chem. Phys. Lett.*, 1983, *100*, 305.
60. Qiu, X.; Mirau, P. *J. Magn. Reson.*, 1999, *142*, 183.

INDEX

Acquisition time
  NMR spectrometry
    data acquisition, 114–115
    multidimensional NMR, 146
    quantitative analysis, 121–122
α-effect
  solid-state NMR polymer characterization, semicrystalline polymers, 256–263
  solution NMR polymer characterization, chemical shift calculations, 173–175
Alternating copolymers, solution NMR polymer characterization, 231–24
Amorphous polymers
  molecular dynamics, solid-state NMR relaxation, wideline deuterium nuclear magnetic resonance, 377–385
  solid-state NMR polymer characterization
    chain conformation, 263–266
    structure and morphology, 295–302
AMX system, spin-spin coupling, 37–39
Angular momentum, nuclear magnetic resonance, 3
Anisotropic shielding, molecular structure, chemical shifts, 24–26
Antiphase magnetization, multidimensional nuclear magnetic resonance, through-bond magnetization, 89–91
Apodization, nuclear magnetic resonance spectrometry
  data processing, 117–119
  multidimensional nuclear magnetic resonance, 153–155
Asymmetry, solid-state nuclear magnetic resonance, chemical shift anisotropy, 67–70
Atactic polymers
  microstructural polymer characterization, stereochemical isomerism, 163
  solution NMR, spectral assignment, 168–169
Autocorrelation function
  molecular dynamics of polymers, chain motion in solution, 339–345
  NMR relaxation, 52

Baseline corrections
  NMR spectrometry, multidimensional NMR, 157
  spectrometer characteristics, solution nuclear magnetic resonance, 117
Bernoullian statistics, solution NMR polymer characterization
  chain architecture, 169–173
  stereochemical resonance assignments, 193–194
Bessel functions, molecular dynamics of polymers, chain motion in solution, 340–341

β-effect
  carbon-13 nuclear magnetic resonance,
    chemical shifts, 31–34
  microstructural polymer characterization,
    regioisomerism, 161–162
  molecular dynamics of polymers, solid-state
    NMR relaxation, chemical shift
    anisotropy, 387–390
  solid-state NMR polymer characterization,
    semicrystalline polymers, 256–263
  solution NMR polymer characterization,
    chemical shift calculations, 173–175
Biexponential correlation function, molecular
  dynamics of polymers, solution-based
  relaxation mechanisms, 355–359
Blended polymers, solid-state NMR
  characterization, 317–332
BLEW-12 pulse sequences, solid-state NMR
  spin diffusion polarization gradients, 292–294
  proton NMR multiple pulse decoupling,
    136–138
Bloch equations
  molecular structure, chemical shifts, chemical
    exchange mechanisms, 26–29
  nuclear magnetic resonance, 9–11
Block copolymers
  solid-state NMR characterization, structure and
    morphology, 302–310
  solution NMR, 234–235
    microstructural polymer characterization,
      167
Boltzmann distribution, nuclear magnetic
  resonance, basic properties, 5
Branching structures
  microstructural polymer characterization,
    stereochemical isomerism,
    164–165
  solution NMR polymer characterization
    chain architecture, 223–225
    chemical shift calculations, 173–175
    defect analysis, 214–218
Butane, γ-effect, chemical shifts, 32–34

Carbon-13 nuclear magnetic resonance
  basic properties, 4–5
  chemical shifts, 30–34
  molecular chemical shifts
    chemical structure, 21–23
    inductive effects, 24
  molecular dynamics of polymers
    solid-state relaxation, 360–367
    solution-based relaxation mechanisms,
      346–359

multidimensional NMR, pulse-field gradients,
  149–152
solid-state NMR
  chemical shift anisotropy, 68–70
  magic-angle sample spinning,
    127–128
  polymer characterization
    amorphous polymers, 263–266
    curing and reactivity, 275–278
    semicrystalline polymer characterization,
      256–263
    spin diffusion coefficients, 286–287
    structure and morphology analysis,
      278–280
  solution NMR polymer characterization
    stereochemical isomerism, 183–190
    stereochemical resonance assignments,
      198–211
  proton decoupling, 116
  quantitative analysis, 121–122
spin-lattice relaxation, 57
  heteronuclear compounds, 59–61
spin-spin coupling, homonuclear interactions,
  insensitive nuclei, 49–51
Carr-Purcell spin-echo sequence
  molecular dynamics of polymers,
    solution-based relaxation
    mechanisms, 348–359
  solid-state NMR, polymer characterization,
    block copolymer structure and
    morphology, 306–310
spin-spin relaxation, 141–142
Chain conformation
  microstructural polymer characterization
    regioisomerism, 161–162
    solution NMR, 165–166
  solid-state nuclear magnetic resonance,
    223–225
    cross polarization, 129–131
    polymer characterization, 250–278
      amorphous polymers, 263–266
      elastomers, 266–273
      reactivity and curing, 273–278
      semicrystalline structures,
        251–263
      solid-solid phase transitions,
        261–263
  solution NMR polymer characterization
    chain statistics, 169–173
    microstructural analysis, 161–162
    solution structure, 237–242
    stereochemical resonance assignments,
      193–194

INDEX

Chain motion in solution, molecular dynamics of polymers, 337–359
  molecular dynamics modeling, 338–345
  relaxation mechanisms, 345–349
  solution relaxation, 349–359
Chemical exchange, molecular structure, chemical shifts, 26–29
Chemical shift anisotropy (CSA)
  molecular dynamics of polymers
    solid-state NMR relaxation, 361–367
      lineshape analysis, 385–390
    solution-based relaxation mechanisms, 347–359
  NMR relaxation, 55
  solid-state nuclear magnetic resonance, 66–70
    magic-angle sample spinning, 71–73, 126–128
Chemical shifts
  NMR spectrometric characteristics, referencing techniques, 119–120
  in nuclear magnetic resonance
    polymer structure, 20–36
      carbon chemical shifts, 30–34
      molecular structure, 21–29
      nuclear properties, 34–36
      proton chemical shifts, 29–30
      product operator formalism, 18–20
  solid-state NMR polymer characterization
    basic principles, 249–250
    chain conformation, 250–278
      amorphous polymers, 263–266
      elastomers, 266–273
      reactivity and curing, 273–278
      semicrystalline structures, 251–263
      solid-solid phase transitions, 261–263
    spin diffusion polarization gradients, 290–294
  solution NMR polymer characterization
    defect analysis, branching structures, 215–218
    microstructure analysis
      copolymers, 166–167
      regioisomerism, 161–162
      spectral assignment, 173–175
    stereochemistry
      isomerism, 183–190
      resonance assignments, 191–193, 194–211
Chemical structure effects, molecular chemical shifts, 21–23

Cole-Cole distribution function, molecular dynamics of polymers
  chain motion in solution, 341–345
  solution-based relaxation mechanisms, 355–359
Coleman-Fox models, solution NMR polymer characterization
  chain statistics, 172–173
  stereochemical resonance assignments, 193–194
Combined rotation and multiple pulse spectroscopy (CRAMPS)
  solid-state nuclear magnetic resonance, dipolar broadening/decoupling, 75
  solid-state proton NMR, 133–138
Compensated pulses, pulsed nuclear magnetic resonance, 15
Composite-pulse decoupling (CPD), nuclear magnetic resonance spectrometry, solution NMR, 115–116
Computer components, nuclear magnetic resonance spectrometry, 108
Cone model, molecular dynamics of polymers, solid-state NMR relaxation, wideline deuterium nuclear magnetic resonance, 372–385
Conformational averaging
  γ-effect, chemical shifts, solution NMR polymer characterization, 177–178
  molecular dynamics of polymers, chain motion in solution, 339–345
  proton-proton coupling, 42–44
  solution NMR polymer characterization, stereochemical resonance assignments, 194–211
  solution structure of polymers, chain conformation, 237–242
Contour plots, multidimensional nuclear magnetic resonance, 82–84
Copolymers
  solid-state NMR characterization
    multiphase polymers, 313–317
    triblock copolymer, 268–273
  solution NMR characterization, 225–235
    alternating copolymers, 231–234
    block copolymers, 234–235
    microstructural polymer characterization, 166–167
    random copolymers, 227–231
Correlated experiments, multidimensional nuclear magnetic resonance, 79–80

Correlation functions, molecular dynamics of
    polymers
  chain motion in solution, 340–345
  solid-state NMR relaxation, 361–367
  solution-based relaxation mechanisms, 353–359
Correlation time
  molecular dynamics of polymers, chain motion
    in solution, 338–345
  spin-lattice relaxation, 59–61
Cross-peak intensity measurements,
    multidimensional nuclear magnetic
    resonance
  solution two-dimensional experiments,
    97–98
  two-dimensional solid-state NMR, 99–100
Cross polarization
  molecular dynamics of polymers, solid-state
    NMR relaxation, 361–367
  solid-state NMR polymer characterization
    amorphous polymers, 264–266
    blended polymers, 325–332
    polymer reactivity and curing, 274–278
    semicrystalline polymers, 251–263
  solid-state nuclear magnetic resonance,
    75–77
    basic principles, 249–250
    cross polarization dynamics, 128–131
Crystallinity measurements, solid-state NMR
    polymer characterization,
    semicrystalline polymers, 252–263
  structure and morphology, 295–302
Curing mechanisms, solid-state NMR polymer
    characterization, 273–278
Cyclodextrins
  molecular dynamics of polymers, solid-state
    NMR relaxation, 366–367
  solid-state NMR polymer characterization,
    325–332
Cylindrical morphology, solid-state NMR,
    polymer characterization, spin
    diffusion and polymer morphology,
    283–287

Data acquisition
  multidimensional (nD) NMR, 145–153
    decoupling, 152–153
    digital resolution and acquisition times,
      146
    inverse detection, 146–147
    phase cycling, 147–148
    pulsed-field gradients, 149–152
    quadrature detection, 148–149
  solution NMR, 114–115

Data processing
  multidimensional nuclear magnetic resonance,
    153–158
    apodization (window functions),
      153–155
    baselines and t_2 noise, 157
    linear prediction and zero-filling, 158
    phasing, 155–157
    quantitative analysis, 121–122
  solution nuclear magnetic resonance,
    117–120
    baseline corrections, 117
    digital resolution and zero-filling, 117
    phasing, 119
    quadrature detection, 119
    referencing, 119–120
    window functions, 117–119
Decoupling techniques
  molecular dynamics of polymers, solid-state
    NMR relaxation, 361–367
  multidimensional nuclear magnetic resonance,
    data acquisition, 152–153
  solid-state NMR
    polymer characterization
      basic principles, 248–250
      semicrystalline polymers, 251–263
    proton NMR, multiple-pulse decouplilng,
      134–138
  decoupler calibration
    pulse width adjustment, 111
    solid-state NMR, 131
    solution NMR, 115–116
Defect analysis, solution NMR polymer
    characterization, 214–223
  branching, 214–218
  endgroups, 218–223
Delays alternating with nutation for tailored
    excitation (DANTE) pulse sequence
  solid-state NMR polymer characterization
    blended polymers, 323–332
    semicrystalline structure and morphology,
      300–302
    spin diffusion polarization gradients,
      291–294
    spin-lattice relaxation, 140–141
δ-effect, microstructural polymer characterization,
    regioisomerism, 162
Dendritic polymers
  microstructural polymer characterization, chain
    architecture, 166
  solution NMR polymer characterization, chain
    architecture, 223–225
Deuterated solvents, solution NMR, 112–114

INDEX

Deuterium nuclear magnetic resonance
  molecular dynamics of polymers
    solid-state relaxation lineshapes, 371–385
    solution-based relaxation mechanisms, 346–359
  solid-state NMR polymer characterization
    multiphase polymers, 310–317
    semicrystalline structure and morphology, 297–302
  solid-state nuclear magnetic resonance quadrupolar interactions, 77–79
  spin-lattice relaxation, 57
  wideline NMR spectrometry, 132–133
Diad structure
  microstructural polymer characterization, stereochemical isomerism, 163
  solution NMR polymer characterization, stereochemical resonance assignments, 204–211
Diamond-lattice model, molecular dynamics of polymers, chain motion in solution, 340–359
Diffusion coefficients, solid-state NMR, polymer characterization, spin diffusion and polymer morphology, 286–287
Diffusion equation, solid-state NMR, polymer characterization, spin diffusion and polymer morphology, 280–295
Digital resolution
  data processing, 117
  multidimensional NMR, 146
Dipolar broadening/decoupling, solid-state nuclear magnetic resonance, 73–75
Dipolar coupling
  defined, 37
  molecular dynamics of polymers
    solid-state NMR relaxation, 361–367
      lineshape analysis, 390–394
    solution NMR relaxation, 346–359
  solid-state NMR
    basic principles, 248–250
    polymer characterization
      semicrystalline polymers, 252–263
      structure and morphology analysis, 278–280
Dipolar dephasing experiments, solid-state nuclear magnetic resonance, polymer characterization
  blended polymers, 327–332
  semicrystalline polymers, 254–263

Dipolar filter pulse sequence, solid-state NMR, polymer characterization, spin diffusion polarization gradients, 289–294
Dipole-dipole interactions
  NMR relaxation, 54
  spin-lattice relaxation, heteronuclear compounds, 58–60
Direct polarization, solid-state NMR polymer characterization, semicrystalline polymers, 251–263
Distortionless enhancement by polarization transfer (DEPT)
  spectral editing, 123–125
  solution NMR polymer characterization
    random copolymers, 228–231
    regioisomerism, 212–214
Distribution functions, molecular dynamics of polymers, chain motion in solution, 341–345
Disyndiotactic polymers, microstructural polymer characterization, stereochemical isomerism, 163
DLM model, molecular dynamics of polymers, solution-based relaxation mechanisms, 355–359
Domain size measurements, solid-state NMR, polymer characterization
  block copolymer structure and morphology, 305–310
  multiphase polymers, 315–317
  semicrystalline structure and morphology, 297–302
  structure and morphology analysis, 279–280
Double bond structures, molecular dynamics of polymers, solution-based relaxation mechanisms, 350–359
Double resonance experiments, nuclear magnetic resonance spectrometry, pulse width adjustment, 111
Dwell time, nuclear magnetic resonance spectrometry, data acquisition, 114–115

Echo delay time, molecular dynamics, solid-state NMR relaxation, wideline deuterium nuclear magnetic resonance, 381–385
Elastomers, solid-state NMR characterization, 266–273
Electron-donating groups, proton chemical shifts, 29–30

Electron-withdrawing groups
    molecular structure, chemical shifts, inductive effects, 23–24
    proton chemical shifts, 29–30
Endgroups
    microstructural polymer characterization, stereochemical isomerism, 164–165
    solution NMR polymer characterization, defect analysis, 218–223
Epimerization, solution NMR polymer characterization, spectral assignment, 168–169
Equilibrium distribution, nuclear magnetic resonance, 6
Erythrodiisotactic polymers, microstructural polymer characterization, stereochemical isomerism, 163
Excitation profile, pulsed nuclear magnetic resonance, 14–15

Fast Fourier transform (FFT), nuclear magnetic resonance, 16
Fermi-contact induction, spin-spin coupling, 37
Fluorine nuclear magnetic resonance
    basic properties, 4–5
    chemical shifts, 34
    molecular structure, chemical shifts, inductive effects, 23–24
    scalar coupling, 46–48
    solid-state NMR
        chemical shift anisotropy, 70
        polymer characterization
            curing and reactivity, 276–278
            elastomers, 269–273
            semicrystalline polymers, 260–263
            semicrystalline structure and morphology, 300–302
            structure and morphology analysis, 279–280
    solution NMR polymer characterization
        stereochemical isomerism, 188–190
        stereochemical resonance assignments, 206–211
Force, nuclear magnetic resonance, 3
Fourier transform (FT)
    molecular dynamics of polymers, chain motion in solution, 338–345
    multidimensional nuclear magnetic resonance, free induction decay, 80–82
    nuclear magnetic resonance, basic principles, 15–16
    pulsed nuclear magnetic resonance, excitation profile, 15
    multidimensional NMR, 154–155
        baseline corrections, 157
        phasing, 119
    solution nuclear magnetic resonance, baseline corrections, 117
Free induction decay (FID)
    multidimensional nuclear magnetic resonance, 80–82
    quadrature detection, 148–149
    nuclear magnetic resonance, Fourier transform, 15–16
    pulsed nuclear magnetic resonance, 12–13
    quadrature detection, 119
    solution nuclear magnetic resonance, baseline corrections, 117
    tuning techniques, gain adjustment, 108–109
    window functions, 117–119
Frequency-dependent phase correction, 119
Frequency-independent phase correction, 119
Frequency labeling, multidimensional nuclear magnetic resonance, 80
Frequency-switched Lee-Goldberg (FSLG) pulse sequence, solid-state NMR
    polymerization characterization, spin diffusion polarization gradients, 293–294
    proton NMR, 137–138
Fuoss-Kirkwood distribution function, molecular dynamics of polymers
    chain motion in solution, 341–345
    solution-based relaxation mechanisms, 355–359

Gain adjustments, nuclear magnetic resonance spectrometer, tuning techniques, 108–109
γ-effect
    carbon-13 nuclear magnetic resonance, chemical shifts, 31–34
    microstructural polymer characterization, regioisomerism, 162
    solution NMR polymer characterization, chemical shift calculations, 173–175
GARP pulse sequence
    composite pulse decoupling, 116
    solution NMR polymer characterization, stereochemical resonance assignments, 209–211
Gauche interactions

INDEX

γ-effect, chemical shifts, 32–34
  solid-state NMR polymer characterization
    amorphous polymers, 263–266
    basic principles, 249–250
    blended polymers, 330–332
    semicrystalline polymers, 255–263
    solution NMR polymer characterization, 175–178
    conformational calculations, 194–196
    proton-proton coupling, 42–44
    solution structure of polymers, 235–236
Gaussian broadening (GB) window functions, 118–119
Geometric isomerism, microstructural polymer characterization, stereochemical isomerism, 163–164
Gill-Mieboom technique, molecular dynamics of polymers, solution-based relaxation mechanisms, 348–359
Glass transition temperature, solid-state NMR, polymer characterization, elastomers, 266–273
Goldman-Shen technique, solid-state NMR, polymer characterization
  multiphase polymers, 310–317
  spin diffusion polarization gradients, 288–294

Hall-Helfand theory, molecular dynamics of polymers
  chain motion in solution, 340–341
  solution-based relaxation mechanisms, 355–359
Hamiltonian equations, nuclear magnetic resonance, product operator formalism, 17–20
Hamming windows, window functions, multidimensional NMR, 154–155
Hartmann-Hahn condition, solid-state nuclear magnetic resonance
  cross polarization, 76–77
    cross polarization dynamics, 129–131
    elastomer polymer characterization, 273
Head-to-head chain structure
  microstructural polymer characterization, regioisomerism, 161–162
  solution NMR polymer characterization
    chemical shift calculations, 174–175
    regioisomerism, 211–214
Head-to-tail chain structure
  microstructural polymer characterization, regioisomerism, 161–162

solution NMR polymer characterization, regioisomerism, 211–214
Helium dewars, superconducting magnetic materials, 106
Heteroatomic structures, molecular dynamics of polymers, solution-based relaxation mechanisms, 350–359
Heterogeneous samples, solid-state NMR, basic properties, 249–250
Heteronuclear correlation (HETCOR) experiments
  molecular dynamics of polymers, solid-state NMR relaxation, 365–367
  multidimensional nuclear magnetic resonance
    solution two-dimensional experiments, 94–95
    through-bond magnetization, 86–91
    two-dimensional solid-state NMR, 101–102
    nuclear Overhauser effect, 64–65
    product operator formalism, 18–20
  proton-carbon coupling, 45–46
  solid-state NMR
    polymerization characterization
      blended polymers, 330–332
      spin diffusion polarization gradients, 292–294
    proton NMR pulse sequences, 136–138
  solution NMR polymer characterization, multidimensional NMR, 181–182
  spectrometry characteristics
    nuclear Overhauser enhancement, 142
    sensitivity enhancement, 122–123
    spin-lattice relaxation, 59–61
Heteronuclear multiple bond correlation (HMBC)
  multidimensional nuclear magnetic resonance, through-bond magnetization, 89–91
  solution NMR polymer characterization
    defect analysis
      branching structures, 217–218
      end groups, 220–223
    stereochemical resonance assignments, 199–211
Heteronuclear multiple quantum coherence (HMQC)
  multidimensional NMR
    baseline corrections, 157
    pulse-field gradients, 151–152
  multidimensional nuclear magnetic resonance, 84
    inverse detection, 147
    solution two-dimensional experiments, 94–95
    through-bond magnetization, 89–91

Heteronuclear multiple quantum coherence (HMQC) (*Continued*)
  solution NMR polymer characterization
    chain architecture, 224–225
    defect analysis
      branching structures, 217–218
      end groups, 220–223
    multidimensional NMR, 181–182
    stereochemical resonance assignments, 198–211
Heteronuclear single quantum coherence (HSQC), multidimensional nuclear magnetic resonance, through-bond magnetization, 89–91
High-field superconducting magnet, 105–106
High-resolution magic-angle spinning (HR-MAS), swollen polymer characterization, 272–273
HNCA experiment, solution NMR polymer characterization, stereochemical resonance assignments, 209–211
Homonuclear experiments
  nuclear magnetic resonance
    nuclear Overhauser effect, 65–66
    product operator formalism, 18–20
  solution NMR polymer characterization, stereochemical resonance assignments, 202–211
  spin-lattice relaxation, 58, 61
  spin-spin coupling, insensitive nuclei, 49–51
Hydrocarbons
  chemical shifts, 31–34
  solution NMR polymer characterization, chemical shift calculations, 173–175
Hydrogen bonding, molecular structure, chemical shifts, inductive effects, 24

Incredible natural abundance double-quantum transfer experiment (INADEQUATE), solution NMR polymer characterization
  regioisomerism, 211–214
  stereochemical resonance assignments, 200–211
Inductive effects
  carbon-13 nuclear magnetic resonance, chemical shifts, 33–34
  molecular structure, chemical shifts, 23–24
Insensitive nuclei enhanced by polarization transfer (INEPT)
  multidimensional nuclear magnetic resonance, through-bond magnetization, 88–91

solution NMR polymer characterization
  regioisomerism, 212–214
  sensitivity enhancement, 122–123
  spectral editing, 123–125
Interfaces, solid-state NMR, polymer characterization, spin diffusion and polymer morphology, 285–286
  polarization gradients, 290–294
  surface-to-volume ratio, 283–287
Intermediate time scales, molecular dynamics, solid-state NMR relaxation, wideline deuterium nuclear magnetic resonance, 380–385
Intermolecular interactions, solution structure of polymers, 242–245
Internuclear distances, solution structure of polymers
  chain conformation, 239–242
  NMR characterization, 235–236
Inverse detection, multidimensional NMR
  data acquisition, 146–147
  decoupling sequences, 152–153
  pulse decoupling, 116
Inversion recovery
  molecular dynamics of polymers
    solid-state NMR relaxation, 361–367
    solution-based relaxation mechanisms, 348–359
  spin-lattice relaxation, NMR spectrometry characteristics, 139–142
Isopentane, γ-effect, chemical shifts, 32–34
Isotactic structures
  microstructural polymer characterization, stereochemical isomerism, 162–163
  molecular dynamics of polymers, solid-state NMR relaxation, 363–367
  solid-state NMR polymer characterization, semicrystalline polymers, 255–263
  solution NMR polymer characterization
    spectral assignment, 168–169
    stereochemical isomerism, 184–190
Isotropic reorientation model, molecular dynamics of polymers
  chain motion in solution, 339–345
  chemical shift anisotropy, 388–390
  solution-based relaxation mechanisms, 349–359
IS scalar coupling, multidimensional nuclear magnetic resonance, through-bond magnetization, 87–91

INDEX

J-resolved nuclear magnetic resonance
multidimensional applications, 98
solution NMR polymer characterization,
multidimensional NMR, 179–182
solution structure of polymers, chain
conformation, 237–242

Karplus relationship
proton-proton coupling, 42–44
solution NMR polymer characterization,
stereochemical resonance
assignments, 208–211
Kramer's theory, molecular dynamics of
polymers, solution-based relaxation
mechanisms, 354–359
KWW distribution function, molecular dynamics
of polymers
chain motion in solution, 344–345
solution-based relaxation mechanisms, 357–359

Laboratory frame of reference, nuclear magnetic
resonance, 7–9
Lamb's equation, nuclear magnetic resonance,
chemical shifts, 20–21
Lamellar systems, solid-state NMR, polymer
characterization, spin diffusion and
polymer morphology, 283–287
Laplace transform, molecular dynamics of
polymers, chain motion in solution,
339–345
Larmor precession
molecular dynamics of polymers, solid-state
relaxation lineshapes, 370–394
nuclear magnetic resonance, 4
Lee-Goldberg pulse sequencing, solid-state
NMR
blended polymer characterization, 330–332
proton NMR, 136–138
Length scale measurements, solid-state NMR,
polymer characterization, blended
polymers, 318–332
Librational correlation function, molecular
dynamics of polymers,
solution-based relaxation
mechanisms, 355–359
Linear prediction, NMR spectrometry,
multidimensional NMR, 158
Lineshape analysis
molecular dynamics of polymers, solid state
relaxation, 359–360, 370–394
chemical shift anisotropy, 385–390
dipolar lineshapes, 390–394
wideline deuterium NMR, 371–385

block copolymer structure and
morphology, 309–310
semicrystalline polymers, 254–263
multidimensional NMR, 146
wideline NMR, 131–133
Linewidth measurements
nuclear magnetic resonance spectrometer
data processing, 117
tuning techniques, homogeneity adjustments,
108
polymer dynamics, 336–337
solid-state nuclear magnetic resonance, polymer
characterization, 250
elastomers, 271–273
spin-spin relaxation, 61–63
Log-X^2 distribution, molecular dynamics of
polymers
chain motion in solution, 341–345
solution-based relaxation mechanisms,
355–359
Longitudinal relaxation, nuclear magnetic
resonance, 6
Bloch equations, 10–11
Lorentz-Gauss window function, nuclear magnetic
resonance spectrometry, data
processing, 118–119

Magic-angle sample spinning
molecular dynamics of polymers, solid-state
NMR relaxation, 363–367
nuclear magnetic resonance spectrometry, probe
requirements, 107
solid-state NMR, 70–73
polymer characterization
amorphous polymers, 263–266
basic principles, 248–250
blended polymers, 325–332
block copolymer structure and
morphology, 305–310
elastomers, 266–273
multiphase polymers, 315–317
reactivity and curing analysis,
274–278
semicrystalline polymers, 251–263
spin diffusion polarization gradients,
293–294
proton NMR, 137–138
Magnetic field inhomogeneity
multidimensional NMR, pulse-field gradients,
149–152
nuclear magnetic resonance spectrometry, shim
coil construction, 106–107
spin-spin relaxation, 62–63

Magnetic field strength
  molecular dynamics of polymers,
    solution-based relaxation
    mechanisms, 351–359
  nuclear magnetic resonance, 4
Magnetic resonance, basic principles,
    3–7
Magnetic susceptibility, molecular structure,
    chemical shifts, anisotropic shielding,
    25–26
Magnetization transfer
  molecular dynamics of polymers,
    solution-based relaxation
    mechanisms, 346–359
  multidimensional nuclear magnetic resonance,
    85–92
    through-bond transfer, 85–91
    through-space transfer, 91–92
    two-dimensional solid-state NMR,
    99–100
  solid-state NMR, polymer characterization,
    250
    spin diffusion, 280–295
    structure and morphology analysis,
    278–280
  spin-lattice relaxation, 57
Magnetogyric ratio, nuclear magnetic resonance,
    3–4
Markov models, solution NMR polymer
    characterization
  chain statistics, 172–173
  stereochemical resonance assignments,
    193–194
Matched filtering principle, nuclear magnetic
    resonance spectrometry, window
    functions, 117–119
McConnell equations, molecular structure,
    chemical shifts, chemical exchange,
    26–29
Meiboom-Gill pulse sequence, NMR spectrometry
    characteristics, spin-spin relaxation,
    141–142
Meso configuration
  microstructural polymer characterization,
    stereochemical isomerism,
    162–163
  solution NMR polymer characterization
    chain statistics, 169–173
    stereochemical resonance assignments,
    196–211
Microstructural polymer characterization, solution
    NMR, 161–167

branching and endgroups, 164–165
chain architecture, 165–166
copolymers, 166–167
geometric isomerism, 163–164
regioisomerism, 161–162
stereochemical isomerism, 162–163
Miscibility properties, solid-state NMR
    blended polymer characterization,
    317–332
MLEV 16/17 decoupling
  multidimensional nuclear magnetic resonance
    solution two-dimensional experiments,
    93–94
  through-bond magnetization transfer, 91
  NMR spectrometry characteristics, 116
Mobile polymers, solid-state NMR, polymer
    characterization
  multiphase polymers, 316–317
  spin diffusion and polymer morphology, 287
Model compounds, solution NMR polymer
    characterization
  basic principles, 168–169
  resonance assignments, 190–193
Model-free technique, molecular dynamics of
    polymers, chain motion in solution,
    339–345
Molecular dynamics of polymers
  chain motion in solution, 337–359
    molecular dynamics modeling, 338–345
    relaxation mechanisms, 345–349
    solution relaxation, 349–359
  solid state relaxation, 359–394
    lineshape analysis, 370–394
      chemical shift anisotropy, 385–390
      dipolar lineshapes, 390–394
      wideline deuterium NMR, 371–385
    NMR relaxation times, 360–367
    spin exchange, 367–370
Molecular structure, chemical shifts, 21–29
  anisotropic shielding, 24–26
  carbon-13 nuclear magnetic resonance, 31–34
  chemical exchange mechanisms, 26–29
  chemical structure effects, 21–23
  inductive effects, 23–24
MREV-8 pulse sequence
  solid-state NMR polymer characterization
    blended polymers, 323–332
    block copolymer structure and morphology,
    304–310
    spin diffusion polarization gradients,
    291–294
    solid-state proton NMR, 134–138

INDEX

Multidimensional (nD) nuclear magnetic resonance, 79–102
  experimental techniques, basic principles, 105
  magnetization transfer, 85–92
    through-bond transfer, 85–91
    through-space transfer, 91–92
  molecular dynamics, solid-state NMR relaxation, wideline deuterium nuclear magnetic resonance, 381–385
  nuclear magnetic resonance spectrometry
    basic requirements, 105
    data acquisition, 114–115
    pulse width adjustment, 110–111
  proton-carbon coupling, 45–46
  pulse profiles, 15–16
  scalar coupling, 41
  solid-state NMR
    polymer characterization
      basic applications, 250
      block copolymer structure and morphology, 309–310
      spin diffusion polarization gradients, 292–294
    proton pulse sequences, 136–138
    two-dimensional experiments, 99–102
      heteronuclear correlation, 101–102
      two-dimensional exchange, 99–100
      wideline separation spectroscopy, 100–102
  solution NMR
    polymer characterization, 179–182
      alternating copolymers, 233–234
      defect analysis, 219–223
      stereochemical resonance assignments, 196–211
    two-dimensional experiments, 92–98
      COSY spectroscopy, 92–93
      heteronuclear quantum coherences, 94–95
      J-resolved NMR, 98–99
      TOCSY spectroscopy, 93–94
      two-dimensional exchange (NOESY), 95–98
    data acquisition, 145–153
      decoupling, 152–153
      digital resolution and acquisition times, 146
      inverse detection, 146–147
      phase cycling, 147–148
      pulsed-field gradients, 149–152
      quadrature detection, 148–149
    data processing, 153–158
      apodization, 153–155

      baselines and t_2 noise, 157
      linear prediction and zero-filling, 158
      phasing, 155–157
Multiphase polymers, solid-state NMR characterization, 310–317
Multiple bond couplings, proton-carbon coupling, 45–46
Multiple-pulse decoupling, solid-state proton NMR, 134–138

Nearest-neighbor structure, carbon-13 nuclear magnetic resonance, chemical shifts, 31–34
90° ($\pi/2$) pulse
  multidimensional nuclear magnetic resonance, through-bond magnetization, 87–91
  pulsed nuclear magnetic resonance, 12–13
Nitrogen nuclear magnetic resonance
  basic properties, 5
  chemical shifts, 36
    molecular structure, inductive effects, 24
  molecular dynamics of polymers
    solid-state relaxation, 360–367
    solution-based relaxation mechanisms, 346–359
  scalar coupling, 49
  solid-state NMR polymer characterization, semicrystalline polymers, 258–263
  spin-spin coupling, homonuclear interactions, insensitive nuclei, 50–51
Nonselective relaxation rate, spin-lattice relaxation, 57–58
Nonspinning shims, magnetic field homogeneity, 106–107
Nuclear magnetic resonance (NMR)
  basic principles, 2–20
    Bloch equations, 9–11
    Fourier transform, 15–16
    magnetic resonance, 3–7
    product operator formalism, 16–20
    pulsed nuclear magnetic resonance, 11–15
    rotating reference frame, 7–9
  chemical shifts and polymer structure, 20–36
    carbon chemical shifts, 30–34
    molecular structure, 21–29
    nuclear properties, 34–36
    proton chemical shifts, 29–30
  multidimensional NMR, 79–102
    magnetization transfer, 85–92
      through-bond transfer, 85–91
      through-space transfer, 91–92

Nuclear magnetic resonance (NMR) (*Continued*)
  solid-state two-dimensional experiments, 99–102
    heteronuclear correlation, 101–102
    two-dimensional exchange, 99–100
    wideline separation spectroscopy, 100–102
  solution two-dimensional experiments, 92–98
    COSY, 92–93
    heteronuclear quantum coherences, 94–95
    J-resolved NMR, 98–99
    TOCSY, 93–94
    two-dimensional exchange (NOESY), 95–98
  relaxation, 51–66
    chemical shift anisotropy, 55
    dipole-dipole interactions, 54
    nuclear Overhauser effect, 63–66
      heteronuclear, 63–65
      homonuclear, 65–66
    paramagnetic relaxation, 55–56
    quadrupolar interactions, 54–55
    spin-lattice relaxation, 56–61
      heteronuclear relaxation, 59–61
      homonuclear relaxation, 61
    spin-spin relaxation, 61–63
  solid-state reactions, 66–79
    chemical-shift anisotropy, 66–70
    cross polarization, 75–77
    dipolar broadening/decoupling, 73–75
    magic-angle sample spinning, 70–73
    quadrupolar nuclei, 77–79
  spin-spin coupling, 36–51
    fluorine couplings, 46–48
    homonuclear coupling, insensitive light, 49–51
    nitrogen, 49
    nomenclature, 38–39
    patterns, 39–41
    phosphorus coupling, 48–49
    proton-carbon coupling, 44–46
    proton-proton scalar coupling, 41–44
    scalar coupling nD NMR, 41
    silicon, 49
    strong coupling, 40
Nuclear Overhauser effect (NOE)
  molecular dynamics of polymers
    solid-state NMR relaxation, 364–367
    solution-based relaxation mechanisms, 347–359
  multidimensional nuclear magnetic resonance, through-space magnetization, 91–92

NMR relaxation, 63–66
  heteronuclear effects, 64–65
  homonuclear effects, 65–66
NMR spectrometry
  quantitative analysis, 121–122
  relaxation mechanisms, 138, 142
  sensitivity enhancement, 122–123
  solid-state NMR polymer characterization, blended polymers, 327–332
  solution NMR polymer characterization, multidimensional NMR, 179–182
Nuclear Overhauser effect spectroscopy (NOESY)
  multidimensional nuclear magnetic resonance, 84
  solution two-dimensional experiments, 95–98
  solid-state NMR, polymer characterization
    elastomers, 267–273
    multiphase polymers, 314–317
  solution NMR polymer characterization
    alternating copolymers, 233–234
    chain architecture, 224
    regioisomerism, 213–214
    solution structure of polymers, 236
      intermolecular interactions, 242–245
    stereochemical resonance assignments, 201–211
  spectrometer characteristics, multidimensional NMR, 157

Off-resonance decoupling, solid-state proton NMR, 136–138
One-dimensional diffusion, solid-state NMR, polymer characterization, polymer morphology, 281–287
One-dimensional nuclear magnetic resonance, data acquisition, 114–115
$180°$ (π) pulse, pulsed nuclear magnetic resonance, 12–13
One-pulse experiments, solid-state NMR, polymer characterization, elastomers, 266–273
Orthogonality, pulsed nuclear magnetic resonance, 12–13

"Pake" powder pattern, solid-state nuclear magnetic resonance
  dipolar broadening/decoupling, 74–75
  multiphase polymers, 312–317
  quadrupolar interactions, 79
Paramagnetic relaxation, nuclear magnetic resonance, 55–56
Pascal's triangle, spin-spin coupling, peak intensity, 37–40

INDEX

Peak intensity
  multidimensional nuclear magnetic resonance, solution two-dimensional experiments, 96–98
  nuclear magnetic resonance spectrometry
    data acquisition, 115
    quantitative analysis, 121–122
  solution NMR polymer characterization, chain statistics, 170–173
  spin-spin coupling, 37–38
Phase cycling, multidimensional nuclear magnetic resonance
  data acquisition, 147–148
  pulse-field gradients vs., 152
Phase separation, solid-state NMR, polymer characterization
  multiphase polymers, 310–317
  relaxation times and polymer morphology, 294–295
  spin diffusion and polymer morphology, 282–287
Phasing, nuclear magnetic resonance spectrometry
  data processing, 119
  multidimensional NMR, 155–157
Phosphorus nuclear magnetic resonance
  basic properties, 5
  chemical shifts, 35–36
  molecular dynamics of polymers
    solid-state relaxation, 360–367
    solution-based relaxation mechanisms, 346–359
  scalar coupling, 48–49
  solid-state NMR
    chemical shift anisotropy, 70
    semicrystalline polymers, 259–263
    structure and morphology, 299–302
π-electron clouds, molecular structure, chemical shifts, anisotropic shielding, 24–26
Planck's constant, nuclear magnetic resonance, 3–4
Polarization gradients, solid-state NMR, polymer characterization
  block copolymer structure and morphology, 304–310
  spin diffusion and polymer morphology, 282–287, 287–294
Polarization transfer, NMR spectrometry characteristics
  sensitivity enhancement, 122–123
  spectral editing, 123–125
Polymer characterization. *See also* Molecular dynamics of polymers
  nuclear magnetic resonance

basic techniques, 1–2
chemical shifts, 20–36
  carbon chemical shifts, 30–34
  molecular structure, 21–29
  nuclear properties, 34–36
  proton chemical shifts, 29–30
relaxation mechanisms, 51–66
  basic principles, 6
  chemical shift anisotropy, 55
  dipole-dipole interactions, 54
  nuclear Overhauser effect, 63–66
    heteronuclear, 63–65
    homonuclear, 65–66
  paramagnetic relaxation, 55–56
  quadrupolar interactions, 54–55
  spin-lattice relaxation, 56–61
    heteronuclear relaxation, 59–61
    homonuclear relaxation, 61
  spin-spin relaxation, 61–63
solid-state nuclear magnetic resonance
  chain conformation, 250–278
    amorphous polymers, 263–266
    elastomers, 266–273
    reactivity and curing, 273–278
    semicrystalline structures, 251–263
    solid-solid phase transitions, 261–263
  structure and morphology, 278–332
    blended polymers, 317–332
    block copolymers, 302–310
    multiphase polymers, 310–317
    semicrystalline polymers, 295–302
    spin diffusion, 280–295
      coefficients, 286–287
      interfaces, 285–286
      polarization gradients, 287–294
      proton relaxation, 294–295
solution NMR
  chain architecture, 223–225
  defect analysis, 214–223
    branching, 214–218
    endgroups, 218–223
  microstructure, 161–167
    branching and endgroups, 164–165
    chain architecture, 165–166
    copolymers, 166–167
    geometric isomerism, 163–164
    regioisomerism, 161–162
    stereochemical isomerism, 162–163
  regioisomerism, 211–214
  solution structure, 235–245
    chain conformation, 236–242
    intermolecular interactions, 242–245

Polymer characterization. *See also* Molecular
 dynamics of polymers (*Continued*)
 spectral assignment, 167–182
  chain statistics, 169–173
  chemical shift calculations, 173–175
  γ-gauche effect, 175–178
  model compounds, 168–169
  multidimensional NMR, 179–182
  spectral editing, 178–179
 spectrometry characteristics, polymer
  solubility and sample preparation,
  112–114
 stereochemical techniques, 182–211
  isomerism, observation of, 183–190
  stereosequence resonance assignments,
   190–211
   chemical shift and conformational
    calculations, 194–196
   model compounds, 190–193
   multidimensional NMR, 196–211
   polymerization statistics, 193–194
Precession, nuclear magnetic resonance, 3
Probe characteristics
 nuclear magnetic resonance spectrometry, 107
  tuning techniques, 109–110
 solid-state nuclear magnetic resonance,
  spectrometry characteristics,
  magic-angle sample spinning,
  126–128
Product operator formalism (POF)
 multidimensional nuclear magnetic resonance,
  through-bond magnetization, 85–91
 nuclear magnetic resonance, 16–20
 nuclear magnetic resonance spectrometry, pulse
  width adjustment, 111
Propane, γ-effect, chemical shifts, 32–34
Proton-carbon coupling, scalar couplings, 44–46
Proton-decoupled heteronuclear spectra,
 composite-pulse decoupling (CPD),
 115–116
Proton densities, solid-state NMR, polymer
 characterization, spin diffusion and
 polymer morphology, 286–287
Proton nuclear magnetic resonance
 chemical shifts, 29–30
 chemical shifts and molecular structure, 21–23
 molecular dynamics of polymers,
  solution-based relaxation
  mechanisms, 346–359
 multidimensional NMR, through-bond
  magnetization, 87–91
 polymer characterization, 250
 solid-state NMR

chemical shift anisotropy, 69–70
polymer characterization
 amorphous polymers, 265–266
 blended polymers, 318–332
 block copolymer structure and
  morphology, 304–310
 elastomers, 266–273
 multiphase polymers, 310–317
 relaxation times and polymer morphology,
  294–295
 semicrystalline polymers, 261–263
 structure and morphology analysis,
  279–280
spin diffusion
 polarization gradients, 290–294
 semicrystalline structure and morphology,
  298–302
 structure and morphology analysis,
  278–280
solution NMR polymer characterization
 basic principles, 235–236
 stereochemical isomerism, 183–190
 stereochemical resonance assignments,
  198–211
 quantitative analysis, 121–122
 solution NMR, sample preparation, 113–114
spin-lattice relaxation, 57
Proton-proton coupling, scalar coupling, 41–44
Pseudoasymmetric structures, microstructural
 polymer characterization,
 stereochemical isomerism, 162–163
Pulsed nuclear magnetic resonance, basic
 principles, 11–15
Pulse-field gradients, multidimensional nuclear
 magnetic resonance, 149–152
Pulse sequences
 molecular dynamics of polymers
  solid-state NMR relaxation, spin-exchange
   experiments, 367–370
  wideline deuterium nuclear magnetic
   resonance, 382–385
 multidimensional nuclear magnetic resonance,
  80
  through-bond magnetization, 87–91
  two-dimensional solid-state NMR, 99–100
 rotating-frame spin-lattice relaxation,
  solid-state NMR, 144–145
 sensitivity enhancement, 122–123
 spectral editing, 123–125
 spin-lattice relaxation
  solid-state NMR, 143–144
  solution NMR, 138–141
 spin-spin relaxation, solution NMR, 141–142

INDEX

nuclear magnetic resonance, product operator formalism, 19–20
solid-state nuclear magnetic resonance
  cross polarization, spectrometry characteristics, 130–131
  MREV-8 pulse sequence, 134–138
  polymer characterization
    blended polymers, 323–332
    semicrystalline polymers, 253–263
    spin diffusion polarization gradients, 289–294
  solid-state proton NMR
    Lee-Goldberg pulse sequencing, 136–138
    WAHUHA sequence, 133–138
  wideline NMR spectrometry, deuterium nuclear magnetic resonance, 132–133
solution NMR polymer characterization
  defect analysis, endgroups, 221–223
  spectral editing, 178–179, 189–190
  stereochemical resonance assignments, 200–211
Pulse tip angle, NMR spectrometry characteristics, quantitative analysis, 121–122
Pulse width adjustment, nuclear magnetic resonance spectrometry, tuning techniques, 110–111

Quadrature detection, NMR spectrometry, 119
  multidimensional NMR, 148–149
  phasing, 156–157
Quadrupolar interactions
  molecular dynamics of polymers
    solid-state NMR relaxation, 361–367
    wideline deuterium nuclear magnetic resonance, 371–385
    solution-based relaxation mechanisms, 347–359
  NMR relaxation, 54–55
  solid-state nuclear magnetic resonance, 77–79
  wideline NMR spectrometry, 131–133
Quantitative analysis, NMR spectrometry, 119–122
Quantum coherence, multidimensional NMR, pulse-field gradients, 150–152

Racemic configuration
  microstructural polymer characterization, stereochemical isomerism, 162–163

solution NMR polymer characterization, stereochemical resonance assignments, 196–211
Radiofrequency (rf) console, nuclear magnetic resonance spectrometry, 107
Radius, nuclear magnetic resonance, 3
Random copolymers
  solution NMR characterization, 227–231
  solution structure, chain conformation, 239–242
Rate constants, solid-state NMR, polymer characterization, spin diffusion and polymer morphology, 286–287
Reactivity mechanisms, solid-state NMR polymer characterization, 273–278
Reference compounds, 119–121
Regioisomerism
  microstructural polymer characterization, 161–162
  solution NMR polymer characterization, 211–214
Relaxation mechanisms
  data analysis, 105
  molecular dynamics of polymers
    chain motion in solution, models for, 337–345
    solid state relaxation, 359–394
      lineshape analysis, 370–394
        chemical shift anisotropy, 385–390
        dipolar lineshapes, 390–394
        wideline deuterium NMR, 371–385
      NMR relaxation times, 360–367
      spin exchange, 367–370
    solution-based polymers, 345–359
  nuclear magnetic resonance, 51–66
    basic principles, 6
    chemical shift anisotropy, 55
    dipole-dipole interactions, 54
    nuclear Overhauser effect, 63–66
      heteronuclear, 63–65
      homonuclear, 65–66
    paramagnetic relaxation, 55–56
    quadrupolar interactions, 54–55
    spin-lattice relaxation, 56–61
      heteronuclear relaxation, 59–61
      homonuclear relaxation, 61
    spin-spin relaxation, 61–63
  solid-state NMR
    basic principles, 249–250
    polymer characterization
      blended polymers, 318–332
      relaxation times and polymer morphology, 294–295

Relaxation mechanisms (*Continued*)
   semicrystalline polymers, 253–263, 295–302
   spin diffusion polarization gradients, 288–294
  spectrometry for NMR, 138–145
   basic requirements, 105
   solid-state NMR, 142–145
    rotating-frame spin-lattice relaxation, 144–145
    spin-lattice relaxation, 143–144
   solution NMR, 138–142
    nuclear Overhauser effect, 142
    sample preparation, 113–114
    spin-lattice relaxation, 138–141
    spin-spin relaxation, 141–142
Resolved experiments, multidimensional nuclear magnetic resonance, 79–80
Response function, solid-state NMR, polymer characterization, spin diffusion and polymer morphology, 280–295
"Ring current" shifts, molecular structure, chemical shifts, anisotropic shielding, 25–26
Ring flips, molecular dynamics of polymers, solid-state NMR relaxation, wideline deuterium nuclear magnetic resonance, 374–385
Rotating-frame nuclear Overhauser effect (ROE)
  multidimensional nuclear magnetic resonance, through-space magnetization, 91–92
  nuclear magnetic relaxation, homonuclear experiments, 65–66
Rotating-frame spin-lattice relaxation time constants, solid-state nuclear magnetic resonance
  cross polarization, 76–77, 129–131
  spectrometric characteristics, 144–145
Rotating reference frame, nuclear magnetic resonance, 7–9
Rotational correlation time, nuclear magnetic resonance, basic principles, 6
Rotational isometric state (RIS) model, solution NMR polymer characterization
  chemical shift and conformational calculations, 194–196
  γ-effect, chemical shifts, 177–178
Rotation geometry, solid-state NMR, magic-angle sample spinning, 71–73

Rothwell-Waugh broadening, molecular dynamics of polymers, solid-state NMR relaxation, 363–367

Sample preparation, nuclear magnetic resonance spectrometry, 104–105
  solution NMR, 112–114
Saturation-recovery pulse sequences, spin-lattice relaxation, 140–142
Scalar couplings
  defined, 37
  magnitude, 38
  multidimensional nuclear magnetic resonance, 41
  product operator formalism, 18–20
  proton-carbon coupling, 44–46
  proton-proton coupling, 41–44
Scalar relaxation, nuclear magnetic resonance, 56
Segmental motion hypothesis, molecular dynamics of polymers, solution-based relaxation mechanisms, 349–359
Selective excitation, pulsed nuclear magnetic resonance, 15
Selective relaxation rate, spin-lattice relaxation, 57–58
Semicrystalline polymers
  molecular dynamics, solid-state NMR relaxation, 366–367
   wideline deuterium nuclear magnetic resonance, 379–385
  solid-state NMR polymer characterization, 251–263
  structure and morphology, 295–302
Sensitivity enhancement, NMR spectrometry, 122–123
Shaped pulses
  pulsed nuclear magnetic resonance, 15
  solid-state nuclear magnetic resonance, cross polarization, 130–131
Shielding anisotropy, solid-state nuclear magnetic resonance, 67–70
Shim coils, magnetic field homogeneity, 106–107
Sidebands, solid-state nuclear magnetic resonance, magic-angle sample spinning, 73
Signal-to-noise ratio (SNR)
  multidimensional nuclear magnetic resonance inverse detection, 146–147
  scalar coupling, 41
  data acquisition, 114–115
  multidimensional NMR, apodization, 153–155

probe tuning, 109–110
pulse width adjustment, 110–111
quantitative analysis, 121–122
sensitivity enhancement, 122–123
window functions, 118–119
proton-carbon coupling, 45–46
pulsed nuclear magnetic resonance, 15
solid-state NMR
  basic principles, 249–250
  polymer characterization, semicrystalline polymers, 252–263
Silicon nuclear magnetic resonance
  basic properties, 5
  chemical shifts, 34–35
  molecular dynamics of polymers
    solid-state relaxation, 360–367
    solution-based relaxation mechanisms, 346–359
  scalar coupling, 49
  solid-state NMR polymer characterization
    semicrystalline polymers, 257–263
    solid-solid phase transition, 261–263
  solution NMR polymer characterization, stereochemical isomerism, 189–190
Single-frequency decoupling, solid-state NMR, spectrometry characteristics, 131
Site populations, molecular dynamics of polymers, solid-state NMR relaxation, wideline deuterium nuclear magnetic resonance, 375–385
Slow exchange regime, molecular structure, chemical shifts, chemical exchange, 27–29
Small-angle X-ray scattering (SAXS), solid-state NMR, polymer characterization, block copolymer structure and morphology, 303–310
Solid-solid phase transition, solid-state NMR polymer characterization, semicrystalline polymers, 261–263
Solid-state nuclear magnetic resonance
  applications, 2
  basic principles, 66–79
    chemical-shift anisotropy, 66–70
    cross polarization, 75–77
    dipolar broadening/decoupling, 73–75
    magic-angle sample spinning, 70–73
    quadrupolar nuclei, 77–79
  experimental techniques, 104–105
  molecular dynamics of polymers, relaxation mechanisms, 359–394

    lineshape analysis, 370–394
      chemical shift anisotropy, 385–390
      dipolar lineshapes, 390–394
      wideline deuterium NMR, 371–385
    NMR relaxation times, 360–367
    spin exchange, 367–370
  polymer characterization
    chain conformation, 250–278
      amorphous polymers, 263–266
      elastomers, 266–273
      reactivity and curing, 273–278
      semicrystalline structures, 251–263
      solid-solid phase transitions, 261–263
    structure and morphology, 278–332
      blended polymers, 317–332
      block copolymers, 302–310
      multiphase polymers, 310–317
      semicrystalline polymers, 295–302
      spin diffusion, 280–295
        coefficients, 286–287
        interfaces, 285–286
        polarization gradients, 287–294
        proton relaxation, 294–295
  basic requirements, 105
  cross-polarization, 128–131
  decoupling, 131
  magic-angle spinning, 126–128
  probe requirements, 107
  proton NMR, 133–138
  referencing techniques, 119
  relaxation mechanisms, 142–145
    rotating-frame spin-lattice relaxation, 144–145
    spin-lattice relaxation, 143–144
  wideline NMR, 131–133
  two-dimensional experiments, 98–102
    heteronuclear correlation, 101–102
    two-dimensional exchange, 99–100
    wideline separation spectroscopy, 100–102
Solution nuclear magnetic resonance
  applications, 1–2
  copolymer characterization, 225–235
    alternating copolymers, 231–234
    block copolymers, 234–235
    random copolymers, 227–231
  experimental techniques, basic principles, 104–105
  molecular dynamics of polymers, chain motion, 337–359
    molecular dynamics modeling, 338–345
    relaxation mechanisms, 345–349
    solution relaxation, 349–359

Solution nuclear magnetic resonance (*Continued*)
  polymer characterization
    chain architecture, 223–225
    defect analysis, 214–223
      branching, 214–218
      endgroups, 218–223
    microstructure, 161–167
      branching and endgroups, 164–165
      chain architecture, 165–166
      copolymers, 166–167
      geometric isomerism, 163–164
      regioisomerism, 161–162
      stereochemical isomerism, 162–163
    regioisomerism, 211–214
    solution structure, 235–245
      chain conformation, 236–242
      intermolecular interactions, 242–245
    spectral assignment, 167–182
      chain statistics, 169–173
      chemical shift calculations, 173–175
      γ-gauche effect, 175–178
      model compounds, 168–169
      multidimensional NMR, 179–182
      spectral editing, 178–179
    stereochemical techniques, 182–211
      isomerism, observation of, 183–190
      stereosequence resonance assignments, 190–211
        chemical shift and conformational calculations, 194–196
        model compounds, 190–193
        multidimensional NMR, 196–211
        polymerization statistics, 193–194
  spectrometer characteristics, 111–125
    basic requirements, 104–105
    data acquisition, 114–115
    data processing, 117–120
      baseline corrections, 117
      digital resolution and zero-filling, 117
      phasing, 119
      quadrature detection, 119
      reference compounds, 119–120
      window functions, 117–119
    decoupling, 115–116
    probe requirements, 107
    quantitative analysis, 120–122
    relaxation mechanisms, 138–142
      nuclear Overhauser effect, 142
      spin-lattice relaxation, 138–141
      spin-spin relaxation, 141–142
    sample preparation, 112–114
    sensitivity enhancement, 122–123
    spectral editing, 123–125

  two-dimensional experiments, 92–98
    COSY spectroscopy, 92–93
    heteronuclear quantum coherences, 94–95
    J-resolved NMR, 98–99
    TOCSY spectroscopy, 93–94
    two-dimensional exchange (NOESY), 95–98
Solution structure of polymers, solid-state polymer properties, 235–245
  chain conformation, 236–242
  intermolecular interactions, 242–245
Solvent viscosity, molecular dynamics of polymers, solution-based relaxation mechanisms, 354–359
Spectral assignment, polymer characterization, solution NMR, 167–182
  chain statistics, 169–173
  chemical shift calculations, 173–175
  γ-gauche effect, 175–178
  model compounds, 168–169
  multidimensional NMR, 179–182
  spectral editing, 178–179
Spectral density function
  molecular dynamics of polymers
    chain motion in solution, 337–359
    research background, 337
  NMR relaxation, 52–54
  spin-lattice relaxation, heteronuclear compounds, 59–60
Spectral editing
  NMR spectrometry characteristics, 123–125
  solid-state NMR polymer characterization, elastomers, 273
  solution NMR polymer characterization, 178–179
Spectrometry for nuclear magnetic resonance
  basic components, 105–108
    computers, 108
    high-field superconducting magnet, 105–106
    probes, 107
    radiofrequency console, 107
    shim coils, 106–107
  multidimensional NMR, 145–158
    data acquisition, 145–153
      decoupling, 152–153
      digital resolution and acquisition times, 146
      inverse detection, 146–147
      phase cycling, 147–148
      pulsed-field gradients, 149–152
      quadrature detection, 148–149
    data processing, 153–158
      apodization, 153–155
      baselines and t_2 noise, 157

INDEX

linear prediction and zero-filling, 158
phasing, 155–157
relaxation mechanisms, 138–145
  in solids, 142–145
    rotating-frame spin-lattice relaxation, 144–145
    spin-lattice relaxation, 143–144
  in solution, 138–142
    nuclear Overhauser effect, 142
    spin-lattice relaxation, 138–141
    spin-spin relaxation, 141–142
solid-state NMR methods, 126–138
  cross-polarization, 128–131
  decoupling, 131
  magic-angle spinning, 126–128
  proton NMR, 133–138
  wideline NMR, 131–133
solution NMR methods, 111–125
  data acquisition, 114–115
  data processing, 117–120
    baseline corrections, 117
    digital resolution and zero-filling, 117
    phasing, 119
    quadrature detection, 119
    referencing, 119–120
    window functions, 117–119
  decoupling, 115–116
  quantitative analysis, 120–122
  sample preparation, 112–114
  sensitivity enhancement, 122–123
  spectral editing, 123–125
  tuning techniques, 108–111
    gain adjustment, 108–109
    homogeneity adjustment, 108
    probe tuning, 109–110
    pulse width adjustment, 110–111
Spherical systems, solid-state NMR, polymer characterization, spin diffusion and polymer morphology, 283–287
Spin diffusion
  molecular dynamics of polymers
    solid-state NMR relaxation, 364–367
      lineshape analysis, 392–395
      spin exchange experiments, 368–370
    solution-based relaxation mechanisms, 346–359
  polymer dynamics, 337
  solid-state NMR, polymer characterization
    blended polymers, 318–332
    multiphase polymers, 310–317
    structure and morphology analysis, 279–295
      block copolymers, 302–310
      coefficients, 286–287
      interfaces, 285–286
      polarization gradients, 287–294
      proton relaxation, 294–295
      semicrystalline structures, 297–302
  spin-lattice relaxation, 57
Spin exchange experiments, molecular dynamics of polymers
  solid-state NMR relaxation, 367–370
    chemical shift anisotropy, 389–390
    wideline deuterium nuclear magnetic resonance, 382–385
Spin-lattice relaxation
  defined, 6
  molecular dynamics of polymers
    solid-state NMR relaxation, 361–367
    solution-based relaxation mechanisms, 346–359
  NMR measurements
    quantitative analysis, 121–122
    solid-state NMR, 143–144
    solution NMR, 138–141
  nuclear magnetic resonance, 56–61
    Bloch equations, 10–11
    heteronuclear relaxation, 59–61
    homonuclear relaxation, 61
  polymer dynamics, research background, 336–337
  pulsed nuclear magnetic resonance, 12–15
  solid-state nuclear magnetic resonance, polymer characterization, 249–250
    semicrystalline polymers, 252–263, 295–302
Spinning shims, magnetic field homogeneity, 106–107
Spinning speed, solid-state NMR, polymer characterization, spin diffusion and polymer morphology, 287
Spin quantum number, nuclear magnetic resonance, 3–4
Spin-rotation mechanism, nuclear magnetic resonance relaxation, 56
Spin-spin coupling, nuclear magnetic resonance, 36–51
  fluorine couplings, 46–48
  homonuclear coupling, insensitive light, 49–51
  nitrogen, 49
  nomenclature, 38–39
  patterns, 39–41
  phosphorus coupling, 48–49
  product operator formalism, 18–20
  proton-carbon coupling, 44–46
  proton-proton scalar coupling, 41–44
  scalar coupling nD NMR, 41

Spin-spin coupling, nuclear magnetic resonance (*Continued*)
  silicon, 49
  strong coupling, 40
Spin-spin relaxation
  defined, 7
  molecular dynamics of polymers
    solid-state NMR relaxation, 363–367
    solution-based relaxation mechanisms, 348–359
  molecular structure, chemical shifts, chemical exchange, 28–29
  NMR measurements, 141–142
  nuclear magnetic resonance
    Bloch equations, 10–11
    mechanisms, 61–63
  pulsed nuclear magnetic resonance, 12–15
  solid-state NMR polymer characterization, semicrystalline polymers, 254–263
  structure and morphology, 295–302
Stacked plots, multidimensional nuclear magnetic resonance, 82–84
Stereochemistry, solution NMR polymer characterization, 182–211
  chain statistics, 171–173
  isomerism, observation of, 183–190
  microstructural isomerism, 162–163
  spectral assignment, 168–169
  stereosequence resonance assignments, 190–211
    chemical shift and conformational calculations, 194–196
    model compounds, 190–193
    multidimensional NMR, 196–211
    polymerization statistics, 193–194
Strong coupling, spin-spin coupling patterns, 40
Superconducting magnetic materials, nuclear magnetic resonance spectrometry, 105–106
Swollen polymer beads, solid-state nuclear magnetic resonance characterization, 272–273
  semicrystalline structure and morphology, 302
Syndiotactic structures
  microstructural polymer characterization, stereochemical isomerism, 163
  solid-state NMR polymer characterization, semicrystalline polymers, 255–263
  solution NMR polymer characterization
    spectral assignment, 168–169
    stereochemical isomerism, 184–190

Tail-to-tail chain structure
  microstructural polymer characterization, regioisomerism, 161–162

solution NMR polymer characterization
  chemical shift calculations, 174–175
  regioisomerism, 211–214
Temperature variables, molecular dynamics of polymers
  solid-state relaxation lineshapes, 371–394
  solution-based relaxation mechanisms, 352–359
Tetramethylsilane, NMR spectrometry, chemical shift referencing, 119
Three-bond vicinal couplings, proton-proton interactions, 41–44
Three-dimensional experiments
  multidimensional nuclear magnetic resonance, 82–84
  solid-state NMR, polymer characterization, spin diffusion and polymer morphology, 281–287
  solution NMR polymer characterization, stereochemical resonance assignments, 208–211
Threodiisotactic polymers, microstructural polymer characterization, stereochemical isomerism, 163
Through-bond magnetization exchange
  multidimensional nuclear magnetic resonance, 85–91
  solution NMR polymer characterization, multidimensional NMR, 179–182
Through-space magnetization exchange
  multidimensional nuclear magnetic resonance, 91–92
  solution NMR polymer characterization, multidimensional NMR, 179–182
Time-proportional phase incrementation (TPPI), multidimensional nuclear magnetic resonance, quadrature detection, 148–149
Tip angle. *See also* Excitation profile
  pulsed nuclear magnetic resonance, 12–13
Torque, nuclear magnetic resonance, 3
Total correlation spectroscopy (TOCSY)
  multidimensional nuclear magnetic resonance, 84
    decoupling sequences, 152–153
    solution two-dimensional experiments, 93–94
    through-bond magnetization, 90–91
  solution NMR polymer characterization
    chain architecture, 224–225
    multidimensional NMR, 181–182
    stereochemical resonance assignments, 202–211
Total supression of sidebands (TOSS), solid-state nuclear magnetic resonance, magic-angle sample spinning, 73

INDEX

TPPI-hypercomplex technique, multidimensional
    nuclear magnetic resonance,
    quadrature detection, 149
Trans conformations
    proton-proton coupling, 42–44
    solid-state NMR polymer characterization,
        semicrystalline polymers, 253–263
Transmission electron microscopy (TEM),
    solid-state NMR, polymer
        characterization, block copolymer
        structure and morphology, 303–310
Transverse relaxation. *See also* Spin-spin
        relaxation
    nuclear magnetic resonance, 6
        Bloch equations, 10–11
Triad structure
    microstructural polymer characterization,
        stereochemical isomerism, 163
    solution NMR polymer characterization,
        stereochemical resonance
        assignments, 204–211
Triblock copolymer, solid-state NMR
    characterization, 268–273
    multiphase polymers, 313–317
Trichlorheptane compound, solution NMR
    polymer characterization, resonance
    assignments, 190–193
Tuning techniques, nuclear magnetic resonance
    spectrometer, 108–111
    gain adjustment, 108–109
    homogeneity adjustment, 108
    probe tuning, 109–110
    pulse width adjustment, 110–111
Two-bond geminal coupling, proton-proton
    interactions, 41–44
Two-dimensional correlated spectroscopy (COSY)
    microstructural polymer characterization,
        regioisomerism, 162
    multidimensional nuclear magnetic resonance,
        80–84
        phase cyclilng, 148
        quadrature detection, 148–149
        solution two-dimensional experiments, 92–93
        spectrometry resolution and acquisition time,
            146
        through-bond magnetization, 90–91
    nuclear magnetic resonance, product operator
        formalism, 16–20
    solution NMR polymer characterization
        regioisomerism, 212–214
        stereochemical resonance assignments,
            197–211
Two-dimensional diffusion, solid-state NMR,
    polymer characterization, polymer
    morphology, 281–287

Two-dimensional exchange experiments,
    molecular dynamics of polymers,
    solid-state NMR relaxation, 367–370
Two-dimensional experiments
    multidimensional nuclear magnetic resonance,
        92–98
        COSY spectroscopy, 92–93
        heteronuclear quantum coherences, 94–95
        J-resolved NMR, 98–99
        TOCSY spectroscopy, 93–94
        two-dimensional exchange (NOESY), 95–98
    solid-state nuclear magnetic resonance, 98–99,
        98–102
        heteronuclear correlation, 101–102
        multidimensional NMR, 99–102
            heteronuclear correlation, 101–102
            two-dimensional exchange, 99–100
            wideline separation spectroscopy,
                100–102
        two-dimensional exchange, 99–100
        wideline separation spectroscopy, 100–102
    solution nuclear magnetic resonance, 92–98
        COSY spectroscopy, 92–93
        heteronuclear quantum coherences, 94–95
        J-resolved NMR, 98–99
        TOCSY spectroscopy, 93–94
        two-dimensional exchange (NOESY), 95–98
Two-phase pulse modulation, solid-state NMR,
    spectrometry characteristics, 131
Two-site jump models, molecular dynamics of
    polymers, solid-state NMR
    relaxation, wideline deuterium
    nuclear magnetic resonance, 375–385

WAHUHA sequence, solid-state proton NMR,
    spectrometric characteristics,
    133–138
WALTZ-16 decoupling
    molecular dynamics of polymers,
        solution-based relaxation
        mechanisms, 348–359
    multidimensional nuclear magnetic resonance,
        152–153
    nuclear magnetic resonance spectrometry,
        115–116
Wideline deuterium nuclear magnetic resonance,
    molecular dynamics of polymers,
    solid-state relaxation lineshapes,
    371–385
Wideline probes, nuclear magnetic resonance
    spectrometry, 107
Wideline separation (WISE) NMR spectroscopy
    molecular dynamics of polymers, solid-state
        NMR relaxation, lineshape analysis,
        392–395

Wideline separation (WISE) NMR spectroscopy (*Continued*)
  multidimensional nuclear magnetic resonance, 100–101
  solid-state NMR, polymerization characterization
    blended polymers, 321–332
    block copolymer structure and morphology, 309–310
    multiphase polymers, 310–317
    spin diffusion polarization gradients, 293–294
Window functions, 117–119
  multidimensional NMR, 154–155
Windowless decoupling sequences, solid-state proton NMR, 136–138
Windowless isotropic magnetization (WIM)
  heteronuclear correlation, two-dimensional solid-state NMR, 101–102
  molecular dynamics of polymers, solid-state NMR relaxation, lineshape analysis, 393–395
  solid-state NMR, polymerization characterization, spin diffusion polarization gradients, 292–294

X nucleus channel, nuclear magnetic resonance spectrometry, pulse width adjustment, 111
X-ray scattering techniques, solid-state NMR, polymer characterization
  multiphase polymers, 317
  spin diffusion and polymer morphology, 287

Zero filling
  data processing, 117
  solution NMR, data acquisition, 114–115